# Lecture Notes in Computer Science 2254

Edited by G. Goos, J. Hartmanis, and J. van Leeuwen

**Springer**
*Berlin
Heidelberg
New York
Barcelona
Hong Kong
London
Milan
Paris
Tokyo*

Murray Reed Little   Laurence Nigay (Eds.)

# Engineering for Human-Computer Interaction

8th IFIP International Conference, EHCI 2001
Toronto, Canada, May 11-13, 2001
Revised Papers

 Springer

Series Editors

Gerhard Goos, Karlsruhe University, Germany
Juris Hartmanis, Cornell University, NY, USA
Jan van Leeuwen, Utrecht University, The Netherlands

Volume Editors

Murray Reed Little
Carnegie Mellon University, Software Engineering Institute
Pittsburgh, PA 15213, USA
E-mail: little@sei.cmu.edu

Laurence Nigay
University of Grenoble 1, Laboratory CLIPS-IMAG, IIHM Research Group
385, rue de la Bibliothèque, 38041 Grenoble Cedex 9, France
E-mail: Laurence.Nigay@imag.fr

Cataloging-in-Publication Data applied for

Die Deutsche Bibliothek - CIP-Einheitsaufnahme

Engineering for human computer interaction : 8th IFIP international
conference ; revised papers / EHCI 2001, Toronto, Canada, May 11 - 13, 2001.
Murray Reed Little ; Laurence Nigay (ed.). - Berlin ; Heidelberg ; New York ;
Barcelona ; Hong Kong ; London ; Milan ; Paris ; Tokyo : Springer, 2001
    (Lecture notes in computer science ; Vol. 2254)
    ISBN 3-540-43044-X

CR Subject Classification (1998): H.5.2, H.5.3, D.2, H.3, H.4, K.4

ISSN 0302-9743
ISBN 3-540-43044-X Springer-Verlag Berlin Heidelberg New York

Springer-Verlag Berlin Heidelberg New York
a member of BertelsmannSpringer Science+Business Media GmbH

http://www.springer.de

© 2001 IFIP International Federation for Information Processing, Hofstrasse 3, A-2361 Laxenburg, Austria
Printed in Germany

Typesetting: Camera-ready by author, data conversion by Steingräber Satztechnik GmbH, Heidelberg
Printed on acid-free paper     SPIN: 10846000     06/3142     5 4 3 2 1 0

# Preface

The papers collected here are those selected for presentation at the Eighth IFIP Conference on Engineering for Human-Computer Interaction (EHCI 2001) held in Toronto, Canada in May 2001.

The conference is organized by the International Federation of Information Processing (IFIP) Working Group 2.7 (13.4) for Interface User Engineering, Rick Kazman being the conference chair, Nicholas Graham and Philippe Palanque being the chairs of the program committee. The conference was co-located with ICSE 2001 and co-sponsored by ACM.

The aim of the IFIP working group is to investigate the nature, concepts, and construction of user interfaces for software systems. The group's scope is:
! ∀to develop user interfaces based on knowledge of system and user behavior;
! ∀to develop frameworks for reasoning about interactive systems; and
! ∀to develop engineering models for user interfaces.

Every three years, the working group holds a working conference. The Seventh one was held September 14-18 1998 in Heraklion, Greece. This year, we innovated by organizing a regular conference held over three days.

Over 50 submitted papers were received, and each of them was reviewed by three WG 2.7 (13.4) members and observers. Twenty-one long papers and four short papers were selected for presentation at the conference. Their authors come from 10 countries, in North America and Europe, reflecting the truly international nature of the conference. The papers are organized into topic areas as follows:
! ∀Software Engineering Methods
! ∀Formal Methods
! ∀Toolkits
! ∀User Interface Evaluation
! ∀User Interface plasticity
! ∀3D User Interfaces
! ∀Input and Output Devices
! ∀Mobile Interaction
! ∀Context Sensitive Interaction

Following each presentation there was a discussion among the participants and the presenter. A transcript of the discussion is found at the end of each paper in these proceedings. Each session was assigned a discussion transcriber, whose responsibility was to collect/transcribe the questions and answers during the session. The original transcripts were distributed to the attendees during the conference, and modifications that clarified the discussion were accepted.

The program committee invited three keynote speakers, David Garlan, Saul Greenberg, and Jeff Raskin. Summaries of their talks are included in these proceedings.

Without the submission of papers, a conference cannot occur. IFIP WG 2.7 (13.4) thanks all those who spent energy and time in writing their papers and preparing their presentations.

July 2001                                    Reed Little and Laurence Nigay

## Program Committee

| | |
|---|---|
| G. Abowd, U.S.A. | C. Gram, Denmark |
| R. Bastide, France | R. Kazman, U.S.A |
| L.J. Bass, U.S.A. | K. Hopper, New Zealand |
| M. Beaudouin-Lafon, France | R. Little, U.S.A. |
| M. Borup-Harning, Denmark | I. Newman, U.K. |
| A. Chabert, U.S.A. | L. Nigay, France |
| S. Chatty, France | P. Palanque, France |
| G. Cockton, U.K. | F. Paterno, Italy |
| J. Coutaz, France | C. Roast, U.K. |
| P. Dewan, U.S.A. | H.G. Stiegler, Germany |
| A. Dix, U.K. | C. Unger, Germany |
| N. Graham, Canada | L. Watts, U.K. |

## Sponsoring Organizations

# Conference Participants

A. Alsumait, Canada
K. Baker, Canada
J. Barnes, U.S.A.
L. Bass, U.S.A.
R. Bastide, France
L. Bergman, U.S.A.
S. Chatty, France
J. Coutaz, France
P. Curzon, U.K.
D. Damian, Canada
W. Dees, Netherlands
R. Eagleson, Canada
D. Garlan, U.S.A.
N. Graham, Canada
P. Gray, U.K.
S. Greenberg, Canada
D. Grolaux, Belgium
M. Borup-Harning, Denmark
S. Henninger, U.S.A.
J. Höhle, Germany
F. Jambon, France
R. Jezek, Austria
R. Kazman, U.S.A.
S. Kettebekov, U.S.A.
R. Little, U.S.A.
H. Lutfiyya, Canada
C. MacGregor, Canada

S. MacKenzie, Canada
M. Mäkäräinen, Finland
B. Miners, Canada
L. Nigay, France
P. Palanque, France
F. Paterno, Italy
G. Phillips, Canada
J. Rapskin, U.S.A.
P. Renevier, France
C. Roast, U.K.
J. Roth, Germany,
N. Roussel, Switzerland
M. Sage, U.K.
D. Salber, U.S.A.
K. Schneider, Canada
P. Smith, Canada
H. Stiegler, Germany
W. Stuerzlinger, Canada
C. Unger, Germany
T. Urnes, Norway
P. Van Roy, Belgium
W. Volz, U.S.A.
J. Willans, U.K.
J. Wu, Canada
C. Yellowlees, Canada
G. Zimmerman, U.S.A.

# Table of Contents

## User Interface Evaluation

## User Interface Plasticity

## 3D User Interfaces

## Input and Output Devices

## Mobile Interaction

## Context Sensitive Interaction

# Aura: Distraction-Free Ubiquitous Computing

David Garlan

School of Computer Science
Carnegie Mellon University
5000 Forbes Avenue
Pittsburgh, PA 15213
garlan@cs.cmu.edu

Technological trends are leading to a world in which computing is all around us – in our cars, our kitchens, our offices, our phones, and even our clothes. In this world we can expect to see an explosion of computational devices, services, and information at our disposal. While this is an undeniable opportunity, currently we are ill-prepared to deal with its implications.

To take a simple example: Today I can easily afford to have 10 PCs in my office. But what can I do with them? Simply keeping them synchronized would be a nightmare. There are few, if any, applications that can exploit them simultaneously. And even if I could harness them all at once, the effort of configuring all of that software, ensuring version compatibility, starting it up, and stopping it in a consistent fashion would be painful. Furthermore, that equipment would not be usable by me in all of the other settings away from my office. Put simply, there is a serious mismatch between the availability of computing resources and our ability to exploit them effectively.

The root of the problem is that the most precious resource in a computer system is no longer its processor, memory, disk or network. Rather, it is a resource not subject to Moore's law: user attention. Today's systems distract a user in many explicit and implicit ways, thereby reducing his effectiveness. Unless we find ways to make technology more "invisible" we will see a widening gap between technological capability and usefulness.

At CMU we are addressing this problem in a new research effort called Project Aura. Aura's goal is to provide each user with an invisible halo of computing and information services that persists regardless of location, and that spans wearable, handheld, desktop, and infrastructure computers. As the user moves from one location to another, and as resources come and go, the system adapts, providing the user with continuous, self-configuring, optimal access to data and computation.

To achieve this goal, Project Aura is attempting to rethink system design, focusing on integrating two broad concepts across all system levels. First, it uses proactivity, or the ability of a system layer to act in anticipation of requests by a higher layer. This is in contrast to today's systems, where each layer is reactive to the layer above it. Second, Aura is self-tuning: layers adapt by observing the demands made on them and adjusting their performance and resource usage characteristics to match demand. This is in contrast to today's systems, where the behavior of a system layer is relatively static.

M. Reed Little and L. Nigay (Eds.): EHCI 2001, LNCS 2254, p. 1, 2001.
© Springer-Verlag Berlin Heidelberg 2001

# Supporting Casual Interaction
# Between Intimate Collaborators

Saul Greenberg

Department of Computer Science, University of Calgary
Calgary, Alberta, CANADA T2N 1N4
saul@cpsc.ucalgary.ca

Over last decade, we have seen mounting interest in how groupware technology can support electronic interaction between intimate collaborators who are separated by time and distance. By intimate collaborators I mean small communities of friends, family or colleagues who have a real need or desire to stay in touch with one another. While there are many ways to provide electronic interaction, perhaps the most promising approach relies on casual interaction. The general idea is that members of a distributed community track when others are available, and use that awareness to move into conversation, social interaction and work. On the popular front, we see support for casual interaction manifested through the explosion of instant messaging services: a person sees friends and their on-line status in a buddy list, and selectively enters into a chat dialog with one or more of them. On the research front, my group members and I are exploring the subtler nuances of casual interaction. We design, build and evaluate various groupware prototypes [1,2,3,4] and use them as case studies to investigate:

- how we can enrich on-line opportunities for casual interaction by providing people with a rich sense of awareness of their intimate collaborators;
- how we can supply awareness of people's artifacts so that these can also become entry points into interaction;
- how we can present awareness information at the periphery, where it becomes part of the background hum of activity that people can then selectively attend to;
- how we can create fluid interfaces where people can seamlessly and quickly act on this awareness and move into conversation and actual work;
- how we can have others overhear and join ongoing conversations and activities;
- how we can make these same opportunities work for a mix of co-located and distributed collaborators; and
- how we balance distraction and privacy concerns while still achieving the above.

## References

1. Boyle, M., Edwards, C., Greenberg, S.: The Effects of Filtered Video on Awareness and Privacy. CHI Letters (CHI 2001 Proceedings), 2(3), ACM Press (2000)
2. Greenberg, S., Fitchet, C.: Phidgets: Incorporating Physical Devices into the Interface. In M. Newman, K. Edwards and J. Sedivy (Eds) Proc. Workshop on Building the Ubiquitous Computing User Experience. Held at ACM CHI'01 (2001)
3. Greenberg, S., Kuzuoka, H.: Using Digital but Physical Surrogates to Mediate Awareness, Communication and Privacy in Media Spaces. Personal Technologies 4(1), Elsevier (2000)
4. Greenberg, S., Rounding, M.: The Notification Collage: Posting Information to Public and Personal Displays. CHI Letters (ACM CSCW 2000 Proceedings) 3(1), (2001) 515-521

M. Reed Little and L. Nigay (Eds.): EHCI 2001, LNCS 2254, p. 3, 2001.
© Springer-Verlag Berlin Heidelberg 2001

# Turning the Art of Interface Design into Engineering

Jef Raskin

8 Gypsy Hill Road
Pacifica, CA 94044
JefRaskin@aol.com

The name of this conference begins with the word "engineering," a skill that I've seen little of in the world of commercial interface design. Here's what I mean by "engineering":

I don't know if my background is typical, but I've enjoyed designing aircraft for some years now. As a child, I built model airplanes, some of which flew. Of necessity, they had many adjustable parts. I could move the wing forward and aft in a slot or with rubber bands, add bits of clay to the nose to adjust the balance, and I'd glue small aluminum tabs to wings and tail surfaces and bend them to correct the flight path.

Once I had gotten past calculus and some college physics, I began to study aerodynamic and mechanical engineering more seriously. I remember with considerable pleasure the first time I was able to design a model (radio-controlled by this time) based on knowledge sound and deep enough so that I knew that --barring accidents--the aircraft would fly, and even how it would fly. I opened the throttle, the plane chugged along the runway and rose into the air, exactly as predicted. It flew and maneuvered, and I brought it back to a gentle landing.

That's engineering. The ability to design from rational foundations, numerically predict performance, and have the result work much as expected without all kinds of ad hoc adjustments.

This is not what I see in interface design. Working in the practical world of the interface designers who produce the commercial products used by millions, even hundreds of millions of people, I find that most of the practitioners have no knowledge of existing engineering methods for designing human-computer interfaces (for example, Fitts' and Hick's laws, GOMS analyses, and measures of interface efficiency). The multiple stupidities of even the latest designs, such as Microsoft's Windows 2000 or Apple's OS X, show either an unjustifiable ignorance of or a near-criminal avoidance of what we do know.

My talk will look at some of the true engineering techniques available to us, and where HCI can go if we allow these techniques -- rather than inertia, custom, guesswork, and fear -- to guide us.

M. Reed Little and L. Nigay (Eds.): EHCI 2001, LNCS 2254, p. 5, 2001.
© Springer-Verlag Berlin Heidelberg 2001

# Towards a UML for Interactive Systems

Fabio Paternò

CNUCE-C.N.R., Via V.Alfieri 1, 56010 Ghezzano, Pisa, Italy
fabio.paterno@cnuce.cnr.it

**Abstract.** Nowadays, UML is the most successful model-based approach to supporting software development. However, during the evolution of UML little attention has been paid to supporting user interface design and development. In the meantime, the user interface has become a crucial part of most software projects, and the use of models to capture requirements and express solutions for its design, a true necessity. Within the community of researchers investigating model-based approaches for interactive applications, particular attention has been paid to task models. ConcurTaskTrees is one of the most widely used notations for task modelling. This paper discusses a solution for obtaining a UML for interactive systems based on the integration of the two approaches and why this is a desirable goal.

## 1 Introduction

The success of UML [9] reflects the success of object-oriented approaches in software development. UML is already composed of nine notations. In practice, software engineers rarely use all of them. In this context, it seems difficult to propose the inclusion of another notation. However, successful approaches have to take into account what the new important problems are, if they want to continue to be successful. The rapidly increasing importance of the user interface component of a software system cannot be denied. Since the early work on UIDE [3], various approaches have been developed in the field of model-based design of interactive applications [7] [8] and they have not affected at all the development of UML. The developers of UML did not seem particularly aware of the importance of model-based approaches to user interfaces: UML is biased toward design and development of the internal part of software systems, and has proven its effectiveness in this. However, despite UML use cases and other notations can effectively capture "functional" requirements or specify detailed behaviours, UML does not specifically - nor adequately - support the modelling of user interfaces aspects.

In the European R&D project GUITARE, we developed a set of tools for task modelling and have proposed their use to software companies. The first issue raised in response was how to integrate their use with their UML-guided design and development practices. The issue of how to integrate model-based approaches to user interfaces with UML was discussed in some CHI workshops where participants agreed that it is conceptually possible to link development practice to HCI practice through the use of object models derived from task analysis and modelling [1].

M. Reed Little and L. Nigay (Eds.): EHCI 2001, LNCS 2254, pp. 7–18, 2001.

Last generation of model-based approaches to user interface design agree on the importance of task models. Such models describe the activities that should be performed in order to reach users' goals. Task models have shown to be useful in a number of phases of the development of interactive applications: requirements analysis, design of the user interface, usability evaluation, documentation and others. Tasks are important in designing interactive applications: when users approach a system, they first have in mind a set of activities to perform, and they want to understand as easily as possible how the available system supports such activities and how they have to manipulate the interaction and application objects in order to reach their goals. Likewise, when modelling human-computer interaction it is more immediate to start with identifying tasks and then the related objects. Conversely, when people design systems (with no attention to the human users) they often first think in terms of identifying objects and then try to understand how such objects interact with each other to support the requested functionality.

ConcurTaskTrees (CTT) [7] is a notation that has been developed taking into account previous experience in task modelling and adding new features in order to obtain an easy-to-use and powerful notation. It is a graphical notation where tasks are hierarchically structured and a rich set of operators describing the temporal relationships among tasks has been defined (also their formal semantics has been given). In addition, it allows designers to indicate a wide set of task attributes (some of them are optional), such as the category (how the task performance is allocated), the type, the objects manipulated, frequency, and time requested for performance. Each task can be associated with a goal, which is the result of its performance. Multiple tasks can support the same goal in different modalities. The notation is supported by CTTE (the ConcurTaskTrees Environment), a set of tools supporting editing and analysis of task models. It has been used in many countries in a number of university courses for teaching purposes and for research and development projects. It includes a simulator (also for cooperative models) and a number of features allowing designers to dynamically adjust and focus the view, particularly useful when analysing large specifications.

In the paper there is first a discussion of possible approaches to integrating task models and UML. Next, the approach selected is presented along with criteria to indicate how to use the information contained in use case diagrams to develop task models. For this part, examples taken from a case study concerning applications supporting help desks are given. Finally, some concluding remarks and indications for further work are presented.

## 2   Integrating UML and Task Models

Aiming at integrating the two approaches (task models represented in CTT and UML) there can be various basic philosophies that are outlined below. Such approaches can exploit, to different extents, the extensibility mechanisms built into UML itself (constraints, stereotypes and tagged values [9]) that enable extending UML without having to change the basic UML metamodel. The types of approaches are:

- *Representing elements and operators of a CTT model by an existing UML notation.* For example, considering a CTT model as a forest of task trees, where CTT operands are nodes and operators are horizontal directed arcs between sibling nodes, this can be represented as UML class diagrams. Specific UML class and association stereotypes, tagged values and constraints can be defined to factor out and represent properties and constraints of CTT elements [6];

- *Developing automatic converters from UML to task models,* for example in [5] there is a proposal to use the information contained in system-behaviour models supported by UML (use cases, use cases diagrams, interaction diagrams) to develop task models.
- *Building a new UML for interactive systems,* which can be obtained by explicitly inserting CTT in the set of available notations, still creating semantic mapping of CTT concepts into UML metamodel. This encompasses identifying correspondences, both at the conceptual and structure levels, between CCT elements and concepts and UML ones, and exploiting UML extensibility mechanisms to support this solution;

Of course there are advantages and disadvantages in each approach. In the first case, it would be possible to have a solution compliant with a standard that is already the result of many long discussions involving many people. This solution is surely feasible. An example is given in [6]: CTT diagrams are represented as stereotyped class diagrams; furthermore constraints associated with UML class and association stereotypes can be defined so to enforce the structural correctness of CTT models. However, I argue that the key issue is not only a matter of expressive power of notations but it is also a matter of representations that should be effective and support designers in their work rather than complicate it. The usability aspect is not only important for the final application, but also for the representations used in the design process. For example, UML state charts and their derivative, activity diagrams, are general and provide good expressive power to describe activities. However, they tend to provide lower level descriptions than those in task models and they require rather complicated expressions to represent task models describing flexible behaviours. In [10] there is another example of using existing UML notations for describing concepts important for designing interactive systems such as task and presentation models. Again, this proves its feasibility but the readability and comprehensibility of the resulting diagrams are still far from what user interface designers expect. Thorny problems emerge also from the second approach. It is difficult to model first a system in terms of objects behaviour and then derive from such models a meaningful task model because usually object-oriented approaches are effective to model internal system aspects but less adequate to capture users' activities and their interactions with the system. The third approach seems to be more promising in capturing the requirements for an environment supporting the design of interactive systems. However, attention should be paid so that software engineers who are familiar with traditional UML can make the transition easily. A possible variant to this solution is to use CTT as a macro-notation for an existing UML notation, such as activity diagrams. This would offer the advantages of being more comprehensible and better able to highlight the elements of interest for user interface designers, while still allowing the constructs to be mapped onto a standard UML notation. In particular, in the following I will outline a proposal aimed at obtaining a UML for interactive systems based on the integration of CTT and current UML.

## 3   The Approach Proposed

In our approach to UML for Interactive Systems the use of task models represented in CTT is explicitly introduced. The approach should also be supported by a defined process and a related tool for integrated use of CTT and the other notations useful for supporting design in this class of software applications.

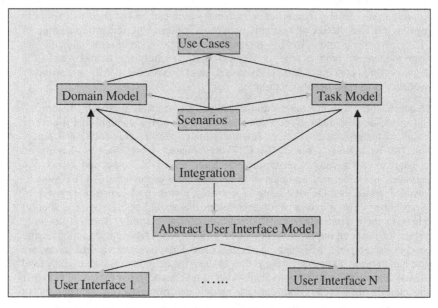

**Fig. 1.** The Representations Involved in the Approach Proposed

Not all UML notations are equally relevant to the design of interactive systems; the most important in this respect appear to be use case, class diagrams and sequence diagrams. In the initial part of the design process (see Figure 1), during the requirement elicitation phase, use cases supported by related diagrams should be used. Use cases are defined as coherent units of externally visible functionality provided by a system unit. Their purpose is to define a piece of coherent behaviour without revealing the internal structure of the system. They have shown to be successful in industrial practise.

Next, there is the task modelling phase that allows designers to obtain an integrated view of functional and interactional aspects. Interactional aspects, related to the ways to access system functionality, in particular cannot be captured well in use cases; so, in order to overcome this limitation they can be enriched with scenarios, informal descriptions of specific uses of the system considered. More user-related aspects can emerge during task modelling. In this phase, tasks should be refined, along with their temporal relationships and attributes. The support of graphically represented hierarchical structures, enriched by a powerful set of temporal operators, is particularly important. It reflects the logical approach of most designers, allows describing a rich set of possibilities, is declarative, and generates compact descriptions.

The decomposition of the task model can be carried on until basic tasks (tasks that involve no problem solving or control structure component) are reached. If this level of refinement is reached it means that the task model includes also the dialogue model, describing how user interaction with the system can be performed. If such a level of description is reached then it is possible to directly implement the task model in the final system to control the dynamic enabling and disabling of the interaction objects and to support features, such as context-dependent task-oriented help.

In parallel with the task modelling work, the domain modelling is also refined. The goal is to achieve a complete identification of the objects belonging to the domain considered and the relationships among them. At some point there is a need for integrating the information between the two models. Designers need to associate tasks with objects in order to indicate what objects should be manipulated to perform each task. This information can be directly introduced in the task model. In CTT it is possible to specify the references between tasks and objects. For each task, it is possible to indicate the related objects, including their classes and identifiers. However, in the domain model more elaborate relationships among the objects are identified (such as association, dependency, flow, generalisation, …) and they can be easily supported by UML class diagrams. There are two general kinds of objects that should be considered: the presentation objects, those composing the user interface, and application objects, derived from the domain analysis and responsible for the representation of persistent information, typically within a database or repository. These two kinds of objects interact with each other. Presentation objects are responsible for creating, modifying and rendering application objects. Given one application object there can be multiple presentation objects (for example, a temperature can be rendered by a bar chart or a numeric value). The refinement of task and objects can be performed in parallel, so that first the more abstract tasks and objects are identified and then we move on the corresponding more concrete tasks and objects. At some point the task and domain models should be integrated in order to clearly specify the tasks that access each object and, vice versa, the objects manipulated by each task.

Thus, we follow a kind of middle-out, model-first [11] approach, by which we identify elements of interest through the Use Cases and then use them to create more systematic models that allow clarifying a number of design issues and then to develop the resulting interactive software system.

Scenarios are a well known technique in the HCI field often used to improve understanding of an interactive applications. They provide informal descriptions of a specific use in a specific context of an application. CTTE, the tool supporting CTT, gives also the possibility of simulating the task model. This means that it is possible to interactively select one task and ask the system to show what tasks are enabled after the performance of that task, and then carry on this activity iteratively. A sequence of interactions with the simulator allows designers to identify an abstract scenario, a specific sequence of basic tasks that are associated with a specific use of the system considered. Figure 2 provides an example describing a scenario derived from a task model for a cooperative ERP application (in the *Scenario to be performed* sub-window). The scenario was created beforehand and now is loaded for interactive analysis within the task model simulator

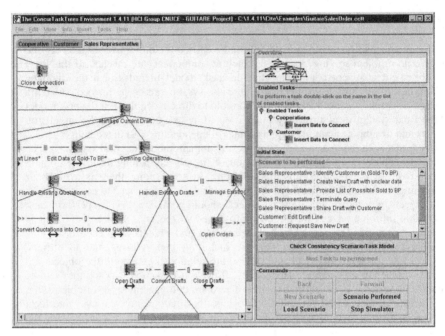

**Fig. 2.** Example of Abstract Scenario Derived from Task Model

A similar concept is supported by sequence diagrams in UML: two-dimensional charts where the vertical dimension is time, while the horizontal dimension shows the classifier roles representing individual objects in the collaboration. Such diagrams provide good representation of limited sequential activities. Their specificity is that they represent the messages sent with arrows, indicating the objects involved and the direction of data flow. In both cases, we can think about them as representations of scenarios that are used to validate the dynamic behaviour of both the domain and task models. In addition, informal scenarios can be used as a technique for preliminary identification of requirements, and thus can give useful input for the development of use cases that provide more structured descriptions.

Task models are useful to support the design of user interfaces. Deriving the concrete user interface can be looked at as a two-step process. First, an abstract user interface model should be identified: it gives a platform-independent description of how the user interface should be structured. Next, the concrete user interface is generated taking into account the specific platform considered (desktop, laptop, handheld device, ...). Note that this framework allows designers to support plasticity [12], the possibility of adapting to a specific device (although this is beyond the scope of this paper).

A number of criteria have been identified [7] to gather information from the task model that is useful to obtain effective user interfaces. The concept of enabled task sets is used to identify the presentations of the user interface. An enabled task set is a group of tasks enabled over the same period of time. Thus it is a good candidate to be supported by a single presentation (however, multiple enabled tasks sets that differ by only a few elements may be supported by a single presentation). Once we have identified the presentations, they can first be structured and then refined according to a number of criteria, for example:

- grouping of tasks can be reflected in the grouping of presentation objects;
- temporal relationships among tasks can be useful to structure the user interface dialogue and indicate when the interaction techniques supporting the tasks are enabled;
- the type of tasks and objects (and their cardinality) should be considered when the corresponding widgets and presentation techniques are selected.

The design process is iterative. So, when a user interface is drafted, then it is still possible to return to the task and domain models, modify them and start the cycle over again.

To summarise, the rationale for the approach proposed is that in current UML notations it is possible to describe activities by activity diagrams or state chart diagrams (which are specialised types of activity diagrams). However, these notations have mainly been conceived of to describe either high-level user work-flow activities, or low-level interactions among software objects. Although they can be used to describe task models or user interactions and some research work has been done in this direction, the results do not seem particularly convincing. Activity diagrams do not seem to provide good support for multiple abstraction levels, which has been found particularly useful in HCI (all successful approaches in task modelling, including HTA, GOMS, UAN share this feature). In addition, they do not seem to scale-up well: when medium-large examples are considered, the graphical representations obtained are often rather difficult to interpret. It is possible to define such diagrams hierarchically, but designers often have difficulties in following their relationships and obtaining an overall view of the specification.

## 4    Moving from Use Cases to Task Models

Use cases have been well accepted in the phase of requirement elicitation and specification. There is some confusion over exactly how to use and structure them (see for example [2] for a discussion of the many ways to interpret what a use case is). Here we will refer to UML use cases that are usually structured informal descriptions accompanied by use case diagrams. Thus, it becomes important to identify a set of criteria helping designers to use the information contained in use cases diagrams as a starting point to develop task models.

Use cases can give useful information for developing the CTT model in a number of ways. They identify the actors involved. Such actors can be mapped onto the roles of the CTT models. Indeed, CTT allows designers to specify the task model of cooperative applications by associating a task model for each role involved, and then a cooperative part describes how tasks performed by one user depend on the tasks performed by other users. In addition, a use case allows the identification of the main tasks that should be performed by users, the system or their interaction. This distinction in task allocation is explicitly represented in CTT by using different icons. Also abstract objects can be identified from an analysis of use cases: during use case elicitation, a number of domain objects are also identified, and collected in term of UML classes and relationships; this enables the incremental construction of the core of the domain model which is a principal result of analysis activities.

In use case diagrams there are actors, use cases and relationships among use cases. Actors can be associated with the roles of the task model and use cases can be considered as high-level tasks. In the use case diagram it is indicated to what use cases actors are involved. This means that they will appear in the corresponding task model. In addition, use cases that involve multiple actors must be included in the cooperative part of the task model. The relationships among use cases are: association (between an actor and a use case), extension (to insert additional behaviour into a base use case that does not know it), generalisation (between a general use case and a more specific one), inclusion (insertion of additional behaviour into a base use case that explicitly describes the insertion).

It is possible to identify criteria to indicate how to consider these operators in the task model. For example, the inclusion relationship can be easily mapped onto the decomposition relation supported by hierarchical tasks, while the generalisation relationship requires both task decomposition and a choice operator between the more specific alternatives. The extension relationship can be expressed as either an optional task or a disabling task performed when some particular condition occurs. These criteria should be interpreted as suggestions rather than formal rules, still giving the possibility of tailoring them to the specific case considered. In Figure 3 there is an example of Use case concerning a case study described in [4]. It is an abstract use case (HD Call) whose goal is to identify and enable access to help desk (HD) services to registered users. It is easy to see that a number of roles are involved: the help desk user, the help desk operator, the expert and the support manager. The relationships used in this diagram are: associations between actors and use cases (for example the File HDU Feedback is associated with the HD User actor); extend (for example, Abort HD Call extend HD Call); include (for example HD Call includes Start HD Call, it is possible also to indicate one-to-many relationships, such as in the case of HD Call and HD Interactions).

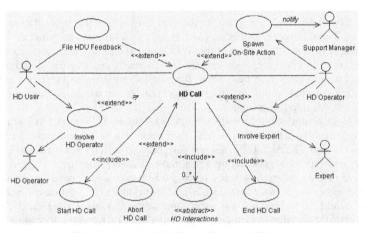

**Fig. 3.** Example of multi-actor Use case Diagram

The use cases included in the above diagram can be further expanded through specific diagram associated with each of them. It is possible then to group all the use cases that involve a certain actor. An example is provided in Figure 4, where the help desk user is considered. We can also find an example of use case generalisation (Identify HD Issue is a generalisation of Init New HDU Issue and Resume HDU Issue). This type of representation (a use case diagram describing all the use cases involving a certain role) can be used as starting point for the development of the corresponding task model.

Figure 5 shows the task model corresponding to our example. A part from the root task (Access HD), this task model just associates a task with each use case. However, the task model already adds further information to the use case diagram because the temporal relationships among the activities are indicated. The Abort HD Call task can interrupt the HD Call task ([> is the disabling operator) at any time. The HD Call is decomposed into a number of tasks: we first have the Start HD Call task, which is further decomposed into Identify HD User followed by Identify HD Issue (>> is sequential operator), and then a set of concurrent cooperating tasks (|[]| is the concurrent with information exchange operator). Then it is possible to perform either Suspend or Close the HDU Issue ([] is the choice operator) and lastly, after the optional performance of File HDU Feedback (optional tasks have the name in squared brackets), the HD Call ends. The model in Figure 5 concerns high-level tasks and represents a good starting point for the design of the user interface.

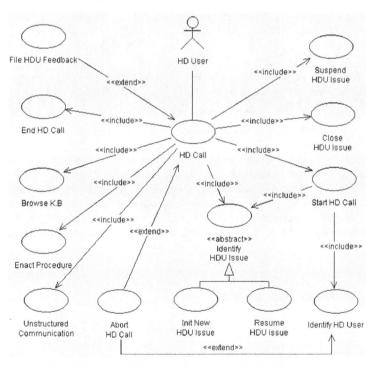

**Fig. 4.** Example of Use case Diagram associated with an Actor

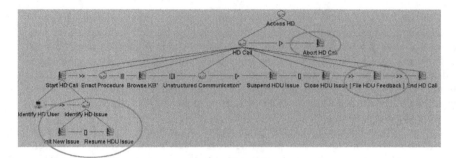

**Fig. 5.** Example of Task Model corresponding to the previous Use Case Diagram

Figure 6 provides a more readable representation of the task model by introducing high-level tasks grouping strictly related sub-tasks that share some temporal constraints (HD Call Session, HD Interactions and Dispose HDU Issue). Thus, we obtain a more immediate representation of the various phases of a HD Call.

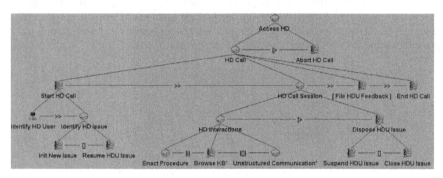

**Fig. 6.** Structured Representation of the Example of Task Model

It is clear that the model obtained at this point is not yet the complete task model. Some tasks need further refinements and should be structured more precisely (for example, by indicating further sub-tasks). In addition, during the final refinement more attention should be paid to designing how activities are supported with the support of the user interface. On the other hand, we have seen how it is possible to obtain a preliminary structure of the task model satisfying the requirements indicated in the use cases.

## 5  Conclusions and Discussion

This paper has discussed an approach to integrating UML and CTT for designing interactive systems based on the use of use cases, Class Diagrams, CTT task models and scenarios. A successful modelling environment for interactive systems must account for the importance of task models and the notations that have proved to be effective in representing them. The logical framework that has been presented gives indications of the types of representations to use in the various phases and their relationships. It supplies clear suggestions on how to leverage recent research work on model-based design of user interfaces to improve UML.

Automatic support for the new method introduced in this paper seems to be a reasonable goal. To this end, an editor of uses cases has been added to the CTTE tool. At this time, this is the most engineered tool for task modelling and analysis and it is publicly available (http://giove.cnuce.cnr.it/ctte.html).

## Acknowledgments

I wish to thank Nando Gallo, Vincenzo Benvenuto, Giovanni Radice (Intecs Sistemi S.p.A.) for useful discussions on the topics of the paper. GUITARE is a R&D Project funded by the European Commission. Support is gratefully acknowledged. More information is available at http://giove.cnuce.cnr.it/guitare.html

# References

1. Artim, J., et al, Incorporating Work, Process and Task Analysis Into Commercial and Industrial Object-Oriented System Development, SIGCHI Bulletin, 30(4), 1998.
2. Cockburn A., Structuring Use cases with Goals, Journal of Object-Oriented Programming, http://members.aol.com/acockburn, 1997.
3. Foley, J., Sukaviriya, N., History, Results, and Bibliography of the User Interface Design Environment (UIDE), an Early Model-based System for User Interface Design and Development, in F. Paterno' (ed.) Interactive Systems: Design, Specification, Verification, pp. 3-14 Springer Verlag, 1994.
4. F.Gallo, A.Paladino, E.Benvenuto, Application Domain Analysis and Requirements for Pilot Applications, GUITARE Deliverable 3.1, 2000.
5. S.Lu, C.Paris, K.Linden, Toward the Automatic Construction of Task Models from Object-Oriented Diagrams, Proceedings EHCI'98, Kluwer Academic Publishers, pp.169-189
6. N.Numes, J.Falcao, Towards a UML profile for user interface development: the Wisdom approach, Proceedings UML'2000, LNCS, Springer Verlag.
7. Paternò, F., Model-Based Design and Evaluation of Interactive Application. http://giove.cnuce.cnr.it/~fabio/mbde.html Springer Verlag, ISBN 1-85233-155-0, 1999.
8. Puerta, A., A Model-Based Interface Development Environment, IEEE Software, pp. 40-47, July/August 1997.
9. J.Rumbaugh, I.Jacobson, G.Booch, The Unified Modeling Language Reference Manual, Addison-Wesley, 1999.
10. P. da Silva, N.Paton, UMLi : The Unified Modeling Language for Interactive Applications, Proceedings UML'2000, LNCS, Springer Verlag
11. M.B. Rosson, Designing Object-Oriented User Interfaces from Usage Scenarios, ACM CHI'97 workshop on Object Models in User Interface Design, http://www.cutsys.com/CHI97/Rosson.html
12. Thevenin D., Coutaz J., Plasticiy of User Interfaces: Framework and Research Agenda, Proceedings INTERACT'99, pp. 110-117, IOS Press, 1999.

## Discussion

*P. Smith:* Why do you say to software developers that believe that use cases + prototypes are sufficient and that task modeling is unnecessary?
*F. Paterno:* These people are making a lot of design decisions that are not captured well in use cases and prototypes alone.

*N. Graham:* I didn't understand how you represent temporal relations in UML
*F. Paterno:* You could use sequence diagram or activity diagrams for instance but we prefer to represent that using the CTT notation.

# An Interdisciplinary Approach for Successfully Integrating Human-Centered Design Methods into Development Processes Practiced by Industrial Software Development Organizations

Eduard Metzker and Michael Offergeld

DaimlerChrysler Research and Technology Center Ulm, Software Technology Lab,
P.O. Box 2360,
D-89013 Ulm, Germany
{eduard.metzker, michael.offergeld}@daimlerchrysler.com

**Abstract.** In a world where competitors are just a mouse-click away, human-centered design (HCD) methods change from a last minute add-on to a vital part of the software development lifecycle. However, case studies indicate that existing process models for HCD are not prepared to cope with the organizational obstacles typically encountered during the introduction and establishment of HCD methods in industrial software development organizations. Knowledge about exactly how to most efficiently and smoothly integrate HCD methods into development processes practiced by software development organizations is still not available. To bridge this gap, we present the experience-based human-centered design lifecycle, an interdisciplinary effort of experts in the fields of software engineering, human-computer interaction, and process improvement. Our approach aims at supporting the introduction, establishment and continuous improvement of HCD processes in software development organizations. The approach comprises a process model, tools, and organizational measures that promote the utilization of HCD methods in otherwise technology-centered development processes and facilitate organizational learning in HCD. We present results of a case study where our approach has been successfully applied in a major industrial software development project.

## 1 Introduction

The relevance of usability as a software quality factor is continually increasing within the industrial software development community. One fact contributing to this trend is the bringing into being of new regulations and standards for software usability that are increasingly influencing national legislation, such as the *EC Directive about Safety and Health of VDU Work* [1, 2]. Contractors tend to demand compliance with these regulations, thus forcing development organizations to adopt human-centered design (HCD) approaches in order to achieve defined usability goals. Another point is the still growing market for commercial web-based services such as on-line banking applications or e-commerce portals: The success of these systems depends even more on usability and user satisfaction than that of traditional desktop applications since

M. Reed Little and L. Nigay (Eds.): EHCI 2001, LNCS 2254, pp. 19–33, 2001.
© Springer-Verlag Berlin Heidelberg 2001

their easy availability via WWW (world wide web) enforces competition between their provider organizations [3]. The very same effects will carry over to the mobile, proactive services embedded in the information appliances of the near future [4].

It is well accepted both among software practitioners and in the human-computer interaction research community that structured approaches and explicit HCD engineering processes are required to build systems with high usability [5-11]. However, a closer look at documented case studies of existing HCD process models shows that these approaches do not provide solutions for the severe organizational obstacles that are encountered in establishing HCD methods in the development processes practiced by industrial software development organizations (see below).

To overcome these shortcomings we developed the experience-based HCD lifecycle, a result of a concerted effort of practitioners and researchers in the fields of software engineering, human computer interaction and process improvement. Our approach is designed to improve the utilization of HCD methods in standard software development processes and facilitate organizational learning concerning HCD activities. We explain the experience-based HCD lifecycle by describing the respective process models, tools and organizational measures that constitute our approach. We describe how our approach was verified during a long-time evaluation where it was successfully applied to improve the development process of an interactive software system. Finally we present the results of this evaluation and the lessons learned.

## 2    Existing Process Models for Human-Centered Design

There is a large body of research and practical experience available on software process models which are used to describe and manage the development process of software systems. Prominent examples are the *waterfall model* [12], the *spiral model* [13], or the *fountain model* [14]. Yet for the shortcomings of traditional process models concerning usability issues, a number of approaches have been developed that take into account the special problems encountered with the development of highly interactive systems [6, 10, 11, 15, 16]. According to ISO13407 these approaches can be embraced by the term 'human-centered design processes' [17]. In this section we focus on those approaches which have been extensively applied to industrial software development projects. We outline their basic principles and discuss some of their drawbacks based on documented case studies.

One of the first approaches used to address usability issues was the soft system methodology (SSM) [15, 18]. SSM was widely applied to capture the objectives, people involved (e.g. stakeholder, actors, and clients), constraints, and different views of interactive systems during development. However, since SSM's origins are in general systems theory, rather than computer science, it lacks many of the specific HCD activities such as construction of user interface mockups or iterative usability testing which are necessary to fully specify interactive systems. These shortcomings limit the utilization of SSM to the early activities of the development process such as requirements or task analysis.

The star lifecycle [16], proposed by Hix and Hartson, focuses on usability evaluation as the central process activity. Around this central task the, activities system / task / functional / user analysis, requirements / usability specifications, design & design representation, rapid prototyping, software production and deployment are placed. The results of each activity such as task analysis are subjected to an evaluation before going on to the next process activity. The bi-directional links between the central usability evaluation task and all other process activities cause the graphical representation of the model to look like a star.

One problem concerning this approach was already outlined by Hix and Hartson [16]: Project managers tend to have problems with the highly iterative nature of the model. They find it difficult to decide when a specific iteration is completed, complicating the management of resources and limiting their ability to control the overall progress of the development process. Furthermore, the star lifecycle addresses only the interactive parts of a software system, leaving open how to integrate the star lifecycle with a general software development method.

The usability engineering lifecycle [10] is an attempt to redesign the whole software development process around usability engineering knowledge, methods, and activities. This process starts with a structured requirements analysis concerning usability issues. The data gathered from the requirements analysis is used to define explicit, measurable usability goals of the proposed system. The usability engineering lifecycle focuses on accomplishing the defined usability goals using an iteration of usability engineering methods such as conceptual model design, user interface mockups, prototyping and usability testing [6]. The iterative process is finished if the usability goals have been met.

As outlined by Mayhew [10], the usability engineering lifecycle has been successfully applied throughout various projects. However, some general drawbacks have been discovered by Mayhew during these case studies: One important concern is that redesigning the whole development process around usability issues often poses a problem regarding the organizational culture of software development organizations. The well established development processes of an organization can not be turned into human-centered processes during a single project. Furthermore, the knowledge necessary to perform the HCD activities is often missing in the development teams, hampering the persistent establishment of HCD activities within the practiced development processes.    How the HCD activities proposed in the usability engineering lifecycle should be integrated exactly and smoothly into development processes practiced by software development organizations, was declared by Mayhew as an open research issue [10].

Usage-centered design [11], developed by Constantine and Lockwood, is based on a process model called activity model for usage-centered design. The activity model describes a concurrent HCD process starting with the activities of collaborative requirements modeling, task modeling, and domain modeling, in order to elicit basic requirements of the planned software system. The requirements analysis phase is followed by the design activities: interface content modeling and implementation modeling. These activities are continuously repeated until the system passes the usability inspections carried out after each iteration. The design and test activities are paralleled by help system / documentation development and standards / style definition for the proposed system. This general framework of activities is supplemented by special methods like essential use case models or user role maps.

Constantine and Lockwood provide many case studies where usage centered design was successfully applied, yet they basically encountered the same organizational obstacles as Mayhew [10] when introducing their HCD approach into software development processes practiced. They emphasize the fact that 'new practices, processes, and tools have to be introduced into the organization and then spread beyond the point of introduction' [11]. A straightforward solution to these problems is training courses for all participants of HCD activities offered by external consultants. However, this solution is regarded as being time consuming and cost intensive in the long run. It tends to have only a limited temporary effect and thus does not promote organizational learning on HCD design methods [11]. Constantine and Lockwood conclude that it is necessary to build up an internal body of knowledge concerning HCD methods, best practices and tools tailored to the needs of the development organization.

As outlined above, the approaches described have some limitations in common which hamper their effective application and establishment in an industrial setting. These shortcomings can be summarized as follows:

- Existing HCD processes are decoupled from the overall system development process.

One common concern relating to these approaches is that they are regarded by software project managers as being somehow decoupled from the software development process practiced by the development teams. It appears to project managers that they have to control two separate processes: the overall system development process and the HCD process for the interactive components. As it remains unclear how to integrate and manage both perspectives, the HCD activities have often been regarded as dispensable and have been skipped in case of tight schedules [19].

- Existing HCD process models are not suitable for tailoring.

Another point that is also ignored by the approaches described is that development organizations are often overwhelmed by the sheer complexity of the proposed HCD process models. The models lack a defined procedure for tailoring the development process and methods for specific project constraints as system domain, team size, experience of the development team or the system development process already practiced by the organization.

- Existing HCD process models assume that HCD methods can be performed ad hoc.

Most approaches also assume that experienced human factors specialists are available throughout the development team and that HCD methods can be performed ad hoc. However, recent research shows that even highly interactive systems are often developed without the help of in-house human factors specialists or external usability consultants [20]. Therefore HCD methods often can not be utilized because the necessary knowledge is not available within the development teams [19, 21].

- Existing HCD process models are not prepared for continuous process improvement.

Almost all approaches do not account for the fact that turning technology-centered development processes into human-centered development processes must be seen as a continuos process improvement task [4]. A strategy for supporting a long-lasting establishment of HCD knowledge, methods, and tools within development organizations is still missing.

# 3   The Experience-Based Human-Centered Design Lifecycle

The experience-based human-centered design lifecycle addresses the shortcomings of the existing HCD process models described above. It aims at promoting the utilization of HCD activities and methods in development processes practiced by software development organizations. It supports the introduction, establishment, and continuous improvement of HCD activities, methods, and tools within software development processes. Our approach is based on two concepts: An *HCD reference model* that contains specific activities for each logical phase of the development process and the *introduction-establishment-improvement (IEI)* model for HCD activities which guides the tailoring and smooth integration of HCD methods in mainstream software development processes.

## 3.1   The HCD Reference Model

The reference model contains HCD activities to be integrated into software development processes for interactive systems. It is organized into virtual process phases based on the framework of the *usability maturity model* (UMM) defined in ISO TR 18529 [22] which is based on ISO 13407 [17] and the Capability Maturity Model (CMM) [33]. The reference model contains various HCD activities from different human-centered design approaches. Currently the model comprises activities adopted from Nielsen [6] and Mayhew [10] as depicted in Fig.1. The shaded blocks represent the UMM framework while the plain blocks represent the respective activities of our reference model[1].

The set of  methods collected in the reference model is continually revised and extended with new methods by using the model in real projects and collecting data on its efficiency. The aim of the reference model is to compile a collection of well-accepted HCD activities in one framework which have been carefully verified for practicability and efficiency together with context information (process context, project context, technology context, domain context, and quality context) that describes in which development context the respective activities are most efficient.

## 3.2   The Introduction-Establishment-Improvement Model for HCD

The set of activities of the reference model must be supplemented with organizational tasks which help to manage and tailor the HCD activities of the reference model according to specific constraints of the respective project and the needs of the development organization. Our introduction-establishment-improvement (IEI) model for human-centered design provides process steps that support the introduction,

---

[1]The graphical representation of the reference model is adopted from the format of ISOTR18529 – 'Human-centred lifecycle process descriptions'. Whilst it is possible to draw a number of simple diagrams which demonstrate the iterative nature of the human- centred lifecycle there are many different versions of lifecycles, depending on the development context of the system planned. It is therefore difficult and may even be confusing to draw one simple diagram which demonstrates how processes are linked.

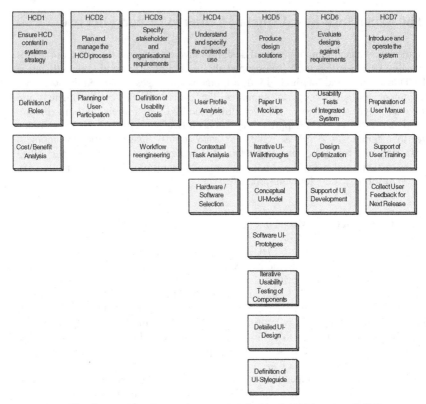

**Fig. 1.** Mapping of the Reference Model to the UMM framework [22]

establishment and continuous improvement of HCD activities throughout the whole development lifecycle of highly interactive systems. These organizational tasks are grouped in the IEI model as depicted in Fig. 2.

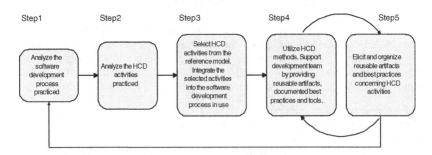

**Fig. 2.** The IEI model

The IEI model consists of the following steps:

*Step 1: Analyze the software development process practiced*
At the beginning, a short analysis of the software lifecycle practiced is performed. The results of this step are, first, a rough description of the process model that is used by the software development organization, and, second, a rough description of its main phases.

*Step 2: Analyze the HCD activities practiced*
The next step comprises an analysis of the practiced HCD process and the related HCD activities, to elicit when, where and how HCD methods are performed within the software development lifecycle in use. The deliverable of this step is a documentation of possible improvements of the HCD process that is currently used.

*Step 3: Select suitable HCD tasks and integrate them into the practiced software development process*
The results of the first steps form the rationale for the selection of HCD activities from the reference model for the improvement of the development process. However, in this step further important factors have to be considered, e.g. the type of system to be developed and  project constraints like budget and schedules.
The HCD activities which have been selected for the improvement of development process have to be integrated in the model of the practiced software development lifecycle and the project planning and form the improved development process, supplemented with appropriate HCD activities.

*Step 4: Support effective performance of the defined HCD activities*
Generally, at this step in the model resources have already been allocated for HCD activities, e.g., a usability engineer was nominated, who is responsible for coordinating and supporting the execution of the various HCD activities of the new process. However, the efficiency and impact of  the proposed HCD methods must be increased by providing information, best practices, tools and reusable process deliverables of past projects (e.g. templates for usability test questionnaires, results of conceptual task analysis or user interface mockups) which facilitate effective performance of the selected HCD activities. This set of information should be easily accessible for all participants of HCD activities.

*Step 5: Collect and disseminate best practices and artifacts concerning HCD tasks*
During the execution of HCD activities, artifacts with a high value for reuse in the same or subsequent projects are generated by the participants of HCD activities, for example, templates for usability tests, reusable code fragments, or best practices on how to most efficiently conduct a user profile analysis. Furthermore observations of gaps in the HCD process should be documented and used as an input for the next improvement cycle. These best practices and artifacts comprise HCD experience and rationale that have to be captured and organized in a way that allows for easy reuse in the same or subsequent projects.

The IEI model contains two cycles: The inner cycle between step 4 and 5 supports the introduction and establishment of HCD activities and methods within the practiced software development process. It supports the effective utilization and improvement of HCD activities and best practices which are tailored to the needs of the development organization. This cycle is continuously iterated during the development process.

The outer cycle which connects step 5 and 1 should be performed in the ideal case at least twice during the development process of large projects as it serves the improvement of the overall HCD process. In this cycle observations and best practices collected in step 5 should be used to integrate needed HCD activities in the process and remove activities that have failed to prove their utility.

## 4  Tool Support

The literature often maintains that there is a lack of tool support for HCD processes [6, 10, 19, 21, 23]. Indeed, few mature tools are commercially available most of them for activities of the late phases of the development lifecycle, e.g. (semi-) automatic GUI (graphical user interface) builders [24-28] or usability test tools [23, 29, 30]. With the increasing relevance of usability as a software quality factor, there will be linked a growing demand for tools which support HCD activities within the industrial software development community. These tools should help the participants of the HCD lifecycle to execute their tasks more effectively.

One important class of HCD tools are systems that support developers by providing the knowledge to perform the various HCD activities and to organize the outputs of iterative design processes [10, 19, 21]. For the lack of available commercial tools, we developed a prototypical tool, the so-called *MMI (Man-Machine-Interaction) hyperbase*. The development and introduction of the system was prepared by a requirements analysis conducted with eighteen professional software developers and six project team leaders [31].

The tool summarizes and presents ergonomic design knowledge and HCD methods for the entire HCD lifecycle using a knowledge base and makes the knowledge available via intranet and web-browser. The ergonomic design knowledge and the HCD methods have been attached to the different HCD activities of our reference model, e.g. 'definition of usability goals' or 'user profile analysis'. The MMI hyperbase supports a software development organization in defining the roles of the project team members responsible for HCD and helps to select the right methods and tools for each development step. Furthermore the MMI hyperbase supports the cooperation between the different participants of the HCD lifecycle with the help of a groupware tool. It enables developers to bundle best practices learned during the performance of HCD activities together with reusable artifacts and attach them to the corresponding activities of the reference model. The aim is to reuse HCD rationale, support organizational learning concerning HCD and to improve the HCD process deployed, and thus supports our experience-based approach outlined in the IEI model.

# 5   Employment of the Experience-Based HCD Lifecycle in an Industrial Setting

In the remainder of this paper we illustrate the application of the methods and tools described above on the basis of a military software development project at DASA[2], called AP3In this project we have supported the introduction, establishment and improvement of HCD activities within the software development lifecycle. The overall aim of our activities in AP3 was to increase the productivity of the user interface development process and to improve the quality-in-use of the system developed. In the following section we briefly describe how each step of the IEI model was applied in AP3:

*Step 1: Analyze the practiced software development process*
The analysis of the practiced software development lifecycle to be used in AP3 showed that the development department used a subset of the V-model [32] as a process model for the development of their software systems. The V-model describes the activities which have to be performed during system development and the deliverables - mainly documents - which are to be created during these activities. The software process model to be used by the development team comprised the following five logical phases: project preparation, requirements analysis, system design, system development and installation/transition into use.

*Step 2: Analyze the practiced HCD process*
The analysis of the V-model with respect to HCD activities showed that there is no explicit HCD process anchored as a sub-process in the V-model. Therefore, HCD activities and their relation to the system development phases had to be defined for all phases of the development process.

*Step 3: Select suitable HCD tasks and integrate them into the practiced software development process*
Based on the results of the first two steps, the type of product to be developed and the project constraints, a set of appropriate HCD activities was selected from the reference model. Some further constraints influenced the selection of HCD activities: Though an old system existed, totally new work flows were to be modeled for the new system, thus making it necessary to integrate the workflow reengineering activity into the development process. Furthermore the proposed system had to comply with legal regulations concerning usability [1], which had to be taken into account in the process activities *definition of usability goals* and *usability testing*.

The HCD activities which have been selected from the reference model were integrated into the respective phases of the practiced development process as elicited in step 1 and formed the process model for the performance of HCD activities in the AP3 project.

The resulting process model is depicted in Fig. 3. The process phases of the overall development process which have been elicited in step 1 are represented as shaded

---

[2]DaimlerChrysler Aerospace AG (DASA) is now partner of the transnational European Aeronautic Defence and Space Company (EADS)

boxes. The plain boxes represent the HCD activities that have been selected from the reference model to supplement the development process. The phases must be understood as logical phases. This means that they do not have to be performed in strict order for the whole system like in the waterfall model, but that they can be iteratively applied to sub-components. However, for reasons of clarity and readability we forgo the drawing of iterative loops between the logical process steps that occur in every real project. The resulting process model served as a basis for allocating human resources and project schedules in AP3.

*Step 4: Support effective performance of the defined HCD activities*
The setup and execution of the HCD activities was coordinated by a usability engineer who was released from other project activities. The impact of the usability engineer was enhanced by the MMI hyperbase tool and the knowledge base included. With this intranet-based tool, the developers had access to practical descriptions of the various tasks of the HCD process set up in step 3, as well as to auxiliary means, e.g. templates, checklists, existing components or documented best practices, at all times - which made the efficient execution of the usability activities possible.

*Step 5: Collect and disseminate best practices and artifacts concerning HCD activities*
During the execution of the various HCD activities, some promising best practices were created. For example streamlined methods for user profile analysis and contextual task analysis together with checklists were developed to facilitate efficient performance of respective HCD activities.

Furthermore, a simple yet effective graphical notation for user-relevant system functionality and work flows was developed during the workflow reengineering sessions. Based on use cases and flow diagrams this notation was both understood and accepted by the contractors, users and software developers thus supporting the communication during the workflow reengineering activity.

These best practices have been documented in the knowledge base of the MMI hyperbase together with information like the HCD activity to which they are related, the role for which they are most useful (e.g. usability engineer, user interface (UI) designer, UI developer ).

Based on the experiences collected in the inner cycle of the model, by iterating steps 4 and 5, we went back to step 2 to improve the practiced HCD process. The improved HCD process was used in the remainder of the AP3 project.

# 6  Lessons Learned

The best methods and tools for developing usable systems cannot achieve full impact if they are not supported by suitable organizational measures. Here are some of the simple measures which contributed to the successful application of usability engineering methods in the AP3 project:

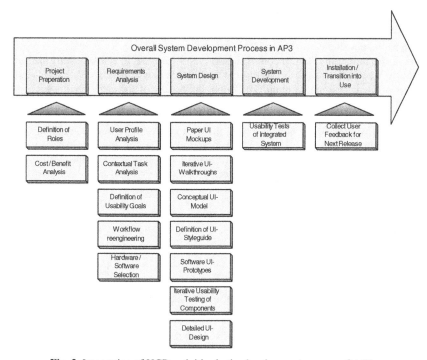

**Fig. 3.** Integration of HCD activities in the development process of AP3

- Nominate personnel which is responsible for HCD during the whole project

  For the successful introduction and establishment of HCD activities it proved to be mandatory to have usability experts involved in both the development organization and the customer organization. In the development organization a usability engineer should be involved right from the beginning of the project and should be responsible for the management of all HCD activities during the whole system development process. In the AP3 project for example the usability engineer was explicitly declared as a member of the core engineering team and was also involved for the requirements engineering phase. On the customer side a usability agent was supervising the system's compliance with usability requirements of the customer organization. In AP3 one usability engineer and one usability agent were nominated who were intensively cooperating.

- Perform regular usability audits with stakeholders

  From our point of view HCD is also always a process of negotiation between different stakeholders, e.g. users, contractor, developers, management and goals of one group often clash with the interests of others. To support communication and facilitate a maximum degree of agreement between these groups on the most important design decisions, usability audits were performed throughout the development process. Each of this audits required at least two days of collaborative work and should be carefully prepared including the needed documents and tools.

During the AP3 project a total number of five usability audits were performed within three years.

- Support the development team with information

One of the most important recommendation is to compile all relevant information and data which is produced during the HCD activities and make it available to all members of the development team. The important results of the early phases such as user profiles, user tasks, or usability goals have to be considered during the whole development process and are vital for the late HCD activities such as testing for compliance with usability goals, designing the user manual by describing major user tasks or supporting user training by explaining and practicing the user tasks at the target system. It has proved to be useful to make the deliverables of the performed usability engineering tasks available in a central repository to all participants of the usability engineering process. In AP3 we used the MMI hyperbase as an HCD support tool.

# 7   Results

There are some important findings to our HCD approach. In comparison to other similar projects in our organization during the last years we identified a lot of different benefits for the AP3 project which was driven with our approach:

## 7.1   General Advantages and Savings

In contrast to previous projects the interaction design and user interfaces were available very early in AP3. Both sides got an early impression of the system's look and feel. The customers felt that they get a system which corresponds to the needs of their users. The suppliers got the certainty that they will deliver a system which satisfies the users. In AP3 the final electronic user interfaces were completed after 60% of the overall development time. In former projects with the same settings usability problems resulted in massive exceeding of deadlines.

## 7.2   Advantages Related to the Experience-Based HCD Lifecycle

One of the most valuable advantages for the development team was that developers and software engineers learned to think in categories of user tasks and not only in terms of system functionality.

In former projects developers only got lists of functions and often did not understand the defects which were communicated to them by the customer. This resulted in user interfaces that rather depicted a collection of single functions than a representation of the users tasks. It was annoying for developers to follow the change orders of the customer.

From the very beginning of the AP3 project the software developers got a deep insight into the user tasks since they participated in performing the contextual task analysis together with the user representatives. Due to this experience they were highly motivated to implement components with a high usability. They were able to understand and solve the problems encountered during usability tests since they knew the user tasks and they realized a mismatch between those tasks and the prototype. Furthermore the whole AP3 project profited from the elicitation and reuse of best practices and artifacts as described above. The development team adapted known HCD activities and methods and streamlined them to fit the development process in used, thus continuously improving the HCD activities practiced.

### 7.3  Synergy Effects Related to Tool Support for HCD Processes

There were immense cost savings since a lot of the material which was produced during the HCD process could be reused after delivery of the system in many ways. One example was the user training material. After delivery the customer organization normally defines an own training team which is responsible for the training of new users. This team got an excellent training material including all the descriptions of user tasks, usability goals, style guide rules, and tests cases for the user interfaces that was compiled using the MMI hyperbase tool. The effort for the training of new users was clearly reduced.

## 8   Conclusions

We presented an experience-based approach to HCD which was designed to overcome the severe organizational obstacles encountered during the application of related approaches. With the help of a case study we showed how to employ our approach in industrial software development organizations and presented organizational measures which contributed to the success of the entire approach. We presented promising results of the case study that showed that applying our approach can facilitate a more effective utilization of HCD methods within industrial software development settings and support the introduction, establishment and continuous improvement of HCD activities in software development organizations.

## References

1.   EC Directive 90/270/EWG: Human-Computer Interfaces, Directive about Safety and Health of VDU-Work (5th Directive Art. 16 Par. 1 of Dir. 89/391/EWG). Official EC-Newsletter, Vol.33, L 156, Minimal Standards, Par.3 (21.6.1990) 18,, 1990.
2.   Verordnung über Sicherheit und Gesundheitsschutz bei der Arbeit an Bildschirmgeräten, Bildschirmarbeitsverordnung - BildscharbV, BGBl. I 1996 pp. 1841, 1996.
3.   J. Nielsen,  Designing Web Usability : The Practice of Simplicity: New Riders Publishing, 2000.
4.   D. A. Norman, »Chapter 10: Want Human-Centered Design? Reorganize the Company.," in The Invisible Computer: MIT Press, 1998.
5.   J. D. Gould and C. Lewis, »Designing for Usability: Key Principles and What Designers Think," Communications of the ACM, vol. 28, pp. 360-411, 1985.
6.   J. Nielsen, Usability Engineering: Morgan Kaufman Publishers, 1994.

7.   J. Karat and T. Dayton, »Practical   Education for Improving Software Usability," presented at Conference on Human Factors in Computing Systems: CHI'95, 1995.
8.   D.R. Wixon J. Ramey (eds.), Field Methods Casebook for Software Design. New York: John Wiley & Sons, 1996.
9.   H. Beyer and K. Holtzblatt, Contextual Design: Defining Customer-Centered Systems: Morgan Kaufmann, 1998.
10.  D. J. Mayhew, The Usability Engineering Lifecycle: A Practioner's Handbook for User Interface Design: Morgan Kaufman, 1999.
11.  L. L. Constantine and L. A. D. Lockwood, Software for Use: A Practical Guide to the Models and Methods of Usage-Centered Design: Addison-Wesley, 1999.
12   W. W. Royce, »Managing the development of large software systems," presented at IEEE WESTCON, San Francisco, USA, 1970.
13.  B. W. Boehm, »A spiral model of software development and enhancement.," IEEE Computer, vol. 21, pp. 61-72, 1988.
14.  B. Henderson-Sellers and J. M. Edwards, »Object-Oriented Systems Life Cycle," Communications of the ACM, vol. 31, pp. 143-159, 1990.
15.  P. B. Checkland, Systems Thinking, Systems Practice: John Wiles & Sons, 1981.
16.  D. Hix and H. R. Hartson, »Iterative, Evaluation-Centered User Interaction Development," in Developing User Interfaces: Ensuring Usability Through Product & Process. New York: John Wiley & Sons, 1993, pp. 95-116.
17.  ISO/TC 159 Ergonomics, »Human-centered Design Processes for Interactive Systems," ISO International Organization for Standardization ISO 13407:1999(E), 1999.
18.  P. B. Checkland and J. Scholes, Soft Systems Methodology in Action: John Wiley & Sons, 1990.
19.  S. Rosenbaum et al., »What makes Strategic Usability Fail? Lessons Learned From the Field.," presented at International Conference on Human Factors in Computing Systems, CHI99, Pittsburgh, PA USA, 1999.
20.  E. Wetzenstein, A. Becker, R. Oed, »Which Support do Developers Demand During the Development of User Interfaces" (in german), to be presented at the 1st interdiciplinary Conference on Human-Computer Issues, M&C2001, Bad Honnef, Germany, 2001.
21.  H. Reiterer, »Tools for Working with Guidelines in Different User Interface Design Approaches," presented at the Annual Workshop of the Special Interest Group on Tools for Working with Guidelines, Biarritz, France, 2000.
22.  ISO/TC 159 Ergonomics, »ISO TR 18529 - The Usability Maturity Model,".
23.  G. Al-Quaimari and D. McRostie, »KALDI: A Computer-Aided Usability Engineering Tool for Supporting Testing and Analysis of Human-Computer Interaction," presented at CADUI99: Computer Aided Design of User Interfaces, Louvain-la-Neuve Belgium, 1999.
24.  Foley et al., »UIDE - An Intelligent User Interface Design Environment," Intelligent User Interfaces, pp. 339-384, 1991.
25.  P. Szekley et. al., »Facilitating the Exploration of Interface Design Alternatives: The HUMANOID Model of Interface Design," presented at Conference on Human Factors in Computing Systems, CHI'92, 1992.
26.  Balzert et.al., »The JANUS Application Development Environment - Generating more than the User Interface," presented at Computer Aided Design of User Interfaces, CADUI'96, 1996.
27.  A.R.Puerta, »A Model Based Interface Development Environment," IEEE Software, vol. 14, pp. 41-47, 1997.
28.  J. Vanderdonckt, »SEGUIA - Assisting Designers in Developing Interactive Business Oriented Applications," presented at internatl. Conference on Human Computer Interaction, HCI'99, Munich, 1999.
29.  M. Macleod and R. Rengger, »The Development of DRUM: A Software Tool for Videoassisted Usability Evaluation," presented at BCS Conference on People and Computers VIII HCI'93, Lougborough, 1993.

30. D. Uehling and K. Wolf, »User Action Graphing Effort (UsAGE)," presented at ACM Conference on Human Aspects in Computing Systems CHI'95, Denver, 1995.
31. H. Wandke, A. Dubrowsky, and J. Huettner, »Anforderungsanalyse zur Einführung eines Unterstüzungssystems bei Software-Entwicklern," presented at Software Ergonomie '99, Walldorf, Germany, 1999.
32. IABG, »Vorgehensmodell zur Planung und Durchführung von IT-Vorhaben - Entwicklungsstandard für IT-Systeme des Bundes," : IABG- Industrieanlagen Betriebsgesellschaft, http://www.v-modell.iabg.de/.
33. Mark C. Paulk, Bill Curtis, Mary Beth Chrissis, and Charles V. Weber, "Capability Maturity Model, Version 1.1," IEEE Software, Vol. 10, No. 4, July 1993, pp. 18-27.

## Discussion

*L. Bass:* How many best practices were incorporated in the project?
*E. Metzker:* Around 30.

*G. Phillips:* How do you control what information regarding best practices make it into the knowledge base?
*E. Metzker:* The project manager and the responsible user interface engineer determined what was stored in the knowledge base.

*F. Paterno:* How did you evaluate the success of the approach in your case study and what elements contributed to it?
*E. Metzker:* We need more case studies to better understand the elements that are more relevant and how they are useful.

*H. Stiegler:* You talked a lot about the UMM which is strongly related to CMM. From my experience it is hard work to arrive at a level beyond 3. Your tool approach even deals with the improvement cycles between level 4 and 5. Did you really achieve that level in such a short time frame?
*E. Metzker:* We did not make a formal assessment yet and we do not claim to have achieved these levels but our overall planning includes the goals of continuous process improvements.

# From Usage Scenarios to Widget Classes

Hermann Kaindl[1] and Rudolf Jezek[2]

[1] Siemens AG Österreich, PSE, Geusaugasse 17,
A-1030 Vienna, Austria
hermann.kaindl@siemens.at
[2] Siemens Business Services GmbH & Co, Erdberger Lände 26,
A-1030 Vienna, Austria
rudolf.j.jezek@sbs.at

In practice, designers often select user interface elements like widgets intuitively. So, important design decisions may never become conscious or explicit, and therefore also not traceable. We addressed the problem of systematically selecting widgets for a GUI that will be built from those building blocks.

Our approach is based upon *task analysis* and *scenario-based design*, assuming that envisaged usage scenarios of reasonable quality are already available. Starting from them, we propose a systematic process for selecting user interface elements (in the form of widgets) in a few explicitly defined steps. This process provides a seamless way of going from scenarios through (attached) subtask definitions and various task classifications and (de)compositions to widget classes. In this way, it makes an important part of user interface design more systematic and conscious.

More precisely, we propose to explicitly assign subtask descriptions to the interactions documented in such a scenario, to the steps of both users and the proposed system to be built. Through combining those subtasks that together make up an interaction, *interaction tasks* are identified. For these, the right granularity needs to be found, which may require task composition or decomposition. The resulting interaction tasks can be classified according to the kind of interaction they require. From this classification, it is possible to map the interaction tasks to a class hierarchy of widgets. Up to this point, our process description is seamless, while the subsequent selection of a concrete widget is not within the focus of this work.

For this mapping, we defined a hierarchy of widget classes according to a functional rather than the usual structural view of classifying widgets. This hierarchy covers the MS Windows style guide. We also defined a hierarchy of interaction tasks that reflects this widget hierarchy.

For an initial evaluation of the usefulness of this approach, we conducted a small experiment that compares the widgets of an industrial GUI that was developed as usual by experienced practitioners, with the outcome of an independent execution of the proposed process. Since the results of this experiment are encouraging, we suggest to investigate this approach further in real-world practice.

Although this systematic process still involves human judgement, we hope that it contributes to a better understanding of the relationship between usage scenarios and user interface elements through the various tasks involved. More generally, it should help to better understand a certain part of user interface design.

M. Reed Little and L. Nigay (Eds.): EHCI 2001, LNCS 2254, pp. 35–36, 2001.
© Springer-Verlag Berlin Heidelberg 2001

## Discussion

*G. Phillips:* Is your work intended to support  - automatic generation of user interfaces - support to allow inexperienced user interface designers to produce better user interfaces
*R. Jezek:* Automatic generation is a long term goal. For now, we are aiming for support to inexperienced designers. Experienced designers can usually with or without our method.

*J Raskin:* How did you choose your widget set?
*R. Jezek:* We used windows 95 widgets.

*J Raskin:* Doesn't that guarantee a poor interface? [laughter]
*R. Jezek:* We had to start somewhere but other sets will be considered in future work.

*J. Coutaz:* You have focused on the mapping between domain objects and widgets. Another strategy could have been the mapping between tasks and widgets?
*R. Jezek:* We have an existing tool where scenarios are defined so we concentrated on scenarios with attached tasks.

*M. Harning:* Have you considered situation where a task like copy maps into multiple widget – a button, a drop-down menu a drag-and-drop operation.
a) do you consider a drag-and-drop action as a widget?
b) Did you consider situations where one task maps into multiple widgets?
*R. Jezek:* We focused on simple user interfaces, without drag-and-drop, but it is good input for future work.

# Evaluating Software Architectures for Usability

Len Bass and Bonnie E. John

Carnegie Mellon University
Pittsburgh, Pa 15213
ljb@sei.cmu.edu, bej@cs.cmu.edu

For the last twenty years, techniques to design software architectures for interactive systems that support usability have been a concern of both researchers and practitioners. Recently, in the context of performing architecture evaluations, we were reminded that the techniques developed thus far are of limited utility when evaluating the usability of a system based on its architecture. Techniques for supporting usability have historically focussed on selecting the correct overall system structure. Proponents of these techniques argue that their structure retains the modifiability needed during an iterative design process while still providing the required support for performance and other functionality.

We are taking a different approach. We are preparing a collection of connections between specific aspects of usability (such as the ability for a user to "undo" or "cancel") and their implications for software architecture. Our vision sees designers using this collection both to generate solutions to those aspects of usability they have chosen to include and to evaluate their system designs for specific aspects of usability. Our contribution is a specific coupling between aspects of usability and their corresponding architecture. We do not attempt to designate one software architecture to satisfy all aspects of usability. Details can be found in [1].

## References

1. Bass, L., John, B. E. and Kates, J. Achieving Usability Through Software Architectural Means, CMU/SEI-2001-TR-005, April 2001, Carnegie Mellon University, Pittsburgh, Pa.

### Acknowledgments

This work is supported by the U.S. Department of Defense.

## Discussion

*F. Jambon:* Do you think that tools can verify some of your checklist items automatically?

*L. Bass:* Yes, some of them can be calculated automatically, but this is not the purpose of this work.

M. Reed Little and L. Nigay (Eds.): EHCI 2001, LNCS 2254, pp. 37–38, 2001.

*J. Coutaz:* Are your patterns expressed at the conceptual level or implementation level?

*L. Bass:* It depends on the scenario; for instance, cancel pattern is pretty much implementational.

*J. Röth:* What (scenario, checklist, etc.) was the most useful to the software engineers?

*L. Bass:* Found that checklist was most useful part of work for software engineers. Did I forget that? Have I thought about it?

*D. Salber:* I like the idea of being proactive versus testing after the fact. But actually you should do both. What support does your approach offer?

*L. Bass:* Approach intended to be complementary to existing software practice.

*D. Salber:* Then what support for coordinating the two phases?

*L. Bass:* Try to have usability engineers plus software engineers talk at beginning of V-cycle not just at the end (what they do actually).

# Interactive System Safety and Usability
# Enforced with the Development Process

Francis Jambon, Patrick Girard, and Yamine Aït-ameur[1]

Laboratory of Applied Computer Science
LISI/ENSMA, BP 40109
86961 Futuroscope cedex, France
jambon@ensma.fr, girard@ensma.fr, yamine@supaero.fr
http://www.lisi.ensma.fr/ihm/index-en.html

**Abstract.** This paper introduces a new technique for the verification of both safety and usability requirements for safety-critical interactive systems. This technique uses the model-oriented formal method B and makes use of an hybrid version of the MVC and PAC software architecture models. Our claim is that this technique –that uses proofs obligations– can ensure both usability and safety requirements, from the specification step of the development process, to the implementation. This technique is illustrated by a case study: a simplified user interface for a Full Authority Digital Engine Control (FADEC) of a single turbojet engine aircraft.

## 1 Introduction

Formal specification techniques become regularly used in the area of computer science for the development of systems that require a high level of dependability. Aircraft embedded systems, the failure of which may cause injury or death to human beings belong to this class.

On the one hand, user-centered design leads to semi-formal but easy to use notations, such as MAD [1] and UAN [2] for requirements or specifications, or GOMS [3] for evaluation. These techniques could express relevant user interactions but they lack clear semantics. So, neither dependability nor usability properties can be formally proved.

On the other hand, adaptation of well-defined approaches, combined with interactive models, gives partial but positive results. Among them, we find the interactors and related approaches [4, 5], model-oriented approaches [4], algebraic notations [6], Petri nets [7] or temporal logic [8, 9]. Thanks to these techniques, some safety as well as usability requirements may be proved.

Nevertheless, theses formal techniques are used in the development process in a limited way because of two constraints:

- Formal techniques mostly depend on ad hoc specification models –e.g. interactors– and do not concern well-known software architecture models as Arch, MVC or

---

[1] Yamine AÏT-AMEUR is now Professor at ENSAE-SUPAERO, ONERA-DTIM, 10 av. Edouard Belin, BP 4032, 31055 Toulouse cedex, France.

M. Reed Little and L. Nigay (Eds.): EHCI 2001, LNCS 2254, pp. 39–55, 2001.

PAC. As a consequence, these unusual models make the specification task hard to use by most user interfaces designers.

• Few of these formal techniques can preserve formal semantics of the requirements from the specification to the implementation steps. Most of them can prove ergonomic properties at the specification level only. So, it cannot be proved that the final software is exactly what has been specified.

This article focuses on the B method [10, 11]. On the one hand, compared to VDM and Z, it makes possible the definition of a constructive process to build whole applications, with the respect of all the rules by the use of a semi-automatic tool [12]. On the other hand, the interactive system can be specified with respect to well-known software architecture models as Arch [13]. In this paper, we will show how the B method can be used to specify a critical system, the FADEC user interface case study, and how dependability as well as user interface honesty can be proved.

This work may be considered as a new step towards the definition of an actual interactive development method based on formal approaches. Our first results [13, 14] focus on low-level interaction mechanisms, such as mouse and window control. We showed that the B method might be used with profit in interactive development. Our aim in this article is to apply the method on critical systems for two main reasons. First, we believe that critical systems are applications of primary importance for safe methods. In addition, critical systems introduce special needs in terms of flow of control. So, we focus on two main points: (1) how can the specification of both critical systems and the B method influence software architecture –e.g. how B method constraints can be interpreted into well known HCI approaches– and (2) what are the benefits of using the B method for the specification process of critical systems.

The paper is organized as follows: in section 2, the B method is presented, and some previous results in applying formal approaches in HCI context are briefly summarized. In section 3, a study upon architecture models suitable for both critical systems and the B method is detailed. Last, the fourth section describes the specification of the case study and explains how the safety and usability requirements can be formally checked.

## 2    The B Method and Interaction Properties [14]

The B method allows the description of different modules, i.e., abstract machines that are combined with programming in the large operators. This combination enables designers to build incrementally and correctly –once all the proof obligations are proved– complex systems. Moreover, the utmost interest in this method, in our case, is the semi-automatic tool it is supported by.

### 2.1    The Abstract Machine Notation

The abstract machine notation is the basic mechanism of the B method. J.-R. Abrial defined three kinds of machines identified by the keywords MACHINE, REFINEMENT and IMPLEMENTATION. The first one represents the high level of specification. It expresses formal specification in a high abstract level language. The second one defines the different intermediate steps of refinement and finally the third one reaches the implementation level. Do note that the development is considered to be correct only when every refinement is proved to be correct with respect to the semantics of the B language. Gluing invariant between the different machines of a development are defined and sets of proof obligations are generated. They are used to prove the development correctness.

A theorem prover including set theory, predicates logic and the possibility to define other theories by the user, achieves the proof of these proof obligations. The proving phase is achieved either automatically, by the theorem prover, or by the user with the interactive theorem prover. The model checking method, which is known to be often overwhelmed by the number of states that are needed to be computed is not used in the present version of the B tool [12].

## 2.2    Description of Abstract Machines

J.-R. Abrial described a set of relevant clauses for the definition of abstract machines. Depending on the clauses and on their abstraction level, they can be used at different levels of the program development. In this paper, a subset of these clauses has been used for the design of our specifications. We will only detail these clauses. A whole description can be found in the B-Book [10]. The typical B machine starts with the keyword MACHINE and ends with the other keyword END. A set of clauses can be defined in between. In our case, these clauses appear in the following order:

- INCLUDES is a programming in the large clause that allows to import instances of other machines. Every component of the imported machine becomes usable in the current machine. This clause allows modularity capabilities.
- USES has the same modularity capabilities as INCLUDES except that the OPERATIONS of the used machines are hidden. So, the imported machine instances cannot be modified.
- SETS defines the sets that are manipulated by the specification. These sets can be built by extension, comprehension or with any set operator applied to basic sets.
- VARIABLES is the clause where all the attributes of the described model are represented. In the methodology of B, we find in this clause all the selector functions which allow accessing the different properties represented by the described attributes.
- INVARIANT clause describes the properties of the attributes defined in the clause VARIABLES. The logical expressions described in this clause remain true in the whole machine and they represent assertions that are always valid.
- INITIALISATION clause allows giving initial values to the VARIABLES of the corresponding clause. Do note that the initial values must satisfy the INVARIANT clause predicate.
- OPERATIONS clause is the last clause of a machine. It defines all the operations – functions and procedures– that constitute the abstract data type represented by the machine. Depending on the nature of the machine, the OPERATIONS clause authorizes particular generalized substitutions to specify each operation. The substitutions used in our specifications and their semantics is described below.

Other syntax possibilities are offered in B, and we do not intend to review them in this article, in order to keep its length short enough.

## 2.3    Semantics of Generalized Substitutions

The calculus of explicit substitutions is the semantics of the abstract machine notation and is based on the weakest precondition approach of Dijkstra [15]. Formally, several substitutions are defined in B. If we consider a substitution S and a predicate P representing a postcondition, then [S]P represents the weakest precondition that establishes P after the execution of S. The substitutions of the abstract machine notation are inductively defined by the following equations. Do notice that we restricted ourselves to the substitutions used for our development. The reader can refer to the literature [10, 11] for a more complete description:

$$[SKIP]P \Leftrightarrow P \tag{1}$$

$$[S1 \parallel S2]P \Leftrightarrow [S1]P \wedge [S2]P \tag{2}$$

$$[PRE\ E\ THEN\ S\ END]P \Leftrightarrow E \wedge [S]P \tag{3}$$

$$[ANY\ a\ WHERE\ E\ THEN\ S\ END]P \Leftrightarrow \forall\ a\ (E \Rightarrow [S]P) \tag{4}$$

$$[SELECT\ P1\ THEN\ S1\ WHEN\ P2\ THEN\ S2\ ELSE\ S3\ END]P \Leftrightarrow \tag{5}$$
$$(P1 \Rightarrow [S1]P) \wedge (P2 \Rightarrow [S2]P) \wedge ((\neg P1 \wedge \neg P2) \Rightarrow [S3]P)$$

$$[x:=E]P \Leftrightarrow P(x/E) \tag{6}$$

The substitution (6) represents the predicate P where all the free occurrences of x are replaced by the expression E. Do notice that when a given substitution is used, the B checker generates the corresponding proof obligation, i.e., the logical expression on the right hand side of the operator "$\Leftrightarrow$". This calculus propagates a precondition that must be implied by the precondition set by the user. If not, then the user proves the precondition or modifies it. For example, if E is the substitution [x+1] and P the predicate $x \neq 2$, the weakest precondition is $x \neq 1$.

## 2.4    Interaction Properties

Proving interaction properties can be achieved by the way of *model checking* or *theorem proving* [16]. Theorem proving is a deductive approach to the verification of interactive properties. Unless powerful theorem provers are available, proofs must be made "by hand". Consequently, they are hard to find, and their reliability depends on the mathematical skills of the designer. Whereas model checking is based on the complete verification of a finite state machine, and may be fully automated. However, one of the main drawbacks of model checking is that the solution may not be computed due to the high number of states [16]. The last sessions of EHCI as well as DSV-IS show a wide range of examples of these two methods of verification.

For instance, *model checking* is used by Palanque et al. who model user and system by the way of object-oriented Petri nets –ICO– [17]. They argue that automated proofs can be done to ensure first there is no cycle in the task model, second a specific task must precede another specific task (*enter_pin_code* and *get_cash* in the ATM example) and third the final functional core state is the final user task (*get_cash* and *get_card*). These proofs are relative to reachability. Furthermore, Lauridsen uses the RAISE formalism to show that an interactive application –functional core, dialogue control and logical interaction– can be built using translations from the functional core adapter specification [18]. Then, Lauridsen shows that his refinement method can prove interaction properties as predictability, observability, honesty, and substitutivity.

In the meantime, Paternó and Mezzanotte check that unexpected interaction trajectories expressed in a temporal logic –ACTL– cannot be performed by the user. The system –a subset of an air traffic control application– is modeled by interactors specified with LOTOS [19]. Brun et al. use the translation from a semi-formal task-oriented notation –MAD– [1] to a temporal logic –XTL– [8] and prove reachability [20].

Our approach in this article –with the B method– deals with the first method, i.e., *theorem proving*. Yet, the method does not suffer from the main drawbacks of theorem proving methods, i.e., proving all the system "by hand". In our former studies [13, 14], about 95% of the proofs obligations, regarding *visibility* or *reachability*, were automatically proved thanks to the "Atelier B" tool. Our present work –the FADEC user interface specification– has been successfully and fully automatically proved. All proof obligations, regarding *safety* and *honesty*, have been distributed in the separate modules of the system specification, as we will see later on.

Moreover, since the specification is incrementally built, the proofs are also incrementally built. Indeed, compositionality in B ensures that the proofs of the whole system are built using the ones of the subsystems. This technique simplifies considerably the interaction property verifications. And then, this incremental conception of applications asserts that the proofs needed at the low-level B-machines of the application, i.e. the functional core, are true at the higher levels, i.e. the presentation. So, the *reliability* is checked by construction.

# 3    The FADEC Case Study and Its Software Architecture

A FADEC (Full Authority Digital Engine Control) is an electronic system that controls all the crucial parameters of aircraft power plants. One of the system roles is to lower the cognitive load of pilots while they operate turbojet engines, and to reduce the occurrence of pilot errors.

Our case study focuses on the startup and the shutdown procedures of a single turbojet engine aircraft. In our scenario, the pilot controls the engine ignition –off, start, run– and the fuel valve –closed, open. The engine states can be perceived via the fuel pressure –low, normal, high– and the engine temperature –low, normal, high. The system interface is composed of lights and push buttons. The interface layout adopts the *dark cockpit philosophy* which minimizes distracting annunciation for pilots, i.e. only abnormal or transition states are visible. So, the normal parameters of the engine during flight do not light up any interface lights.

In this section, we start with an analysis of constraints imposed by the B language over architecture design. Secondly, we explain why "pure" MVC and PAC models fail against B requirements. Lastly, we describe our hybrid model, named CAV.

## 3.1    Rules for B Architecture Design

Our case study belongs to the safety-critical interactive-system category. More precisely, as in common interactive systems, the user controls the software system, but, as a reactive system, a third part, that evolves independently from both the software system and the user, must also control the system. In first approximation, the software may be modeled as a unique view that displays some functional core under a specific interactive control. Our first idea for designing such a system was to use a well-known multi-agent model, such as MVC or PAC, because acceptability of formal methods is greatly influenced by using domain standard methods.

The interactive system specifications must however stay in the boundaries of the B language constraints. We selected three kinds of constraints that relate to our purpose. These main constraints are:

1. Modularity in the B language is obtained from the inclusion of abstract machine instances –via the INCLUDES clause– and, according to the language semantics, all these inclusions must form a tree.
2. The substitutions used in the operations of abstract machines are achieved in parallel. So, two substitutions –or operations– used in the same operation cannot rely on the side-effects of each other. So, they are not allowed on the abstract machines specifications.
3. Interface with the external world, i.e. the user actions as well as the updates of system state must be enclosed in the set of operations of a single abstract machine.

## 3.2    Classical Multi-agent Architecture Models

As we explained in the upper section, our first impulse was to apply directly a classical multi-agent approach to our problem. Nevertheless, we discovered rapidly that none of them could be used without modification. In this section, we briefly describe MVC and PAC architecture models and relate how they are inappropriate.

**MVC** is an acronym for "Model, View, Controller". It is the default architecture model for the Smalltalk language [21], and became the first agent-based model for HCI. This model splits the responsibility for user interface into autonomous agents that communicate by messages, and are divided into three functional perspectives: *Model* stands for application data, and their access. It is the only object that is allowed to communicate with other *Model* objects. *View* is in charge of graphical outputs. It gives the external representation of the domain data, using *Model* objects services to extract the data to be presented. It also presents the perceivable behavior of the agent. This separation between *Model* and *View* allows a *Model* to own several *Views*. Lastly, *Controller* is responsible for inputs, and for making perceivable the behavior of the agent. It also manages the interactions with the other *Controllers*.

When the user gives some input, the associated *Controller* triggers a *Model* function. Then, the *Model* sends a message to all its *Views* to inform them for a change. Each *View* may request for the new state of the *Model*, and can refresh the graphical representation if needed.

The main problem with MVC is the double link between the *View* and the *Model*. This point violates the first rule we identified, which concerns B abstract machine inclusion order. More precisely, the *Model* cannot access the *View* –for sending it a message– if the *View* must ask the *Model* for data to be visualized.

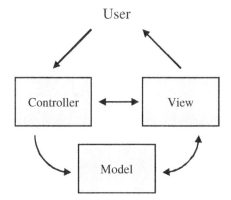

**Fig. 1.** The three components of the Model-View-Controller software architecture model

The **PAC** model, for "Presentation, Abstraction, Control" was proposed in 1987 by J. Coutaz [22]. Opposed to MVC, it is absolutely independent from languages. Agents are made of facets, which express complementary and strongly coupled computational services. The *Presentation* facet gives the perceivable input and output behavior of the object. The *Abstraction* facet stands for the functional core of the object. As the application itself is a PAC agent, no more component represents the functional core. There is no application interface component. The *Control* facet insures coherency between *Presentation* and *Abstraction*. It solves conflicts, synchronizes the facets, and refreshes the states. It is also takes charge of communication with other agents –their *Control* facets. Lastly, it controls the formalism transformations between abstract and concrete representations of data.

The PAC model gives another dimension as interactive objects are organized in hierarchies. Communication among the *Control* facets in the hierarchy is precisely defined. Modification to an object may lead its *Control* facet to signal this modification to the *Control* facet of parent object, which may in turn communicate this modification to its siblings. This allows all the parts of the application to correctly refresh. We believe that this precise point –i.e. the honesty property– may be directly addressed in a B development. It is the basis for the refinement steps we intend to conduct.

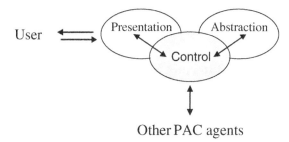

**Fig. 2.** The three facets of the Presentation-Abstraction-Control software architecture model

PAC solves the problem of MVC, because the *Control* facet is the only responsible for synchronizing the other two facets. Unfortunately, PAC does not respect the third rule on a unique entry point. In classical interactive systems, the unique entry point is the user. So, the *Presentation* facet may be considered as the unique entry point of the program. But, in safety critical systems, the reactive system itself is another external entity that must be taken into account. For that purpose, the presentation facet does not seem to be a good candidate.

### 3.3    The Hybrid CAV Model (Control-Abstraction-View)

We propose an hybrid model from MVC and PAC to solve this problem. The model uses the external strategy of MVC: the outputs of the system are devoted to a specific abstract machine –the *View*– while inputs are concerned by another one –the *Control*– that also manages symmetrical inputs from the reactive system which is directed by the third abstract machine –the *Abstraction*. The *Control* machine synchronizes and activates both *View* and *Abstraction* machines in response to both user and aircraft events, though assuming its role of control.

To limit exchanges between control and the two other components, a direct link is established between the *View* and the *Abstraction*, to allow the former to extract data from the latter. This point is particularly important in the B implementation of this model, because of the second rule we mentioned upper, that enforces to pay particular attention to synchronization problems between machines. This last point is mainly discussed in the following section.

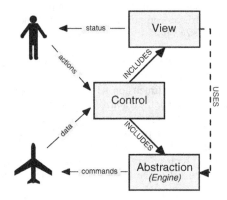

**Fig. 3.** The three components of the Control-Abstraction-View software architecture model

## 4    The FADEC User Interface Specification in B

In this section, we detail the development process we used for the case study, and we focus on HCI requirement satisfaction. The first step consists in modeling the FADEC, using the CAV architecture model we described in the previous section. The second step concentrates on safety requirements, that concerns inputs. The third step pays attention to honesty property, mainly outputs. Lastly, we illustrate the expressive power of the B method with iterative resolution of side effects on our specification.

Two kinds of requirements must be fulfilled:
- The system must be safe, i.e. the pilot must not be able to damage the engine. For example, the fuel pressure must be *normal* in order to begin the startup sequence.
- The system must be honest, i.e. the user interface lights must reflect the exact engine parameters –pressure and temperature– at any moment.

## 4.1    Modeling the FADEC User Interface

Applying our architecture model to the FADEC case study is straightforward. Each B machine encapsulates few attributes.

The *Engine* abstract machine –the *Abstraction*– models the functional core of the FADEC, e.g. the engine control parameters in the VARIABLE clause, and the variation sets of them in the SETS and INVARIANT clauses. The SETS are defined in respect to the logical description of the system, and a unique "SetProbeData" is defined for both the fuel prestsure and the engine temperature:

```
MACHINE
       Engine
SETS
       SetIgnition = {off, start, run} ;
       SetFuelValve = {open, closed} ;
       SetProbeData = {low, normal, high}
VARIABLES
       Ignition , FuelValve , FuelPress , EngineTemp
INVARIANT
       Ignition ∈ SetIgnition ∧
       FuelValve ∈ SetFuelValve ∧
       FuelPress ∈ SetProbeData ∧
       EngineTemp ∈ SetProbeData ∧ ...
```

The *View* abstract machine models what the pilot can perceive from the user interface, e.g. the lights and their status (TRUE for on, and FALSE for off). Because of the dark cockpit philosophy, we chose to use two lights for reflecting either ignition state, pressure or temperature that have three different states, and only one for fuel valve status which has only two different states. Moreover, the *View* abstract machine uses an instance of the *Engine* abstract machine in order to be aware of its sets of variables. It is expressed by a USES clause:

```
MACHINE
       View
USES
       engine.Engine
VARIABLES
       IgnitionOff, IgnitionStart, FuelValveClosed,
       EngineTempLow, EngineTempHigh, FuelPressLow, FuelPressHigh,
       StartButtonEnabled
```

**INVARIANT**

$IgnitionOff, IgnitionStart \in BOOL \times BOOL \land$
$FuelValveClosed \in BOOL \land$
$EngineTempLow, EngineTempHigh \in BOOL \times BOOL \land$
$FuelPressLow, FuelPressHigh \in BOOL \times BOOL \land$
$StartButtonEnabled \in BOOL \land \ldots$

The *Control* abstract machine is the centralized control of the system. So, it does not need to define any functional core nor presentation variables which are already defined in the *Engine* and *View* abstract machines respectively. On the other hand, it must include the sets, the variables and the operations of both instances of *Engine* and *View*:

**MACHINE**
    *Control*
**INCLUDES**
    *engine.Engine, view.View*

The sets, the variables and some of the invariants of the three abstract machines are now precisely defined. We can focus on the INVARIANT clauses that ensure safety.

## 4.2   Safety Requirements

The first requirement of the FADEC is safety. For instance, the start mode must not be used if the fuel pressure is not *normal*. This property must always be satisfied. In B, this requirement may be enforced with an INVARIANT clause that applies on the variables *Ignition* and *FuelPress*. The *Engine* abstract machine which represents the system functional core is responsible for it, with the following B expression:

$\neg (Ignition = start \land FuelPress \neq normal)$

We do not pay attention to what action is done. We only focus on the fact that never abnormal fuel pressure may be observed when startup is processing. In the semantics of B, the invariant must equal true at the initialization of the abstract machine, at the beginning and at the end of any operation. Note that the substitutions of the initialization as well as the operations are assumed to be executed in parallel.

Of course, the startup operation of the engine must satisfy this invariant, so the operation is guarded by an ad-hoc precondition PRE that ensures the operation will never be used if the fuel pressure is different from *normal*:

**startup =**
    PRE      *FuelPress = normal*
    THEN     *Ignition := start*
    END ;

For HCI, what is interesting now is: *Is the user able to make an error?* Whatever the user does, the B specification of the engine machine ensures that it will not be possible to violate the invariant. What about user actions and user interface state now? In our architecture model, the *Control* abstract machine is responsible for user inputs and for functional core actions activation. As a consequence, it must include an instance of the *Engine* abstract machine, and must use its operations within their specifications. Using B allows propagating the conditions. A new guard can/must be set for the operation used when the pilot presses the startup button. We do not give the engine the responsibility for controls, we propagate this semantic control to the user interface. One basic solution is to guard the activation of the startup button in the *Control* abstract machine by the precondition *engine.FuelPress = normal*:

```
start_button_pressed =
     PRE      engine.FuelPress = normal
     THEN     engine.startup ||
                  ...
     END ;
```

This basic solution suffers from two main drawbacks:
- The redundancy of preconditions is needed by modular decomposition of the B abstract machines
- The pilot's action on the system is blocked without proactive feedback, i.e. the pilot can press the startup button and nothing happens.

A more clever design is to delegate the safety requirements of the functional core to the user interface. In this new design, the user interface startup button is disabled while the startup sequence cannot be initiated for safety reasons. Now the user interface is in charge with the safety requirements of the system. Two modifications are needed:
- The *startup* operation of the *Control* abstract machine is now guarded by the state of the startup button:

  > *view.StartButtonEnabled* = **TRUE**

- The invariant must ensure that the startup button is disabled when the fuel pressure is not normal and enabled otherwise:

  > ( (*engine.FuelPress* = *normal* ∧ *view.StartButtonEnabled* = **TRUE**) ∨
  > (*engine.FuelPress* ≠ *normal* ∧ *view.StartButtonEnabled* = **FALSE**) )

The B semantics –and the *Atelier B* tool– checks for the validity of these assertions, and ensures for the compatibility of all abstract machines operations. Our software is now assumed not to allow the pilot doing anything wrong that can damage the turbojet engine.

## 4.3    Usability Requirements

Honesty is a well known property in user interfaces [23]. In safety critical systems, system honesty is crucial because user actions depend on the user capacity to evaluate the state of the system correctly. This point assumes the displayed state **is** the state of the actual system. Ensuring user interface honesty requires the specification to prove that the system state –represented by *Engine* variables– is always reflected in the pilot interface –represented by *View* variables. Like safety requirements, this requirement stands for an *always true* invariant.

It seems conspicuous that the honesty property must be stipulated in the INVARIANT clause of the *View* abstract machine. However, updates of the *Engine* and *View* variables are achieved in parallel by the operations of the *Control* abstract machine, because of the B semantics constraints –quoted in the §3.1. As a result, it is impossible to get the *Engine* variables update **before** the *View* variables update. As a consequence, the honesty property must be stipulated in the INVARIANT clause of the *Control* abstract machine only. For example, the light *FuelValveClosed* must be *on* only when the fuel valve is closed. We can express it by an exhaustive invariant that gives the two right solutions:

( (*engine.FuelValve* = *closed* ∧ *view.FuelValveClosed* = **TRUE**) ∨
(*engine.FuelValve* ≠ *closed* ∧ *view.FuelValveClosed* = **FALSE**) )

Our software architecture model assumes that the *Control* abstract machine really acts. So, the operation of the *Control* abstract machine, which is used when the pilot presses the close button, i.e. the *close_fuel_button_pressed* action, must update both the *Engine* state and the *View* state:

**close_fuel_button_pressed =**
    BEGIN
              engine.close_fuel ‖
              view.update_ui(    *engine.Ignition, closed,*
                                         *engine.FuelPress, engine.EngineTemp,*
                                       *view.StartButtonEnabled*    )
    END ;

Another consequence is that the operation of the *View* abstract machine must properly update the variable:

**update_ui (ignition, fuel_valve, fuel_press, engine_temp, start_button_enabled) =**
    PRE
              ignition ∈ SetIgnition ∧
              fuel_valve ∈ SetFuelValve ∧
              engine_temp ∈ SetProbeData ∧
              fuel_press ∈ SetProbeData ∧
              start_button_enabled ∈ **BOOL**
    THEN
              ANY fvc WHERE
                      fvc ∈ **BOOL** ∧
                      ( (fuel_valve = closed) ⇒ (fvc = **TRUE**) ) ∧
                      ( (fuel_valve ≠ closed) ⇒ (fvc = **FALSE**) )
              THEN
                      FuelValveClosed := fvc
              END ‖ ...
    END

## 4.4    Specificity of Asynchronous Systems

In critical systems such as the FADEC, some parts of the system change without any interaction with the user. For example, a probe whose control is obviously outside of the system updates the fuel pressure. A real-time kernel is in charge of interrogating every probe and sending responses to the whole system. In our analysis, the real-time kernel is out of our scope. Nevertheless, its entry point into our system must be defined. We propose to manage the aircraft in a way that is symmetric to the user. As stated in figure 1, it is controlled by the *Engine* abstract machine. When events are initiated by the aircraft power plant, their entry point is the *Control* abstract machine. So doing, the *Control* is completely in charge of updating the internal state – functional core– and the external state of the application –view. More, it can also ensure that the state of interaction is also correct.

Because we were focusing on HCI, we did not really pay attention to this side of the application. In our sense, one of the most interesting result of our study is the following: the B method helped us to discover a hazardous but important side-effect. The automatic prover detected a problem with fuel pressure invariant as soon as we introduced the action that updates this pressure into the functional core: on the one hand, the fuel pressure must be normal during the startup sequence, otherwise, the pilot cannot press the start button. On the other hand, if the fuel pressure falls down when the *Engine* abstract machine is in start mode, the turbojet engine must stop. We did not take this case into account. Fortunately, the B semantics does not allow this. Therefore, the *Engine* abstract machine must be enhanced:

```
update_fuel_press (fuel_data) =
     PRE      fuel_data ∈ SetProbeData
     THEN
               FuelPress := fuel_data ||
               SELECT Ignition = start ∧ fuel_data ≠ normal
               THEN    Ignition := off
               END
     END ;
```

As a result, the *Control* abstract machine must update the *View* accordingly:

```
fuel_press_event (data) =
     PRE data ∈ SetProbeData
     THEN
               SELECT engine.Ignition = start ∧ data ≠ normal THEN
                       engine.update_fuel_press (data)         ||
                       view.update_ui(   off, engine.FuelValve,
                                         data, engine.EngineTemp,
                                         FALSE  )
               ELSE
                       engine.update_fuel_press (data) ||
                       view.update_ui(   engine.Ignition, engine.FuelValve,
                                         data, engine.EngineTemp,
                                         TRUE   )
               END
     END ;
```

# 5    Conclusion

A previous work [13, 14] show that the B language can be used to specify WIMP interactive systems and ensure usability properties. This work shows that the B method can enforce safety and usability in a process-control interactive system with asynchronous behavior. Moreover, this study covers specification and design topics:

we define a new software architecture model, that allows an actual instantiation with the B method, and we describe an empiric method for modeling safety-critical interactive systems in B. Four points may be enlightened:
- safety and usability are ensured by using invariant that are relatively easy to find from the non formal description of the system behavior,
- incremental development can be achieved with this method, which is particularly suitable in HCI domain,
- using the B tool is very helpful for ensuring specification completeness, as we discovered during our analysis,
- the modules and then the whole specification may be completely validated thanks to the prover, in a whole automated way.

At the end of the last step, the specification must be considered as safe respect with to the requirements –safety and usability.

This work is a second step towards a real method for specifying, designing and implementing interactive systems with the B method. The perspectives are numerous. First, we need to use the B refinement theory to implement the specifications we realized. This step, which has already been initiated, will lead us to pay a particular attention to the connections with user interface toolkits. Then, a more exhaustive study must be accomplished to evaluate what properties may be enforced using the B language, and what properties cannot. This will then allow us to design a safety critical system with the collaboration of a set formal methods to avoid limitations among each methods.

### Acknowledgements

We thank Hervé BRANLARD and Joël MITARD for their documentation about operational FADEC procedures, and Sylvie FERRÉS for the English review.

# References

1. Scapin, D.L. and Pierret-Golbreich, C. Towards a method for task description : MAD in *Working with display units, edited by L. Berliguet and D. Berthelette.* Elsevier Science Publishers, North-Holland, 1990. pp. 371-380.
2. Hix, D. and Hartson, H.R. *Developping user interfaces: Ensuring usability through product & process.* John Wiley & Sons, inc., Newyork, USA, 1993.
3. Card, S., Moran, T. and Newell, A. *The Psychology of Human-Computer Interaction.* Lawrence Erlbaum Associates, 1983, 280 p.
4. Duke, D.J. and Harrison, M.D. Abstract Interaction Objects. *Computer Graphics Forum.* 12, 3 (1993), pp. 25-36.
5. Paternò, F. A Theory of User-Interaction Objects. *Journal of Visual Languages and Computing.* 5, 3 (1994), pp. 227-249.
6. Paternò, F. and Faconti, G.P. On the LOTOS use to describe graphical interaction in . Cambridge University Press, 1992. pp. 155-173.
7. Palanque, P. *Modélisation par Objets Coopératifs Interactifs d'interfaces homme-machine dirigées par l'utilisateur.* PhD Université de Toulouse I, Toulouse, 1992, 320 p.

8.  Brun, P. XTL: a temporal logic for the formal development of interactive systems in *Formal Methods for Human-Computer Interaction, edited by P. Palanque and F. Paternò.* Springer-Verlag, 1997. pp. 121-139.
9.  Abowd, G.D., Wang, H.-M. and Monk, A.F. A Formal Technique for Automated Dialogue Development, in *Proc. DIS'95, Design of Interactive Systems* (Ann Arbor, Michigan, August 23-25, 1995), ACM Press, pp. 219-226.
10. Abrial, J.-R. *The B Book: Assigning Programs to Meanings.* Cambridge University Press, 1996, 779 p.
11. Lano, K. *The B Language Method: A guide to practical Formal Development.* Springer, 1996.
12. Steria Méditerranée. Atelier B in 1997.
13. Aït-Ameur, Y., Girard, P. and Jambon, F. A Uniform approach for the Specification and Design of Interactive Systems: the B method, in *Proc. Eurographics Workshop on Design, Specification, and Verification of Interactive Systems (DSV-IS'98)* (Abingdon, UK, 3-5 June, 1998), Proceedings, pp. 333-352.
14. Aït-Ameur, Y., Girard, P. and Jambon, F. Using the B formal approach for incremental specification design of interactive systems in *Engineering for Human-Computer Interaction, edited by S. Chatty and P. Dewan.* Kluwer Academic Publishers, 1998. Vol. 22, pp. 91-108.
15. Dijkstra, E. *A Discipline of Programming.* Prentice Hall, Englewood Cliff (NJ), USA, 1976.
16. Campos, J.C. and Harrison, M.D. Formally Verifying Interactive Systems: A Review, in *Proc. Eurographics Workshop on Design, Specification and Verification of Interactive Systems (DSV-IS'97)* (Granada, Spain, 4-6 June, 1997), Springer-Verlag, pp. 109-124.
17. Palanque, P., Bastide, R. and Sengès, V. Validating interactive system design through the verification of formal task and system models, in *Proc. IFIP TC2/WG2.7 Working Conference on Engineering for Human-Computer Interaction (EHCI'95)* (Grand Targhee Resort (Yellowstone Park), USA, 14-18 August, 1995), Chapman & Hall, pp. 189-212.
18. Lauridsen, O. Systematic methods for user interface design, in *Proc. IFIP TC2/WG2.7 Working Conference on Engineering for Human-Computer Interaction (EHCI'95)* (Grand Targhee Resort (Yellowstone Park), USA, 14-18 August, 1995), Chapman & Hall, pp. 169-188.
19. Paternò, F. and Mezzanotte, M. Formal verification of undesired behaviours in the CERD case study, in *Proc. IFIP TC2/WG2.7 Working Conference on Engineering for Human-Computer Interaction (EHCI'95)* (Grand Targhee Resort (Yellowstone Park), USA, 14-18 August, 1995), Chapman & Hall, pp. 213-226.
20. Brun, P. and Jambon, F. Utilisation des spécifications formelles dans le processus de conception des Interfaces Homme-Machine, in *Proc. Journées Francophones sur l'Ingénierie de l'Interaction Homme-Machine (IHM'97)* (Poitiers-Futuroscope, 10-12 septembre, 1997), Cépaduès Éditions, pp. 23-29.
21. Goldberg, A. *Smalltalk-80: The Interactive Programming Environment.* Addison-Wesley, 1984.
22. Coutaz, J. PAC, an Implementation Model for the User Interface, in *Proc. IFIP TC13 Human-Computer Interaction (INTERACT'87)* (Stuttgart, September, 1987), North-Holland, pp. 431-436.
23. Gram, C. and Cockton, G. *Design Principles for Interactive Software.* Chapman & Hall, 1996, 248 p.

## Discussion

*J. Williams:* An observation: your manner of modeling seems similar to Jose Campos at The university of York. He uses a combination of theorem proving and model checking.
*F. Jambon:* The use of B offers a highly automated approach to verification. The approach taken by Campos will be investigated further.

*J. Roth:* Seems to need a lot of specification for even a simple user interface.
*F. Jambon:* Full specification for example system from the paper is six pages. We need to choose between simplicity of expression and reasoning power; here we err on the side of reasoning power. However, some real systems have been successfully specified in B including the Paris Metro control system. In addition, the verified design has proven safety properties.

*P. Van Roy:* In the specification you have an abstraction of the engine. How do you ensure correspondence between the abstraction and the engine itself?
*F. Jambon:* That is the subject of another talk. Briefly, there is an interface between the secure verified code and the insecure elements not coded in B. One approach is to have a state machine at either end of each engine operation. This is unverified code so it can't be effectively reasoned about and we test it instead.

*N. Graham:* Your CAV architecture looks similar to some very early architectures for interactive systems, having inputs trigger outputs directly. Is the design of the architecture motivated by the idea that this is appropriate to the problem domain, or to make it easy or possible to model the system using B?
*F. Jambon:* It was forced on us by the B theorem prover. The architecture is problematic because the control component tends to acquire all the functionality of a PAC control and an MVC control so it's quite large. However the one advantage of the architecture is that it is fully symmetrical.

*L. Bass:* The only way the control passes information to the user is through the aircraft? Can the abstraction communicate directly with the user?
*F. Jambon:* This is forced on us by the B modeling notation. It limits the kind of systems we can model but gives us provable safety.

*L. Bass:* If you replace the airplane with a database for example this model wouldn't work, would it?
*F. Jambon:* No, not unless the database actively provided status updates. There's an inherent assumption that the abstraction is dynamic.

*J. Coutaz:* It appears we could have an input component between the user and the control and a functional core adapter between the plane and the control, then you'd have sort of a "micro-arch".
*F. Jambon:* Yes, there are multiple interfaces so it sort of looks like that.

*S. Chatty:* What does the 'uses' line between the view and the abstraction mean, exactly?
*F. Jambon:* Visibility of data types.

*S. Chatty:* What is the flow of information?
*F. Jambon:* None --- the view cannot read the abstraction. This would introduce a situation that the theorem prover could not deal with because of the parallelism in the system. All dynamic information goes through the control.

*H. Stiegler:* The command and status arrows are not part of the model? Or are they?
*F. Jambon:* The arrows are not part of the formal model at the specification level because they represent non-formal components of the system about which we cannot reason rigorously.

*H. Stiegler:* But wouldn't they break the hierarchy of your formal model?
*F. Jambon:* Yes, if we included them.

*P. Van Roy:* If there really is a cyclic dependency in the system, can this be modeled in B? This seems a serious limitation in real life, since many (most?) systems seem to end up with cycles at run time.
*F. Jambon:* B does not allow this is there is no way to rigorously prove that a cycle would not bottom out at run time. Also, B cannot prove temporal behaviour, just state evolution. Temporal reasoning is required to address this problem and B does not provide this.

# Detecting Multiple Classes of User Errors

Paul Curzon and Ann Blandford

Interaction Design Centre, School of Computing Science, Middlesex University, UK
{p.curzon,a.blandford}@mdx.ac.uk

**Abstract.** Systematic user errors commonly occur in the use of interactive systems. We describe a formal *reusable* user model implemented in higher-order logic that can be used for machine-assisted reasoning about user errors. The core of this model is a series of non-deterministic guarded temporal rules. We consider how this approach allows errors of various specific kinds to be detected and so avoided by proving a single theorem about an interactive system. We illustrate the approach using a simple case study.

## 1    Introduction

In this paper, we present an approach to the verification of interactive systems that allows the detection of systematic user errors. The approach extends standard hardware verification techniques based on machine-assisted proof to this new domain. Human error in interactive systems can be just as disastrous as errors in the computer component of the system. Whilst it is impossible to eradicate all human error without trivializing system functionality, there are whole classes of persistent user errors whose presence is predictable due to their distinct cognitive cause [12]. Their possibility can be removed completely with appropriate design [9,2]. If methods for detecting a range of such errors in a systematic way are available, system reliability can be improved. Designing interactive systems so that a single class of error is absent is relatively straightforward. However, there are many different reasons for users making mistakes. Design principles used to avoid such errors can conflict, so that without care, in eliminating one class of error, other errors are introduced.

People do not normally behave randomly. Neither do they behave completely logically. However, it is a reasonable, and useful, approximation to say that they behave *rationally*. They enter an interaction with goals and some knowledge of the task they wish to perform. They act in ways that, given their knowledge, seem likely to help them achieve their goals. It is precisely because users are behaving rationally in this way that they make certain kinds of persistent errors with certain interactive system designs. For example, with the early designs of cash machines, users would frequently forget to take back their card. Most current cash machines have been redesigned so that this no longer occurs.

We investigate a method based on a generic user model specified in higher-order logic. This contrasts, for example, with an approach based on formulating properties corresponding to user errors and checking that those properties do not hold of the system (as might be done in a model checking approach). In our approach *rational* user behaviour is specified within a *reusable*, generic user model. This model is based

M. Reed Little and L. Nigay (Eds.): EHCI 2001, LNCS 2254, pp. 57–71, 2001.

on theory from the cognitive sciences and requires validating only once, not for each new interactive system considered. The user model is treated as a component of the system under verification. A single theorem is proved using the HOL proof system [8] that the task under consideration will be completed by this combined system. Systematic user errors occur as a side effect of the user behaving rationally. Our reasoning is based directly on the underlying behaviour that causes the problems to arise. We then check a single positive property (that the task is completed) not a whole range of negative properties (that various situations do not arise). We are concerned here with the detection of persistent user errors. It is not our aim that our user model explains other aspects of behaviour.

Our approach is a formal verification approach and as such is used by the designer of a system at the design stage without direct user involvement. However, the formal models used are based on information collected with the involvement of users. The method does not explicitly identify design improvements. However, in conducting a verification a very detailed understanding of the design is obtained. Thus when errors are detected the verifier also gains a detailed understanding of why they occur and, as has been demonstrated in hardware verification applications, can also suggest improvements to designs as a result of the verification [6].

We build here on our previous work. In [4] we demonstrated the feasibility of our approach using a generic user model to detect the possibility, or prove the absence of, systematic user errors (under the assumptions of the model). We used a very simple user model. It took user goals into account in only a limited way and so could only detect one class of error: the post completion error [2]. This is the situation where once the goal has been achieved other outstanding tasks are forgotten. In [5] we introduced a more accurate generic user model that took user knowledge into account and allowed for a wider range of rational behaviour. We demonstrated how this more accurate model can be used to detect in isolation a second class of user errors related to communication goals. It includes order errors and errors where the system design assumes the user knows they must perform a device[1] specific (as opposed to task specific) action.

In this paper we extend this work by demonstrating that a single theorem can be used to detect the presence or absence of both classes of error previously considered together with a third class of user error related to device delay. This ability to simultaneously detect multiple classes of errors is important because design guidelines to avoid such errors can contradict. For example post-completion errors can be eliminated if the order of user actions is carefully dictated by the computer system. However, computer systems dictating the order of user actions is precisely what causes the second class of errors we examine.

Formal user modeling is not new. Butterworth *et al* [1] use TLA to describe behaviour at an abstract level that supports reasoning about errors. However, it does not support re-use of the user model between devices. Moher and Dirda [10] use Petri net modeling to reason about users' mental models and their changing expectations over the course of an interaction; this approach supports reasoning about learning to use a new computer system but focuses on changes in user belief states rather than proof of desirable properties. Paterno' and Mezzanotte [11] use LOTOS and ACTL to specify intended user behaviours and hence reason about interactive behaviour.

---

[1] We use the term *device* throughout this paper to refer to the machine component of an interactive system whether implemented in hardware or software.

Because their user model describes how the user is intended to behave, rather than how users might actually behave, it does not support reasoning about errors. Duke *et al* [7] express constraints on the channels and resources within an interactive system; this approach is particularly well suited to reasoning about interaction that, for example, combines the use of speech and gesture. Our work complements these alternative uses of formal user modelling. It also complements traditional hardware and software verification approaches where an implementation is verified to meet a machine-centred specification, or where properties of a device are checked. Such approaches are concerned with the detection of errors in the computer system, rather than user errors as here. This is discussed further in [4].

**Table 1.** Higher-order Logic Notation

| | |
|---|---|
| a **AND** b | both a and b are true |
| a **OR** b | either a is true or b is true |
| a **IMPLIES** b | a is true implies b is true |
| **FOR ALL** n. P(n) | for all n, property P is true of n |
| **EXISTS** n. P(n) | there exists an n for which property P is true of n |
| f n | the result of applying function f to argument n |
| a = b | a equals b |
| (a, b) | a pair with first element a and second b |
| [a; b; c] | a list with elements a, b and c |
| **IF** a **THEN** b **ELSE** c | if a is true then b is true, otherwise c |
| **Theorem** P | P is a definition or theorem |

## 2    The HOL Theorem Prover

Our work uses the HOL system [8]. It is a general purpose, interactive theorem prover that has been used for a wide variety of applications. A typical proof will proceed by the verifier proving a series of intermediate lemmas that ultimately can be combined to give the desired theorem. Proofs are written in the meta-language of the theorem prover: SML. Each proof step is a call to an SML function. The proof script is developed by calling such functions interactively. The resulting ML proof script can later be rerun in batch mode to subsequently regenerate the theorems proved. If modifications are made to the system under verification then much of the proof script is likely to be reusable to verify the new design.

The HOL system provides a wide range of definition and proof tools, such as simplifiers, rewrite engines and decision procedures, as well as lower level tools for performing more precise proof steps. The architecture of the system means that new, trustable proof tools for specific applications can easily be built on top of the core system. Such proof tools are just SML functions that call existing proof functions.

All specifications and goals in HOL are written in higher-order logic. Higher order logic treats functions as first class objects. Specifications are thus similar to functional programs with logical connectives and quantification. The notation used in this paper is summarised in Table 1.

# 3    A Generic User Model and Task Completion Theorem

Our user model is based on a series of non-deterministic (disjunctive) temporally guarded action rules. The user model specifies each rule as a possibility that may be acted upon but does not specify which is actually chosen. The model therefore does not specify a single behaviour but a range of possible behaviours in any situation. Each rule describes an action that a user *could* rationally make. The rules are grouped corresponding to the user performing actions for specific related reasons. Each such group then has a single generic description. The model can not be used to detect errors that occur as a result of users acting randomly when there are obvious and rational behaviours available. A user who acts randomly because there are no rational options available is treated by the user model as erroneous behaviour. The model does not describe what a user *does* do, just what a user *could* do rationally. Our model makes no attempt to describe the likelihood of particular actions: a user error is either possible or not. Since we consider classes of error that can, by appropriate design, be eliminated, this strict requirement is appropriate. If it is possible for errors to arise from rational behaviour, they are liable to occur persistently, if not predictably. Their eradication will thus improve the reliability and usability of the system greatly.

Full details of the model are given elsewhere[2] [5]. Here, for clarity of explanation we use a semi-formal higher-order logic notation to give an overview.

## 3.1    Reactive Behaviour

The first group of rules we consider is that of *reactive* behaviour, where a user reacts to a stimulus, that clearly indicates that a particular action should be taken. For example, if a light next to a button is on, a user might, if the light is noticed, react by pressing the button. All the rules have the basic form below. If at a time t some condition is true (here `light`) then the NEXT action taken by the user out of the list of possible actions *may be* the given action (here pressing `button`).

```
(light t) AND NEXT user_actions button t
```

Note that the relation NEXT does *not* require that the action is taken on the *next cycle*, but rather that it is taken at an unspecified later time but before any other user action. A relation is recursively defined that, given a list of such pairs of inputs and outputs, asserts the above rule about them, combining them using disjunction so that they are non-deterministic choices. To target the generic user model to a particular interactive system, it is applied to a concrete version of this list containing specific signals:

```
[(light₁, button₁);...;(lightₙ, buttonₙ)]
```

People do not interact with interactive systems purely in a reactive way. They may ignore reactive stimuli for very rational reasons. In subsequent sections we show how the model is extended to take into account some such rational behaviour.

---

[2] See also www.cs.mdx.ac.uk/staffpages/PaulCurzon/

## 3.2    Communication Goals

People enter an interaction with some knowledge of the task that they wish to perform. In particular, they enter the interaction with communication goals: a task dependent mental list of information the user knows they must communicate to the computer system. For example, on approaching a vending machine, a person knows that they must communicate their selection to the machine. Similarly, they know they must provide money before the interaction is terminated. While inserting coins is not strictly a communication goal in cognitive science terms, for the purposes of this paper we treat it in the same way. Communication goals are important because a user may take an action as a result not of some stimulus from the machine but as a result of seeing an apparent opportunity to discharge a communication goal. For example, if on approaching a rail ticket machine the first button seen is one with the desired destination on, the person may press it, irrespective of any guidance the machine is giving. No fixed order can be assumed over how communication goals will be discharged if their discharge is apparently possible and the task as opposed to any particular device does not force a specific order.

Communication goals can be modelled as guard-action pairs. The guard describes the situation under which the discharge of the communication goal can be attempted. The action is the action that discharges the communication goal. They form the guard and action of a temporally guarded rule. We include an additional guard to this rule, stating that the action will only be attempted if the user's main goal has not yet been achieved. Strictly a similar guard ought to be added to the reactive rules. Currently they describe purely reactive behaviour.

**NOT**(goalachieved t) **AND** guard t **AND** NEXT user_actions action t

As for reactive behaviour a list of guard-action pairs is provided as an argument to the user model rather than the rules being written directly. The separate rules are combined by disjunction with each of the other non-deterministic rules.

As the user believes they have achieved a communication goal, it is removed from their mental list. This is modelled by a daemon within the user model. It monitors the actions taken by the user on each cycle, removing any from the communication goal list used for the subsequent cycle.

## 3.3    Completion

In achieving a goal, subsidiary tasks are often generated. Examples of such tasks include replacing the petrol cap after filling a car with petrol, taking the card back from a cash machine and taking change from a vending machine [2]. One way to specify these tasks would be to explicitly describe each such task. Instead we use the more general concept of an *interaction invariant*. This terminology is based on that of a *loop invariant* from program verification where a property holds at the start of each iteration of a loop. It does not necessarily hold during the body of the loop but must be restored before the next iteration commences. An interaction invariant similarly holds at the start of an interaction and must be restored before the next interaction commences, but is perturbed whilst the interaction progresses. The underlying reason why the subsidiary tasks of an interaction must be performed is that in interacting with the system some part of the state must be temporarily perturbed in order to achieve the desired task. Before the interaction is completed such perturbations must be undone. For example, to fill a car with petrol the petrol cap must be removed, and later restored. We specify the need to perform these completion tasks indirectly by supplying this interaction invariant as a higher-order argument to the user model.

We assume that a user, on completing the task in this sense, will terminate the interaction, irrespective of any other possible actions. Rather than specifying it as a non-deterministic rule we model it using an if-then-else construct, so that it overrides all other actions. A special user action, finished, indicates that the user has terminated the interaction. If the interaction had been terminated previously then it remains terminated.

```
IF invariant t AND goalachieved t OR finished(t-1)
THEN NEXT user_actions finished t
ELSE non-deterministic rules
```

Cognitive psychology studies have shown that users also sometimes terminate interactions when only the goal itself has been achieved [2]. This can be modeled as an extra non-deterministic rule.

```
goalachieved t AND NEXT user_actions finished t
```

The model also assumes an interaction finishes when no rational action is available. This rule acts as a final default case in the user model. Its guard states that none of the other rules' guards are true. In practice a user may behave randomly in this situation – the model assumes that if this occurs before the task is completed then a preventable user error occurs.

The user model is a relation that combines the separate rules. It takes a series of arguments corresponding to the details relevant to a specific machine: the list of possible user actions, the list of communication goals for the task, the list of reactive stimuli and actions they might prompt, the relevant possessions, the goal of the user and the interaction invariant. By providing these specific details as arguments to the relation, a user model for the specific interaction under investigation is obtained automatically. The important point is that the underlying cognitive model does not have to be provided each time, just lists of relevant actions, etc, specific to the current interaction. The way those actions are acted upon by the model is modelled only once.

## 3.4    Correctness Theorem

We now consider the theorem we prove about interactive systems. The usability property we are interested in is that if the user interacts rationally with the machine, based on their goals and knowledge, then they are guaranteed to complete the task for which they started the interaction. As noted earlier, task completion is more than just goal completion. In achieving the goal, other important sub-tasks may result that must then be done in addition to completing the goal. The property required is that eventually the goal has been achieved and all other sub-tasks have been completed (i.e. the interaction invariant restored).

The user model and the device specification are both described by relations. The device relation is true of its input and output arguments if they describe consistent input-output sequences of the device. Similarly, the user model relation is true if the inputs (observations) and outputs (actions) are consistent sequences that a rational user could perform. The combined system can then be described as the conjunction of the instantiated user model and the specification of the system. The task completion theorem we wish to prove thus has the form:

**Theorem**

> **FOR ALL** *state traces.*
> > *initial state* **AND**
> > *device specification* **AND**
> > *user model* **IMPLIES**
> > > **EXISTS** t. *invariant* t **AND** *goalachieved* t

If a theorem of this form can be proved then even if a user is capable of making the rational errors considered, that potential will not affect the completion of the task: the errors will never manifest themselves.

The theorem is generic and so reusable in the same way as the user model. The same information must be provided: notably the user's goal and the interaction invariant, together with the device specification and specialised user model.

## 4    User Errors Detected by the User Model

Though the user model is simple, it describes a user who is capable of making a range of persistent but rational errors. The model does not imply that mistakes will always be made, just that the potential is there. The errors are consequences of describing the way results from cognitive science suggest people act in trying to achieve their goals. The errors are detected by attempts to prove the task completion theorem. If an interactive system design is such that users can make errors then it will be impossible to prove that the task can be completed in all situations. It should be noted that we define classes of errors by their cognitive cause *not* by their effects. We do not claim that in proving the absence of an error that a similar effect might not happen due to some other cause such as a fire alarm ringing in the middle of an interaction.

### 4.1    Post-Completion Errors

One kind of common, persistent user error that emerges from the user model is the post-completion error [2]. This is the situation where a user terminates an interaction with completion tasks outstanding. For example, with old cash machines users persistently, though unpredictably, took cash but left their card. Even in laboratory conditions people have been found to make such errors [2]. This behaviour emerges as a consequence of the rule in the model allowing a user to stop once the goal has been achieved. If a system is to be designed so that such errors are eliminated, then the goal must not be achievable until after the invariant has been restored. If this is so, the rule will only become active in the safe situation when the task is fully completed.

Such errors could still occur (less frequently) if the system is designed so that the goal is achieved first but that warning messages are printed or beeps sounded to remind the user to do the completion tasks. Such designs do not remove all possibility of the error being made; they just reduce its probability. In our framework such an interactive system is still considered erroneous.

## 4.2    Communication-Goal Errors

A second class of error that can be detected is based on communication goals. Where there is no task-related, rather than device-related, restriction on the order that communication goals must be discharged, different users may attempt to discharge them in different orders. This will occur even in the face of the device using messages to indicate the order it requires. As with post-completion errors this problem is persistent but occurs unpredictably. It can be avoided if the interactive system does not require a specific order for communication goal actions. In the model this error is a consequence of the communication goal rules activating in any order provided their guard is active, and that if the action is taken the communication goal is removed. This means that the user model may be left at a later time with no rational action to take. The abort rule is then activated and the user model terminates the interaction before the task is completed. Similarly if a design assumes device-specific knowledge of a task that is not a communication goal without giving reactive stimuli, then the user model will abort: a user error occurs.

## 4.3    Device Delay Errors

The user model can also detect some errors related to device delay. If there is no feedback during delays users often repeat the last action. In the user model if there is no light to react to and the user has no outstanding communication goals then only the abortion rule is active. If such a situation can occur, then the task completion theorem cannot be proved. If outstanding communication goals are active, the model would force one of those actions to be scheduled. The action could be taken before the device is ready, thus having no effect. The communication goal would be removed from the communication goal list however, so would not necessarily be repeated. At a later point this would lead to only the abortion rule being possible. Again it would not be possible to prove task completion. The device would need to be redesigned so that the delay occurred when the user had no opportunity to discharge outstanding communication goals, and a "wait" message of some form displayed. This would be reactive in the sense that whilst displayed the user would react by doing nothing. Erroneous systems could still escape detection, however. In particular, if the light indicating the previous action remained lit during the computation time, then task completion could be proved though the reasoning would require the user repeating an action. The problem here is that our current user model is not sufficiently accurate. In particular, humans do not react to stimulus unless they believe that it will help them achieve their goal. In future work we will add additional guards to the reactive rules to model *rational reactive* behaviour.

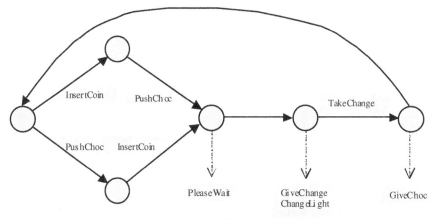

**Fig. 1.** Finite State Machine Specification of a Chocolate Machine

## 5    Case Study

To illustrate the use of the model and demonstrate the approach we consider a vending machine example. A finite state machine description of the machine is given in Figure 1. This example is simple: however, any interactive system that could be described in terms of a finite state machine and for which specific user actions and goals can be formulated could in principle be treated in the same way. In fact, we use a relational description of the finite state machine: our approach is not restricted to finite state machine specifications but could be based on any form of relational specification. This includes not only walk-up-and-use machines but also safety critical systems such as Air Traffic Control Systems.

The vending machine we consider requires users to supply a pound coin and gives change. The user must insert the coin and make a selection (we simplify this here to the pressing of a single button). Processing time is needed once the money is inserted before change is released. We assume for the sake of simplicity that the chocolate machine always contains chocolate. Despite its simplicity, without careful design, such a machine has the potential for users making communication goal errors - the specific order that the coin is inserted and the selection made is not forced by the task so a user could do them in any order. On the other hand if chocolate is given out before the change, a post-completion error could be made. Delay errors might also occur due to the processing time. Indeed, vending machines with such design problems are widespread. Here we consider a design that overcomes these problems.

Our design accepts coin and selection in any order. Once both have been completed it releases the change. A light flashing by the change slot indicates this. However this only occurs after a delay. A "wait" light indicates to the user that the machine is busy. Note that this processing is scheduled after all communication goals are fulfilled. A sensor on the change slot flap releases chocolate when the change is taken.

We must target the generic user model for the interactive system in question. This involves supplying concrete values for each of the model's arguments. We must provide information about the device inputs and outputs, and the user's internal state. This involves defining tuple types with each field corresponding to traces of inputs, outputs, states, etc. Accessor functions to the fields are then used to represent that event.

The first argument that must be supplied to the user model is a list of the actions a user could ever take that affect the interaction. This is used in the rules to specify that all the other actions do not occur when we specify that a particular event happens next. For our example the possible actions are represented by an InsertCoin field, a PushChoc field corresponding to the user pushing the selection button, a UserFinished field indicating the termination of the interaction, a TakeChange field and a Pause action which means the user is actively waiting:

    [InsertCoin; PushChoc; UserFinished; TakeChange; Pause]

The second piece of information that must be supplied is the initial list of communication goals together with their guards. Here there are two communication goals specified. When a user approaches the machine they know they must insert a coin and that this can only be done if they possess the coin. They also know they must make a selection. There are no task enforced conditions on when this can occur, so its guard is TRUE. It can happen at any time. The communication goal list thus has the form:

    [(HasCoin, InsertCoin); (TRUE, PushChoc)]

We must provide a list of reactive signals. This is a list pairing observations with actions that they prompt the user to make. In our design, there are two such signals: the ChangeLight prompts the user to take the change and so trigger the change flap sensor; the PleaseWait light prompts the user to wait (i.e. intentionally do nothing).

    [(ChangeLight, TakeChange); (PleaseWait, Pause)]

We must also indicate the possessions of the user. A relation POSSESSIONS in the generic user model gathers this information into an appropriate form. We supply it with the details of the user having chocolate and coins, the machine giving chocolate, the user inserting a coin and counts of the number of coins and chocolate bars possessed by the user.

Finally we must specify the goal of this interaction and the interaction invariant. Both are also used to create the concrete task completion theorem. The goal is to obtain chocolate. Its achievement is given by the field of the user state: UserHasChoc. For vending machines the interaction invariant, VALUE_INVARIANT, can be based on the value of the user's possessions. After interacting with a vending machine the value of the user's possessions should be at least as great as it was at the start (time 1). The value of a user's possessions is calculated from possession count and value fields using a relation VALUE.

    VALUE_INVARIANT possessions state t =
        (VALUE possessions state t ≥ VALUE possessions state 1)

This relation will not hold throughout the interaction. When the coin is inserted the value will drop and will only return to its initial value once both chocolate and change are taken. It is an invariant in the sense that it must be restored by the end of the interaction.

### 5.1    Proving the Task Completion Theorem

The task completion theorem we proved about this interactive system has the form:

```
Theorem
    FOR ALL state COINVAL CHANGEVAL CHOCVAL.
      COINVAL = CHANGEVAL + CHOCVAL AND
      DeviceState state 1 = RESET AND
      UserHasCoin state 1 AND
      NOT(UserHasChoc state 1) AND
      MACHINE_USER state AND
      MACHINE_SPEC state IMPLIES
        EXISTS t.
        (UserHasChoc state t) AND
        (VALUE_INVARIANT
          (POSSESSIONS possessions CHOCVAL COINVAL CHANGEVAL)
          state t)
```

MACHINE_SPEC is the behavioural specification of the vending machine. MACHINE_USER is the instantiated user model: notice that its only argument is the state. All the other details required such as communication goals have been provided in the instantiation. The theorem also contains assumptions about the initial system state and that the user has a coin but no chocolate at time 1.

An advantage of our approach is that proofs can be generic. The correctness theorem we proved is generic with respect to the value of coins, change and chocolate, for example. They are represented in the predicate POSSESSIONS by variables COINVAL, CHANGEVAL and CHOCVAL rather than by fixed integers. The correctness theorem contains an assumption that restricts the values concerned:

```
COINVAL = CHANGEVAL + CHOCVAL
```

The correctness theorem holds for any triple of values that satisfy this relation. This contrasts with model checking approaches based on binary decision diagrams where fixed concrete values would have to be provided. The verification could similarly be made generic with respect to a range of selections that could be made.

We proved the task completion theorem using symbolic simulation by proof within the HOL theorem prover [8]. Our verification is fully machine-checked. An induction principle concerning the stability of a signal is used repeatedly to step the simulation over periods of inactivity between a rule activating and the action happening.

For example, the theorem below states that if the machine is in the CHOC state at some time $t_1$ greater than 0, then eventually state WAIT is entered. It requires that at time $t_1$ the user has a coin but no chocolate yet, that the interaction was not terminated on the previous cycle and that inserting a coin is still an undischarged communication goal. Once the new state is entered (at some time $t_2$) the user will still not have terminated the interaction, the communication goal will have been discharged, the count of the user coins will have been decremented but the counts of chocolate and change will be unchanged.

```
Theorem
   0 < t₁ AND
   DeviceState state t₁ = CHOC AND
   UserHasCoin state t₁ AND
   NOT(UserHasChoc state t₁) AND
   NOT(UserFinished state (t₁-1))AND
   UserCommgoals state t₁ = [(InsertCoin, UserHasCoin)] AND
   MACHINE_USER state AND
   MACHINE_SPEC state IMPLIES
      EXISTS t₂.
         t₁ ≤ t₂ AND
         DeviceState state t₂ = WAIT AND
         NOT(UserHasChoc state t₂) AND
         NOT(UserFinished state (t₂-1)) AND
         UserCommgoals state t₂ = [] AND
         CountChoc state t₂ = CountChoc state t₁ AND
         CountCoin state t₂ = CountCoin state t₁ - 1 AND
         CountChange state t₂ = CountChange state t₁
```

A similar theorem is proved about each finite state machine state. The final theorem is proved by combining these separate theorems.

These theorems are currently proved semi-automatically. A series of lemmas must be proved for each. They are formulated by hand, but then proved automatically by a set of specially written proof procedures. As the lemmas are of a very standard form it should be straightforward to automate their formulation. Furthermore, as design changes are made, many of the lemmas proved are reusable.

## 6    Conclusions and Further Work

We have presented a verification approach that allows multiple classes of user errors to be detected or verified absent from designs by proving a single task completion theorem. The approach is based on the use of a *generic* user model specified as a higher-order function. By using higher-order logic it can be done elegantly and flexibly – for example the goal and interaction invariant are instantiated with relations.

We specify rational behaviour, not erroneous properties. This means that errors that are the result of that rational behaviour are detected. This is less restrictive than verifying that nothing bad can possibly happen, whatever the user's actions. An advantage of the approach is that the informal and potentially error-prone reasoning implicitly required to generate appropriate properties is not needed. To do this reasoning formally would need a formal user model. By using a generic user model directly, the cognitive basis of the errors is specified and validated once rather than for each new design or task. It also does not need to be revalidated when errors are found and the design subsequently modified.

To illustrate our approach, we described a small case study. We considered the verification of a vending machine design free of the classes of user errors covered. With a faulty design the correctness theorem would not be provable. For example, suppose the design released the chocolate first. We would not be able to prove from the user model that the change was taken. Instead, we would only be able to prove that either the change was taken or that the user finished without change. This is because for this design the completion rule becomes active before the task has been completed. The rule's guard is that the goal has been completed and this is achieved as soon as the user takes the chocolate. Thus rather than proving a step theorem such as that given, we would only be able to prove a conclusion that one of two situations arise, only one of which leads to full task completion. A case study discussing in more detail the attempted verification of a faulty design using our approach is given in [5].

We intend to carry out more complex case studies to test the utility of the methodology. In particular we are currently working on an Air Traffic Control case study. The main difficulty in verifying more complex systems is the time taken to develop the proof. This problem will be eased as we automate the proofs. We used interactive proof in the HOL system to prove the correctness theorem presented here. We intend to continue to develop tactics to increase the automation in doing this. Currently tactics have been written which automate the proofs of the main lemmas. Further work will automate the formulation of the lemmas and their combination. The lemmas and proofs are very formulaic, so much of this task is likely to be straightforward. The use of a common user model makes such automation easier.

By using an interactive proof system, theorems about generic interactive systems can also be proved. For example, in our case study we prove a correctness theorem for machines for all possible prices of chocolate and coins. Changing the price of the chocolate does not require the re-verification of the machine. Similarly we could prove a single correctness theorem that holds for all possible choices of item available in the vending machine. It is thus not a correctness theorem about one interactive system, but about a whole family of systems. A new correctness theorem does not need to be proved if the design is changed within the family.

The example considered here involves a hard-key interface. The approach can also be used with soft-key interfaces where the meaning of particular inputs such as buttons can be different at different points in the interaction. Such interfaces can be modelled using signals that explicitly model the interface indicating when each "virtual" button is available.

The general approach is suitable for finding systematic errors that arise from rational user behaviour: that is, errors that arise as a direct consequence of the user's goals and knowledge of the task. It cannot be used to avoid non-systematic errors or malicious user behaviour. Such errors can never be completely eliminated from interactive systems unless the functionality of the system is severely limited.

Our approach requires traditional informal analysis of the task and of the system's interface to be performed. This is needed to gather the information upon which to instantiate the formal model. If the information so gathered is inaccurate then errors could be missed. As with all formal modelling approaches, the theorems proved are only as good as the assumptions upon which they are based.

The current user model by no means covers all aspects of rational user behaviour. We are building on it to improve its accuracy. In doing so we will increase the number of classes of error detectable. For example, the rules concerning reactive behaviour need to be made rational so that users only react when it appears to help them achieve their goals. There is also a delay between a person committing to an action and actually taking that action. This can be modelled by linking each external action with an additional internal "commit" action. This will allow user errors resulting from such delays to be detected.

We do not claim our methodology can prevent all user errors. However, by providing a mechanism for detecting a series of classes of systematic errors, the usability of systems in the sense of absence of user errors is improved.

### Acknowledgements

This work is funded by EPSRC grant GR/M45221. Matt Jones and Harold Thimbleby made useful comments about an early version of this paper.

## References

1. R. Butterworth, A. Blandford and D. Duke. Using formal models to explore display based usability issues. *Journal of Visual Languages and Computing*, 10:455-479, 1999.
2. M. Byrne and S. Bovair. A working memory model of a common procedural error. *Cognitive Science*, 21(1): 31-61, 1997.
3. F. Corella, Z. Zhou, X. Song, M. Langevin and E. Cerny. Multiway Decision Graphs for automated hardware verification. *Formal Methods in System Design*, 10(1): 7-46, 1997.
4. P. Curzon and A. Blandford. Using a verification system to reason about post-completion errors. Presented at Design, Specification and Verification of Interactive Systems 2000. Available from http://www.cs.mdx.ac.uk/puma/.
5. P. Curzon and A. Blandford. Reasoning about order errors in interaction. Supplementary Proceedings of the International Conference on Theorem Proving in Higher-order Logics, August 2000. Available from http://www.cs.mdx.ac.uk/puma/.
6. P. Curzon and I. Leslie. Improving hardware designs whilst simplifying their proof. *Designing Correct Circuits*, Workshops in Computing, Springer-Verlag 1996.
7. D.J. Duke, P.J. Barnard, D.A. Duce, and J. May. Syndetic modelling. *Human-Computer Interaction*, 13(4): 337-394, 1998.
8. M.J.C. Gordon and T.F. Melham. *Introduction to HOL: a theorem proving environment for higher order logic*. Cambridge University Press 1993.
9. W-O Lee. The effects of skills development and feedback on action slips. In Monk, Diaper, and Harrison, editors, *People and Computers VII*. CUP, 1992.
10. T.G. Moher and V. Dirda. Revising mental models to accommodate expectation failures in human-computer dialogues. In *Design, Specification and Verification of Interactive Systems'95*, pp 76-92. Wien: Springer, 1995.
11. F. Paterno' and M. Mezzanotte. Formal analysis of user and system interactions in the CERD case study. In *Proceedings. of EHCI'95: IFIP Working Conference on Engineering for Human-Computer Interaction*, pp 213-226. Chapman and Hall, 1995.
12. J. Reason. *Human Error*. Cambridge University Press, 1990.

## Discussion

*L. Bass:* I lost the big picture somewhere. The aim is to have a model of the user together with the interactive system and....

*P. Curzon:* Yes, we model both together and then try to prove that the system as a whole has particular properties, in this case the absence of systematic errors.

*L. Bass:* So the work is in generating the model?

*P. Curzon:* The work is in three parts: generating the system model, generating the specific user model from the generic one, and performing the property proofs. Right now performing the proofs is the most time consuming aspect, although we are trying to improve this by writing tools to automate proof steps and reusing HOL scripts from previously completed proofs wherever possible.   Having a generic user model removes most of the work in generating a user model - you just supply arguments to the existing model.

*N. Graham:* For large, feature-rich programs it seems impossible to evaluate their usability strictly by hand. But is this approach really any more tractable?

*P. Curzon:* All case studies to date have been have been fairly trivial systems and proofs have taken a couple of afternoons. On the other hand, a second proof in a given application domain goes much faster because you can reuse lemmas and proof scripts in the tool. Scaling up to a large complex system is currently beyond what we can reasonably accomplish with the current level of proof automation. But in principle this is possible using layered abstractions.

*L. Bergman:* When you can't prove a theorem does the tool help you?

*P. Curzon:* In a sense you just get stuck, but because proving the theorem is interactive you generally get a good sense of where the problem must be and this helps you to find solutions.  In doing a proof you get a very detailed understanding of the design and why it does or does not work.

*J.Williams:* Can you solve scalability by decomposing a large system into smaller subsystems?

*P. Curzon:* Yes, that appears possible but the sub-theorems may have a different form from the main theorems. It may be better to break the system down along task-oriented lines.  Then each task's theorem would be just a simpler version of that for the full task.  Ideally the design would be structured along the same lines.

*J. Hohle:* Do you need a Ph.D. in mathematics to use this system or is it more accessible?

*P. Curzon:* Just at the moment the Ph.D. would be advisable. But ultimately the proof mechanisms are expected to become more automated; This is a major area of work. However a certain degree of skill will always be required. Because the approach is based on a standard user model automation is likely to be easier than if a completely new model was written for each verification.

# Exploring New Uses of Video with VideoSpace

Nicolas Roussel

DIUF, Université de Fribourg
Chemin du Musée, 3
1700 Fribourg, Switzerland
nicolas.roussel@unifr.ch

**Abstract.** This paper describes videoSpace, a software toolkit designed to facilitate the integration of image streams into existing or new documents and applications to support new forms of human-computer interaction and collaborative activities. In this perspective, videoSpace is not focused on performance or reliability issues, but rather on the ability to support rapid prototyping and incremental development of video applications. The toolkit is described in extensive details, by showing the architecture and functionalities of its class library and basic tools. Several projects developed with videoSpace are also presented, illustrating its potential and the new uses of video it will allow in the future.

## 1 Introduction

Although almost forty years have passed since AT&T's first PicturePhone, most commercial systems and applications dealing with video communication today are still based on a phone paradigm. This paradigm usually implies a formal and explicit setup of the link by both parties and the display of the images "as-is" on a monitor or in a window on a computer screen. Over the last twenty years however, the HCI and CSCW research communities have proposed a number of innovative uses of video including support for informal communication [1] as well as highly focused collaborative activities [2], or image processing for human-computer interaction [3].

In this paper, I present videoSpace, a software toolkit designed to facilitate the use of image streams to support such new forms of human-computer interaction and computer-supported collaborative activities. VideoSpace is motivated by the desire to focus on the *uses* of video, rather than the *technologies* it requires. In this perspective, the toolkit is not focused on performance or reliability issues, but rather on the ability to support rapid prototyping and incremental development of video applications. This approach contrasts with many of the research themes usually associated to video in the Multimedia or Network communities such as compression, transport or synchronization. VideoSpace is not aimed at these topics. It is rather intended to help HCI and CSCW researchers who want to explore new uses of the images.

VideoSpace is designed after A. Kay's famous saying: "simple things should be simple, complex things should be possible". It provides users and developers with a set of basic tools and a class library that make it easy to integrate image streams within existing or new documents and applications. The tools, for example, allow

M. Reed Little and L. Nigay (Eds.): EHCI 2001, LNCS 2254, pp. 73–90, 2001.
© Springer-Verlag Berlin Heidelberg 2001

users to display image streams in HTML documents in place of ordinary static images or to embed these streams into existing X Window applications. Creating a video link with the library requires only a few lines of code; managing multiple sources and including video processing is not much more complicated. Since the image streams managed by videoSpace often involve live video of people, the toolkit also provides a flexible mechanism that allows users to monitor and control access to their own image.

The paper is organized as follows. After introducing some related work, I describe videoSpace by showing the architecture and functionalities of its library and basic tools. I then present several projects based on the toolkit that illustrate its potential and the new uses of video it will allow in the future. Finally, I discuss some lessons learned from this work and conclude with directions for future research.

## 2  Related Work

Prototyping has been recognized as an efficient means of developing interactive applications for some time [4]: iterative design promotes the refinement and optimization of the envisioned interaction techniques through discussion, exploration, testing and iterative revision. However, exploring new uses of video through prototyping is hard. Researchers are faced with multiple difficult problems such as the need for digitizing hardware as well as specialized encoding/decoding algorithms and communication protocols. Moreover, evaluating a solution to any of these problems usually means having some solution to all of them.

Many innovative works on the use of image streams overcome these problems by using specific video hardware. Early Media Spaces, for example, were based on analog audio/video networks [5, 6, 7]. ClearBoard [2] also uses an analog video link and dedicated video overlay boards to superimpose two image streams in real-time. Likewise, Videoplace [8] relies on dedicated hardware for image processing. Although these specific hardware solutions allow researchers to focus on the interactions and not the technology required to implement them, they are usually expensive, hard to setup, hard to maintain and sometimes even hard to reproduce.

Specific video hardware allows to create fully functional, high-fidelity prototypes. But high-fidelity prototyping is not good for identifying conceptual approaches, unless the alternatives have already been narrowed down to two or three, or sometimes even one [4]. Low-fidelity prototypes have proved useful to narrow these alternatives. Pen and paper, painting programs or other simple electronic tools can be used to get feedback from potential users from the very early stages of product development. Software toolkits such as Tcl/Tk or GroupKit [9] have long been used for rapid prototyping and iterative development of graphical interfaces and groupware applications. The more recent advent of Web-based applications also promotes the use of quickly hacked HTML interfaces that can be easily modified to explore alternative designs.

The motivation for creating videoSpace resides in the lack of such flexible software tools for exploring new uses of video through the creation of high-fidelity as well as low-fidelity prototypes. Most modern operating systems provide software libraries to manipulate image streams, such as Apple QuickTime, Microsoft DirectX or SGI

Digital Media Libraries. These libraries have all their own advantages and disadvantages, but they are usually platform-dependent and incompatible with each other. These characteristics make them difficult to use for building the complex - and usually distributed - applications required to explore new uses of video. Although I realize the importance of the low-level services offered by these libraries, I believe that the HCI and CSCW communities would benefit from a higher-level software platform for image streams manipulation.

The Mash streaming media toolkit [10] and some other platforms developed by the Multimedia and Network research communities offer high-level video digitizing and transmission services. However, these platforms naturally tend to focus on the transmission techniques and usually rely on the idea that images are to be displayed "as-is", as big and as fast as possible. Although this conception is well suited to applications such as videoconferencing, tele-teaching or video-on-demand, it is too restrictive for more innovative applications that might involve image processing or composition at any point between its production and final use.

The Java Media Framework (JMF) is probably the existing platform closest to videoSpace. It provides programmers with a set of high-level classes to digitize, store, transmit, process and display images. However, it is a closed product of a commercial organization. At the time of this writing, for example, JMF supports video digitizing on Microsoft Windows and Sun Solaris platforms only. Since its source code is not publicly available, implementing this feature on other platforms requires writing separate extensions and possibly rewriting some existing code. Moreover, correcting bugs or implementing new features is a privilege of a few people whose interests might differ from those of the HCI and CSCW research communities.

VideoSpace grew out of my numerous experiences and frustrations in writing applications that digitize, synthesize, transmit, store, retrieve, display, modify or analyze image streams. By moving the common elements of these applications into a library and providing tools such as a network video server, I can now quickly create video applications whose complexity shrunk from several thousand lines of code to only a few hundred lines. This, in turn, facilitates the exploration of a wider range of uses of the image streams to support human-computer interaction and distant collaboration.

## 3   The VideoSpace Library

The videoSpace library was initially developed in C++ on SGI workstations. In its current state, it consists of less than a hundred classes and 15000 lines of code. The hardware and system dependent code is clearly separated from the rest of the code, which allowed us to easily port it to Linux ans Sun Solaris and should facilitate port to other systems. The library and several videoSpace applications were indeed successfully ported to Apple MacOS in September 1999, although this development branch was later abandoned. In addition to the primary C++ library, some of the core services of videoSpace are also available from Python scripts, through a dynamic extension, and some others have been re-implemented in pure Java. Integration with Tcl/Tk is also provided through a specificity of the X Window system that will be detailed in the next section.

## 3.1 General Overview

The videoSpace library is built around the concept of *image*: it provides developers with classes and functions to *produce*, *process*, *transmit*, *display* and *record* images (Fig. 1) and to *multiplex* these basic operations.

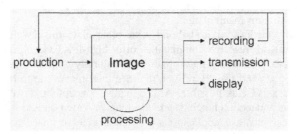

**Fig. 1.** Typical life-cycle of a videoSpace image

VideoSpace images are produced by local or network-accessible *image sources*. An image source may digitize or synthesize in real-time the images it produces. It may also retrieve them from a pre-recorded stream. All image sources are described by a URL that defines them in a unique way. The operation mode of a particular source can be adjusted through the addition of query string parameters to its URL. Such parameters can be used, for example, to specify the frame-rate, the size or the quality of the images produced.

VideoSpace supports real-time processing of images through *filters* that transform or analyze them. Transformation filters modify the dimensions, the data or the encoding of the images whereas analysis filters only extract valuable information from them. The library provides a number of filters for both transforming and analyzing images. Some of them allow to convert images between the different encodings available and to resize them. Others are used for gamma correction or convolution by a 3x3 kernel. Others to paint rectangular regions, to insert an image in another one or to superimpose two images. More complex filters such as chroma-keying, image difference and basic motion detection are also provided.

*Image sinks* are used to transmit images on the network, to display them on the computer screen or on an analog video output (e.g., a separate monitor) and to record them into files. Like image sources, image sinks and their operation mode are described by URLs. As we will see in the following sections, this use of text-based descriptions for sources and sinks allows to specify them at run-time, making it possible to create simple yet powerful applications for manipulating the image streams.

Combining an image source with a sink may result in complex execution flows. For example, if an image coming from a remote source has to be displayed on an analog video output, the application has to listen to the network connection to get the image data and, at the same time, it has to make sure that the hardware is ready to display it when it is completely decoded. The use of multiple image sources or sinks makes this execution flow more complex. Interactive user interfaces add even more event sources to monitor, increasing the complexity further. In order to reduce this complexity, the videoSpace library provides simple mechanisms based on class inheritance and method overriding for multiplexing objects that deal with files, network connections or hardware devices.

## 3.2 Implementation Details

**Image, pixel encodings and memory management.** The `Image` class implements simple data structures that describe the width, height and pixel encoding of a rectangular bitmap and point to the memory location of the corresponding data. The following encodings are supported: L - for luminance -, RGB, ABGR, RGBA, Y'CbCr 4:2:0 and JPEG. In order to facilitate memory management between successive manipulations, the memory location of an `Image` can not be modified without explicitly stating what should happen to the new buffer when it is lately replaced. This information, stored along with the pointer, allows a single `Image` object to successively use different memory areas that are automatically de-allocated when not needed anymore.

**Network support.** The videoSpace library provides several classes for sending and receiving UDP datagrams - unicast or multicast -, creating TCP servers and clients and decoding HTTP messages. These classes allow videoSpace applications to exchange any kind of data, including images. Four network protocols for image streaming have been implemented on top of them: Netscape's HTTP Server-Push extension [11] applied to a series of JPEG images, the client side of the RFB protocol[1], and VSMP and VSTP, two proprietary protocols.

VSMP is based on UDP and can be used for one-to-one (unicast) or one-to-many (multicast) video transmissions. Images are JPEG-encoded so they can fit in a single datagram and a "best effort" strategy is used: lost datagrams are not retransmitted. VSTP was designed for client-type applications that request video from a server. It combines a VSMP transmission of images with an HTTP connection used as a signaling channel: the client sends on the HTTP connection the information required to start the VSMP transmission, i.e. the local host name and a UDP port number, and closes it to stop the transmission.

Although VSMP and VSTP were designed for simplicity and not performance or reliability, their performance level is quite acceptable: over the last three years, they have been used to transmit video streams between France, The Netherlands, Denmark, Germany, Switzerland and Austria at up to 20 QCIF images per second (176x144 pixels) and 10 CIF images per second (320x240 pixels) with a latency of less than half a second and no perceivable image loss.

---

[1] The Remote Frame Buffer (RFB) protocol was developed by AT&T for their Virtual Network Computing (VNC) project [12]. It allows thin client applications to display real-time images of a remote X Window, Apple MacOS or Microsoft Windows desktop server and to send mouse and keyboard events to this server.

**Image production.** The videoSpace library defines an abstract `ImageSource` class with methods for starting the image production, getting the next or the most recent image available[2] and stopping the image production. Several classes derived from `ImageSource` implement the actual source types supported by the library. A factory function allows the run-time creation of a generic `ImageSource` object from the appropriate derived class, given a URL and the desired encoding.

The URL describing a hardware video input on the local machine specifies the device and the actual input used on this device (e.g., analog input number two on digitizing board one). Two generic names, `anydev` and `anynode`, can be used as default values for portability. Three optional query string parameters (`zoom`, `length` and `pause`) control the size of the images, their number and the time between two subsequent ones. A fourth one allows users to indicate whether the hardware resource should be locked (`locked=1`) or whether other applications can preempt it when needed.

URLs corresponding to network sources specify the remote host name or address (or group address, in the case of multicast), possibly followed by a TCP or UDP port number, some path information and some query string parameters. In addition to the four image streaming protocols we mentioned, videoSpace supports to some extent the real-time capture of a window on the computer screen through the X protocol. It also supports two types of pre-recorded image sources, one for single image JPEG files, and the other for files containing JPEG image streams produced by the Server-Push protocol. The following URLs illustrate and summarize the image sources currently implemented:

| | |
|---|---|
| `videoin:/anydev/camera?zoom=3` | local digitizing hardware |
| `http://host:5555/push/video` | HTTP server using the Push extension |
| `rfb://host:1` | VNC server |
| `vsmp://host:9823` | videoSpace app. using VSMP unicast |
| `vsmp://225.0.0.252:5557` | videoSpace app. using VSMP multicast |
| `vstp://host/video?pause=1` | videoSpace app. using VSTP |
| `xwindow://localhost:0/0x1c0000e` | X window (experimental) |
| `file:/tmp/test.jpg` | single JPEG image |
| `file:/tmp/demo.vss` | JPEG stream in the Server-Push format |

**Image processing.** Image filters can be implemented as functions, or as classes that associate data and possibly state-based transitions to the processing algorithm. In the latter case, a `SimpleFilter` class can be used as a base class to share a common syntax and allow the run-time specification of filters. Processing algorithms are usually implemented for a subset of the available encodings (e.g., RGB, RGBA and ABGR). Consequently, encoding conversion might be required before and/or after applying a filter.

---

[2] The most recent image might not be the next one if the source implements some buffering algorithm.

**Image transmission, display and recording.** A pointer to the memory location of an `Image` data can be obtained through a `getData` method. Since the encodings supported by videoSpace are also supported by many low-level graphical or video libraries, its images can be easily manipulated with them. OpenGL, GTK or the Linux SVGA library, for example, can be used to display the images on the computer screen. On SGI O2 and Octane workstations with proper hardware, the Digital Media Libraries can also be used to send them to an external analog video device such as a monitor or a VCR. Recent additions to the library also provide some preliminary support for creating MPEG-1 and RealPlayer compatible video streams.

The videoSpace library defines an abstract `ImageSink` class and several derived classes to facilitate the transmission of images using VSMP, their display using several graphical toolkits, and their recording in several formats. As for image sources, a factory function allows the run-time creation of a generic `ImageSink` object from given a URL. The following URLs illustrate and summarize the image sinks currently implemented:

| | |
|---|---|
| `vsmp://localhost:9823` | videoSpace app. using VSMP unicast |
| `vsmp://225.0.0.252:5557` | videoSpace app. using VSMP multicast |
| `glxwindow://localhost:0` | OpenGL display in an X Window |
| `gtkwindow://localhost:0` | GTK window |
| `svga:/640x480?centered=1` | SVGA full-screen display (Linux only) |
| `videoout:/anydev/anynode` | analog video output (SGI only) |
| `file:/tmp/capture.jpeg` | single JPEG image |
| `file:/tmp/capture.vss` | JPEG stream in the Server-Push format |
| `file:/tmp/capture.mpeg` | MPEG-1 stream (experimental, Linux only) |
| `file:/tmp/capture.rm` | RealPlayer stream (experimental, Linux only) |

**Multiplexing.** The videoSpace library provides a `MultiplexNode` class for multiplexing low-level operations on files, network connections and hardware devices. The `multiplex` method of this class, based on the UNIX `poll` system call, suspends the execution of the application until a timer expires or some descriptor associated to a file, connection or device becomes readable or writable. `multiplex` relies on two other methods: `prepare`, that specifies the set of descriptors to watch and the time limit, and `check`, that defines how the object reacts to low-level state changes. Many classes of the library, including all image sources and sinks, derive from `MultiplexNode` and override these two methods to implement high-level *reactive objects*.

Every `MultiplexNode` object maintains a list of other `MultiplexNode` instances associated to it by the application developer. Calling the `multiplex` method of one object automatically calls the `prepare` and `check` methods of all associated instances in addition to those of the primary object. This allows developers to describe hierarchical structures of high-level reactive objects and to multiplex them in a single call. The following code example illustrates this by showing how to blend a local and a remote image stream and display the resulting images on an analog video output:

```
ImageSource *src1 = createImageSource(Image::RGB,
                   "videoin:/anydev/camera?zoom=2") ;

ImageSource *src2 = createImageSource(Image::RGB,
                   "vstp://remotehost/video") ;

ImageSink *dst =
createImageSink("videoout:/anydev/anynode");

dst->addNode(src1) ; // Associate the two image sources
dst->addNode(src2) ; //     to the image sink

src1->start() ;
src2->start() ;
dst->start() ;

Image img1, img2, composite ;
bool newComposite = false ;

while (dst->isActive()) {

    dst.multiplex() ; // Multiplex the sink and the
sources

    if (src1->getNextImage(&img1)
        || src2->getNextImage(&img2)) {
        // At least one image has changed
        //    update the composite
        blendImages(&img1, &img2, &composite) ;
        newComposite = true ;
    }

    if (newComposite) {
        // The composite stays "new" until handled
        newComposite = !dst->handle(&composite) ;
    }

}
```

## 4  VideoSpace Basic Tools

In addition to the C++ library, aimed at developers, the videoSpace toolkit provides end-users with a number of tools that can be used off-the-shelf, with no or little programming, and serve as building blocks for more complex applications.

## 4.1 VideoServer

The main tool provided by videoSpace is videoServer [13]. VideoServer is a personal Web server run by the user of a workstation and dedicated to video: it is the unique point of access to that person's video sources. The three services it provides are:
1. creating a one-way live video connection;
2. retrieving a pre-recorded video file;
3. acting as a relay for another image source.

VideoServer services are mapped to resource names that are accessible through the HTTP protocol and can be described by simple URLs. Video data itself can be transmitted with the HTTP Server-Push or VSTP. The following URL, for example, requests a live video stream with 1 frame every 60 seconds to be sent with HTTP Server-Push (5555 is videoServer's default port number):
`http://host:5555/push/video?pause=60`

This URL requests a pre-recorded video file to be sent with VSTP to a client listening on `desthost` on port 9257:
`http://host:5555/vstp/movie/demo.vss?host=desthost&port=9257`

This third example illustrates the use of videoServer as a relay for another source, in this case an RFB server. The URL describing the source relayed, `rfb://srchost:1`, has to be encoded in order to be used in the query string of the request:
`http://host:5555/push/relay?src=rfb%3A%2F%2Fsrchost%3A1`

Using custom HTTP servers to provide video services is not new. A number of Webcams available on the Internet work this way, and in fact videoSpace users often use them as image sources. However, videoServer differs from these Webcams on a major issue: it provides its user (i.e., the person who runs it) with access control and notification mechanisms to support privacy. For every request it receives, videoServer executes an external program, the *notifier*, with arguments indicating the name of the remote machine, possibly the remote user's login name, the resource that led to the server - the HTTP referrer - and the requested service. In response, the notifier sends back to the server the description of the service to execute, which can differ from the one the client requested.

The default notifier, a UNIX shell script, allows users to easily define access policies. Low quality or pre-recorded video, for example, can be sent to unidentified users while known people get a high quality live video stream. In addition to computing the service to execute, the notifier can trigger several auditory or graphical notifications to reflect some of the information available, such as the identity of the remote user or the service requested. When a live video request does not specify the number of images, videoServer limits it to 5000, that is, up to three minutes. This ensures that constant monitoring cannot take place without periodically asking permission and thus triggering notifications. All these elements facilitate the acceptance of videoServer by the users and helps finding a trade-off between accessibility and privacy.

## 4.2 VideoClient

VideoClient started as a simple lightweight application designed to display videoSpace streams on the computer screen, the video being scaled to match the window size [13]. As the class library evolved, it became a more generic tool for easily filtering video streams and displaying, recording, or sending them on the network. The current implementation of videoClient is about 100 lines of C++ and allows to specify at run-time an `ImageSource`, a `SimpleFilter` and an `ImageSink`. For example, videoClient can be used for:

1. recording images from local hardware into a file
   videoClient -i videoin:/anydev/anynode \

   -o file:demo.vss

2. multicasting the recorded sequence after applying a difference filter
   videoClient -i file:demo.vss \

   -f difference \

   -o vsmp://225.0.0.252:5557

3. displaying the multicasted stream in an X window using OpenGL
   videoClient -i vsmp://225.0.0.252:5557 \
       -o glxwindow://localhost:0

In addition to top-level windows which can be manipulated through the window manager, the `glxwindow` image sink can take advantage of the architecture of the X Window system to display video in a new subwindow of an existing one. This can be done by simply specifying the id of the parent window:

```
videoClient -o glxwindow://localhost:0?parent=0x120000e
```

This feature of `glxwindow` can be used to "augment" existing X Window applications with video streams. Since the video window is a child of the host window, it is moved, raised, lowered and iconified with it. This way for example, one can easily add video to a text-based chat application. This approach can also be used with user interface toolkits that explicitly support widgets that host external applications. For example, the frame widget of the Tcl/Tk toolkit can host a separate application by setting the widget's *container* property to true. The sample code below illustrates the use of videoClient to embed a video stream in a Tk frame:

```
# create a frame widget for the video
frame .video -container true -width 160 -height 120
set id [winfo id .video]
# launch a videoClient inside the frame
exec videoClient \
```

```
        -i vstp://host/video \
        -o glxwindow://localhost:0?parent=$id &
# make the frame visible
pack .video
```

## 4.3 Other Video Tools

In addition to videoServer and videoClient, videoSpace provides users with a number of other tools to manipulate, display or convert video streams. One of these tools, for example, allows to change the width and height of a stream. Another one can combine several streams in a single one using a mosaic placement. A simple UNIX Shell script also allows users to "stick" a videoClient on any existing X Window using the parent parameter of the glxwindow image sink.

# 5  Exploring New Uses of Video

In this section, I present several projects that were developed with videoSpace and benefitted from its flexibility and extensibility. The details of these projects unfortunately fall beyond the scope of this paper. Some have been published and some others will be. What I want to illustrate here is how the variety of image sources provided by videoSpace combined with its facilities for transmitting, displaying and recording video streams support the rapid prototyping and incremental development of video applications. How videoSpace facilitates the exploration of new uses of video.

## 5.1  Using Documents as Interfaces to Awareness and Coordination Services

Over the last few years, I have been exploring the use of Web-based video environments to provide awareness and coordination services to distributed groups of people [13]. Inspired by previous work on Media Spaces, I have used videoSpace to promote the development of collaborative environments in which video communication facilities are embedded into the existing environment of the users, i.e. their documents and applications, rather than provided as separate applications.
Most Web browsers can display JPEG images and some of them can also display an HTTP Server-Pushed JPEG stream in place of an ordinary image without any plug-in. Combined with these browsers, videoServer allows to include live or pre-recorded video streams in HTML documents by using code such as:

```
<img src="http://host:5555/push/video">
```

By including such references to videoServers, one can easily create dedicated interfaces to the video communication space, such as the awareness view of Fig. 2, but also "augment" existing documents by embedding video services in the existing content. For example, one can include a video link in an e-mail message so the receiver will see the sender's office when he reads the message. When cooperatively editing an HTML document, the authors can also include live views of their offices in the document so they can see if their co-authors are present when they work on it.

**Fig. 2.** HTML-based awareness view showing several videoServers

HTML-based interfaces can be easily shared and exchanged. As more and more people are getting familiar with HTML authoring, they are also easily tailorable. Using HTML and videoServers to create a Web-based video communication environment emphasizes the design principle that "simple things should be simple". VideoServer is indeed in daily use in several places around the world. End users have created HTML interfaces to it and have developed usage patterns without any knowledge of the architecture of the system.

## 5.2  Designing a New Communication Device to Support Teleconviviality

Videoconferencing systems are getting closer to technical perfection every year, making them more and more pleasant to use. However, the usual setting of these systems favors a face to face between all the participants of each site, which makes cross-site informal conversations difficult: before and after a formal meeting or during pauses, people tend to discuss with some of their co-located partners and often ignore the remote people. I strongly believe that one of the keys to informal communication in these situations is in the ability for people to move away from the central focus point of the formal meeting and gather in small cross-site groups isolated from each other. Starting from this idea, several research partners and I designed a new communication device, *le puits* [14] (the well), that combines an SGI O2 workstation with microphones, cameras, speakers and a horizontal video projection system to establish an audio and video link between distant groups of up to 6 people (Fig. 3).

**Fig. 3.** *Le puits*: hardware prototype and sample display configurations. The two rightmost configurations add pre-recorded simulated views to the real views from the prototype

The design of *le puits* required several iterations from the initial sketches to fully functional prototypes. What is interesting to mention here is that videoSpace allowed to develop the software used to compose the projected images several hundred kilometers away from the place where the hardware parts were actually assembled. Pre-recorded video streams showing a rough view of what the cameras would see were used to experiment with different composition methods even before the first prototype was built. Network sources were later used to simulate analog video cables and proprietary codecs between two prototypes. The mosaic composer application was also used to simulate its analog equivalent when it was decided to use such a device to put all the cameras of a prototype on a single video stream. In addition to supporting rapid prototyping of the software and simulation of various hardware configurations, videoSpace allowed us to develop the software required for *le puits* on a laptop running Linux and to later recompile it without a single modification and run it on the SGI O2s of the prototypes.

### 5.3 Using the Hand as a Telepointer

Gesturing is a natural means of communication for humans. Hand gestures in particular are often used to express ideas, to refer to objects, to attract attention or to signal turn taking. To recreate this communication channel over distance, real-time groupware systems usually display telepointers that participants can move over the shared view. However, standard telepointers usually lack semantic information. At best, they are chosen among a predefined set of shapes and/or colors, which makes it hard to draw attention, to designate several objects at the same time or to express an idea. A mouse cursor is a very poor substitute for the hand for gesture communication, and some colleagues and I thought that the image of the hand itself, captured in real-time, would do a better telepointer. We set up a camera above a desk covered by a large blue sheet of paper and recorded several video sequences showing some hand gestures over this solid-color background. We then developed a chroma-keying filter to extract the image of the hands from these sequences and several prototypes to display them over other image streams or running applications [15].

The first prototype superimposed the chroma-keyed video stream over a screen-shot of a supposedly shared application. This first implementation gave us some interesting hints for further tests. We realized, for example, that the chroma-keying process allowed us to annotate the shared view with real world objects. We implemented a second prototype in which the user could control the size, the position and the transparency of the chroma-keyed overlay with the mouse and the keyboard. This second prototype used the same "parent window" trick as videoClient to overlay the chroma-keyed video stream on a running application. This solution, however, couldn't be applied to more than one application at the same time and was too specific to the X Window system. Instead of trying other complex ways to superimpose our chroma-keyed video stream over a traditional computer desktop, we simply used an RFB image stream as the background of a third prototype (Fig. 4).

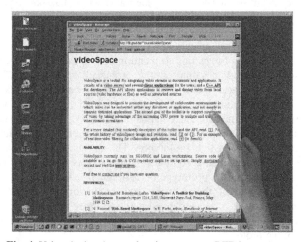

**Fig. 4.** Using the hand as a telepointer over an RFB image stream

It is still too early to claim that the image of the hands are indeed better than a traditional telepointer. However, again, this project emphasizes the rapid prototyping and iterative design paradigm supported by videoSpace, from the development of the chroma-keying filter using pre-recorded sequences to the implementation of several alternative designs using different image sources. As the chroma-keying filter was integrated into the videoSpace library, the complexity of the prototypes was reduced to the minimum: each of them is only 200 lines of C++ code.

### 5.4 Exploring New Desktop Interactions

As illustrated by the previous example, using the images of computer desktops in videoSpace applications offers some interesting perspectives for exploring new interactions or metaphors for these desktops. I am currently developing videoDesktop, a new videoSpace application that acts as a window manager for an X Window server accessible through RFB (an Xvnc server). As a window manager application, videoDesktop knows the dimensions of all the application windows. By using a simple tiling placement algorithm (Fig. 5), it is able to extract the images of individual applications from the RFB image stream and use them as textures to compose a new desktop view.

**Fig. 5.** Sample view of the tiled desktop image stream

Figure 6 shows an example of graphical re-composition of the applications shown in Fig. 5. This composition scales, rotates and translates every application and adds some transparency and a simple shadow. In its current state, videoDesktop allows to send keyboard events to the applications based on a click-to-focus policy. Mouse event coordinates would have to be transformed before sending the event to the Xvnc server. However, this transformation is not currently implemented.

**Fig. 6.** View composed by videoDesktop from the tiled desktop image stream of Fig. 5

The technique used by videoDesktop is quite similar to the *redirection mechanism* of the Task Gallery [16]. However, whereas Task Gallery requires a modified version of Windows 2000, videoDesktop is only 800 lines of code and requires only a standard Xvnc server. Although this project is still in a very early stage, it already allows to use a modern graphical toolkit such as OpenGL to experiment new window managing techniques that were until now relying on graphical models from the 80s or 90s.

## 6   Discussion

The library and tools composing videoSpace are the result of an iterating process of design and use. As the library was getting more complex, the tools were getting simpler. As the level of abstraction of the classes was increasing, the number of lines of code of each tool was decreasing. In this section, I discuss several aspects of the toolkit that contributed to this evolution and its success.

One of the early key choices was the separation between the Image class, used as a simple data container, and the sources, filters and sinks that operate on the data. The implementation of these image operators as independent classes instead of methods of the Image class increases the flexibility of the toolkit and facilitates its extension. Every application can easily define its own operators that can be created and modified at run-time. If these new operators prove useful for other applications, they can then be integrated into the core library.

The JPEG encoding and the HTTP Server-Push protocol were initially chosen for their ease of use and implementation, not their performance. The same desire of simplicity inspired the design of VSMP and VSTP. However, videoSpace is not restricted to these simple choices. New encoding or protocols can be easily added to the library to overcome performance problems or to communicate with other applications. Adding an encoding simply consists in adding a constant to the list of supported encodings in the Image class and providing a few conversion filters. Similarly, as illustrated by the rfb image source, adding a protocol simply consists in adding classes derived from ImageSource and/or ImageSink and changing the associated factory functions.

Another early key choice was the use of URLs to name video sources. When a new source type is added to the toolkit, the factory function is modified and all applications benefit from it without any other change. As explained in the previous section, local sources or pre-recorded streams are often used instead of network sources during development to test an application, to compare alternative designs or to evaluate performance. The use of URLs make it possible to switch between sources without even recompiling the application; therefore, applications can be developed on machines that do not have any video hardware or even a network connection. Another advantage of using URLs is the seamless integration with Web applications, most notably Web browsers.

# 7  Conclusion and Future Work

VideoSpace is a software toolkit designed with one specific goal in mind: facilitate the exploration of new uses of video. In this paper, I have described the class library and basic tools it provides. I have also presented several projects that illustrate its flexibility and its ability to support rapid prototyping and incremental development. VideoSpace has now reached a stable state and some of its applications are in daily use in several research labs. Its source code is publicly available for download[3]. Short-term development plans include the integration of additional video encodings and protocols such as MPEG-4. I would also like to experiment with filters for detecting, identifying and tracking objects or people, based on the techniques presented in [3]. Another direction for future work is the addition of high level functionalities related to session management, such as explicit support for bi-directional or multiple connections.

### Acknowledgments

VideoSpace grew out of the many video applications I wrote while I was working at LRI - Université Paris-Sud on the Telemedia project, funded by France Télécom R&D. Thanks are due to Michel Beaudouin-Lafon, who contributed to many ideas presented in this paper. I am also grateful to Jacques Martin, Jean-Dominique Gascuel and the other people from CSTB and iMAGIS who contributed to the design and prototyping of *le puits*. Current work on videoSpace is supported by ERCIM, the European Research Consortium for Informatics and Mathematics, and the Swiss National Science Foundation.

# References

1.  W. Mackay. Media Spaces: Environments for Informal Multimedia Interaction. In M. Beaudouin-Lafon, editor, Computer-Supported Co-operative Work, Trends in Software Series. John Wiley & Sons Ltd, 1999.
2.  H. Ishii, M. Kobayashi, and K. Arita. Iterative Design of Seamless Collaboration Media. Communications of the ACM, 37(8):83–97, August 1994.
3.  J.L. Crowley, J. Coutaz, and F. Bérard. Perceptual user interfaces: things that see. Communications of the ACM, 43(3):54–64, March 2000.
4.  J. Rudd, K. Stern, and S. Isensee. Low vs. high-fidelity prototyping debate. ACM interactions, 3(1):76–85, 1996.
5.  M. Olson and S. Bly. The Portland Experience: a report on a distributed research group. International Journal of Man-Machine Studies, 34:211–228, 1991.
6.  C. Cool, R.S. Fish, R.E. Kraut, and C.M. Lowery. Iterative Design of Video Communication Systems. In Proc. of ACM CSCW'92 Conference on Computer-Supported Cooperative Work, Toronto, Ontario, pages 25–32. ACM, New York, November 1992.
7.  W. Buxton and T. Moran. EuroPARC's Integrated Interactive Intermedia Facility (IIIF): Early Experiences. In Multi-User Interfaces and Applications, pages 11–34. S. Gibbs and

---

[3] See http://www-iiuf.unifr.ch/~rousseln/projects/videoSpace/

A.A. Verrijn-Stuart, North-Holland, September 1990. Proceedings of IFIP WG8.4 Conference, Heraklion, Greece.

8.  M. Krueger. Artificial Reality II. Addison-Wesley, 1991.

9.  M. Roseman and S. Greenberg. Building Real Time Groupware with GroupKit, A Groupware Toolkit. ACM Transactions on Computer-Human Interaction, 3(1):66–106, March 1996.

10. S. McCanne, E. Brewer, R. Katz, L. Rowe, E. Amir, Y. Chawathe, A. Coopersmith, K. Mayer-Patel, S. Raman, A. Schuett, D. Simpson, A. Swan, T-L. Tung, D. Wu, and B. Smith. Toward a Common Infrastructure for Multimedia-Networking Middleware. In Proc. of 7th Intl. Workshop on Network and Operating Systems Support for Digital Audio and Video (NOSSDAV 97), May 1997.

11. An Exploration of Dynamic Documents. Technical report, Netscape Communications, 1995.

12. T. Richardson, Q. Stafford-Fraser, K.R. Wood, and A. Hopper. Virtual Network Computing. IEEE Internet Computing, 2(1):33–38, Jan-Feb 1998.

13. N. Roussel. Mediascape: a Web-based Mediaspace. IEEE Multimedia, 6(2):64–74, April-June 1999.

14. N. Roussel, M. Beaudouin-Lafon, J. Martin, and G. Buchner. Terminal et système de communication. Patent submitted for approval by France Télécom R&D (INPI n°00-08670), July 2000.

15. N. Roussel and G. Nouvel. La main comme télépointeur. In Tome 2 des actes des onzièmes journées francophones sur l'Interaction Homme Machine (IHM'99), Montpellier, pages 33–36, Novembre 1999.

16. G. Robertson, M. van Dantzich, D. Robbins, M. Czerwinski, K. Hinckley, K. Risden, D. Thiel, and V. Gorokhovsky. The task gallery: a 3D window manager . In Proc. of ACM CHI 2000 Conference on Human Factors in Computing Systems, pages 494–501. ACM, New York, April 2000.

## Discussion

*D. Salber:* This is more of a comment than a question - you mentioned the privacy provided for the person in front of the camera, but what about the person connecting to the camera their address is revealed. There is privacy, but it is only one way.

*N. Roussel:* The user identity is hidden, but one cannot hide connecting machine identity. I'm not highly concerned with this level of privacy, I do not think it is a major issue - if someone wants to connect to me, then they should not be concerned with me knowing who it is. That level of privacy is OK.

*J. Coutaz:* A remark: Because of the lack of reciprocity, I stopped looking at your web page. I felt I was a "voyeur", that is, looking at you while you were not aware of me.

*N. Roussel:* Experience with the system so far has not shown this to be a common reaction.

*N. Graham:* Where did you use VNC protocol in the example showing a hand over a desktop. I did not realize this was possible.

*N. Roussel:* VNC was used because the example is not showing screen sharing, it is showing a remote display, the window shown is that of different system.

# Prototyping Pre-implementation Designs
# of Virtual Environment Behaviour

James S. Willans and Michael D. Harrison

Human-Computer Interaction Group
Department of Computer Science, University of York
Heslington, York, Y010 5DD, U.K.
{James.Willans,Michael.Harrison}@cs.york.ac.uk

**Abstract.** Virtual environments lack a standardised interface between the user and application, this makes it possible for the interface to be highly customised for the demands of individual applications. However, this requires a development process where the interface can be carefully designed to meet the requirements of an application. In practice, an *ad-hoc* development process is used which is heavily reliant on a developer's craft skills. A number of formalisms have been developed to address the problem of establishing the behavioural requirements by supporting its design prior to implementation. We have developed the Marigold toolset which provides a transition from one such formalism, Flownets, to a prototype-implementation. In this paper we demonstrate the use of the Marigold toolset for prototyping a small environment.

## 1  Introduction

One of the characteristics of virtual environments is the lack of a standard interface between the user and system. With virtual environments it is necessary to construct interfaces that support the specific requirements of individual applications [3]. Consider a flight simulator. The components that constitute its interface (including the devices, interaction techniques and objects rendered to the user) are all concerned with simulating the effect of flying the real aircraft. Another application, such as medical training, could not reuse the interface component of the airplane successfully, even though the application may share common goals such as training. This generic lack of standardisation is not surprising considering that virtual environments often seek to imitate the real world. For instance, compare interfaces for driving a car, flying a plane, opening a tin or opening a carton of milk. Each interface matches the requirements of its application and is quite different in terms of the information communicated and physical actions.

The lack of standardisation in virtual environment interfaces contrasts with the dominant style of WIMP (windows, icons, mice and pointers) interaction. WIMP interfaces, such as Microsoft Windows, reuse a consistent interface regardless of application. The devices (mouse and keyboard), interaction techniques (point and click) and components (buttons) are all standardised. This consistent style forms the basis of the success of WIMP applications, because users are aware of how to interact

M. Reed Little and L. Nigay (Eds.): EHCI 2001, LNCS 2254, pp. 91–108, 2001.
© Springer-Verlag Berlin Heidelberg 2001

with new applications because of their knowledge of previous applications. However, the success of virtual environment applications relies on the ability to recreate real world, or novel, interfaces particular to the needs of individual applications. An important side effect of the inconsistent nature of virtual environments is that the work required to build an application interface is vastly increased compared to that of a WIMP. A developer must consider carefully how the requirements of a particular application determine the design of the interface for that application.

There are two parts of a virtual environment application that form the human-computer interface. Firstly, the visual appearance and geometry of the renderings including representations of the user within the environment and, secondly, the behaviour of the interface, including the mapping of the user onto the environment via interaction techniques, and the behaviour of the environment itself in response to user interaction. The visual renderings are usually constructed in a 3D-modeller such as 3Dstudio [2]. These tools model the renderings as they will appear in the environment allowing easy verification of whether they meet the requirements or not (often these are compared to photographs [14]). By contrast, the behaviour of the interface is defined using program code within a virtual environment development application such as [5, 7] in a form that is difficult to analyse in terms of the initial requirements. Consequently, rather than dealing with abstract requirements such as the door must open half way when the user clicks the middle mouse button, the developer must deal with implementation oriented abstractions such as the low-level data generated from the input device(s) and complex mathematical transformations of 3D co-ordinate systems. Additionally, design decisions are embedded within the program code without any higher level documentation. This makes inspection and maintenance difficult. These issues are discussed in more detail in [26].

A strategy which has found some success addressing similar problems within software engineering and human-computer interaction is behavioural design (see [28, 11], for instance). Behavioural design is the process of constructing abstract representations of the behaviour of a system. The aim of this is to describe characteristics of the behaviour in a way that is independent of unnecessary implementation concerns. These descriptions are also used to discuss the design, perform reasoning (formal and informal) and inform the programmer of the precise requirements of the implementation. In addition, such designs facilitate the documentation and subsequent maintenance of the system. The motivation for incorporating this style of behavioural design into the development of virtual environments has recently been recognised and a number of formalisms which attempt at various levels of rigour to support this process have been developed (see [9, 12, 16, 23, 25, 31]).

Despite the clear advantages of pre-implementation behavioural design, it is unrealistic to expect this approach to determine the behaviour of a virtual environment interface completely, particularly in view of the diversity of possible designs. As noted by Myers `the only reliable way to generate quality interfaces is to test prototypes with users and modify the design based on their comments' [18], a view strongly supported in [30]. Thus, a more realistic situation is where behavioural design is closely integrated into a prototyping process such that designs can be tried out and tested with users and the designs refined to reflect their feedback. In order to address this goal we have developed the Marigold toolset [33, 34]. This toolset supports the rapid transition from pre-implementation designs of virtual environment interface behaviour (using the Flownet specification formalism [23, 25]) to fully working implementation-prototypes. Marigold also provides a means of exploring different configuration of devices and behaviours [34] early in the design cycle. Currently the code-generation module of Marigold is for the Maverik toolkit [1], but the approach is independent of any specific environment. A number of rudimentary model-checking facilities are also provided [35]. In this paper we demonstrate how Marigold supports the prototyping of Flownet specifications by incrementally building a small kitchen environment.

The remainder of the paper is structured as follows. In section 2 we give an overview of the Flownet specification formalism and, in order to strengthen motivation for the use of this representations, we compare a specification to its equivalent program code. In sections 3, 4 and 5 we exemplify Marigold by building specifications and prototypes incrementally for a virtual kitchen. In section 6 we examine related work. Finally, in section 7 we summarise our conclusions.

## 2    Flownets

A number of formalisms have been used for the description of virtual environment interface behaviour. In [22, 31], process algebra is used to describe interaction techniques, and in [24] the use of state based notations such as Statecharts [10] are investigated. However, more recently there has been a general opinion that virtual environment interface behaviour is better considered as a hybrid of discrete and continuous data-flow components [12, 36]. With this in mind, a number of further formalisms have been investigated [32, 36]. The formalism developed and presented in [13, 17] and Flownets presented in [23, 25] were both developed specifically for the description of virtual environment interaction techniques. This make it possible to abstract from low-level mathematical transformations and data-structures which may confuse a design. One of the major differences between the two notations are that Flownets use Petri-nets [20] to describe the discrete abstractions rather than state-transition diagrams. This enables the description of multi-modal (concurrent) interaction using Flownets. Moreover, although virtual environment behaviour may only consist of interaction techniques in simple environments (walk-throughs), more advanced environments contain world objects with their own behaviour. The user interacts with these world objects using appropriate interaction techniques. In [27] it is demonstrated how Flownets are able to model world objects in addition to interaction techniques. The relation between the user, interaction technique behaviour and world object behaviour is shown in figure 1.

In addition to using Petri-nets to describe the discrete elements, Flownets use constructs from a notation originally intended to model system dynamics [8] for the description of the continuous data flow and transformation processes. We will describe the formalism by way of an example of an interaction technique. The mouse based flying interaction technique enables the user to navigate through a virtual environment using the desktop mouse to control the direction and speed. Variations of this technique are used in many virtual environment packages (see for example VRML [4]). One variation works as follows. The technique is initiated by pressing of the middle mouse button and moving the mouse away from the clicked position. Once the mouse is a threshold distance away from the clicked position, the user's movement through the environment is directly proportional to the angle between the current pointer position and the point where the middle mouse button was pressed. The distance between these two positions determines the speed. A second press of the middle mouse button deactivates flying.

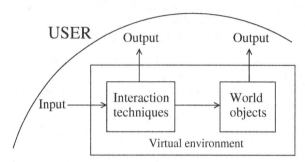

**Fig. 1.** The relation between the user the behaviour of virtual environments

The Flownet specification of the mouse based flying interaction technique is shown in figure 2. The technique has two plugs to the external environment: one input *mouse*, and one output *position*. When the middle mouse button is pressed the middle *m/butt sensor* is activated and the *start* transition fired (1). The *start* transition enables the flow control, which enables the transformer (*update origin*), which updates the value of *origin* held in a store with the current mouse position (2) (taken from the *mouse* plug). A token is then placed in the *idle* state. When the *out origin* sensor detects that the mouse has moved away from the *origin* position, transition (3) is fired which moves the token from the *idle* state to the *flying* state. A token in the *flying* state enables the corresponding flow control which enables the transformer (*update position*) to update the *position* data in the store using the current *mouse* position and the *origin* position (4). This is then output to the *position* plug. Whenever the *flying* state is enabled, the inhibitor connecting this state to the *start* transition implies that the *start* transition cannot be re-fired. When the *in origin* sensor detects that the *mouse* has moved back to the *origin* position, a transition is fired which returns the token from the *flying* state to the *idle* state closing the flow control and halting the transformation on *position*. Regardless of whether the technique is in the *idle* or the *flying* state, it can be exited by the *middle m/butt* sensor becoming true and firing either one of the two *exit* transitions (5 or 6).

The argument for using this formalism is that there is clarity about the implementation of requirements and about the characteristics of behaviour. This point can be illustrated by comparing the Flownet representation of the mouse based flying interaction technique in figure 2 with the equivalent implementation code in Appendix A written in Maverik. It is not important to understand the code, rather to appreciate how the behaviour of the technique is clearer in the Flownet representation of figure 2. This is because the behavioural structure (what happens when and why) of the technique is explicit in the Flownet but implicit in the code. Additionally, although the code contains more detail than the Flownet representation (low level data state and transformations, for instance), it is difficult to relate the abstractions to the requirements.  If we treat the informal description given earlier in this section as the requirements, then it can be seen that the Flownet captures these concepts, for instance *idle*, *flying* and *middle mouse button*.  More importantly, the Flownet captures a precise relation between these concepts. For instance, that it is necessary to be in the state of *flying* in order to *update position,* and that *update position* requires positional data *(origin)* created during the transition to the *idle state.*

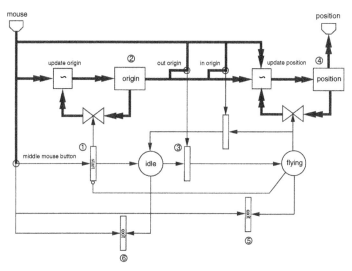

**Fig. 2.** Flownet specification of the mouse based flying interaction technique

## 3  Interaction Techniques

The Marigold toolset supports a design process which closely integrates virtual environment behaviour specification using the Flownet formalism, and the prototyping of that behaviour. In the next sections we exemplify this design process by building elements of a virtual kitchen. Within this section we provide a method for the user to navigate around the kitchen. For this we will use the mouse based flying technique introduced in the previous section. Prototyping a Flownet description of an interaction technique using Marigold is a two stage process. The first stage takes place in the hybrid specification builder (HSB) which supports the specification of a Flownet using direct manipulation (figure 3). At this point, a small amount of code is added to some of the nodes of the specification. This code describes the semantics of some of the Flownet components more precisely. There are three types of code that can be added:

Variable code. This is placed in the plugs of the specification. It describes what kind of information flows in and out of the plugs and, hence, around the specification. Illustrated in figure 4 (a) is the code added to the *mouse* plug. An integer variable represents the state of the mouse buttons and a vector represents the mouse position. Variable code is also used to define data which reside in the stores.

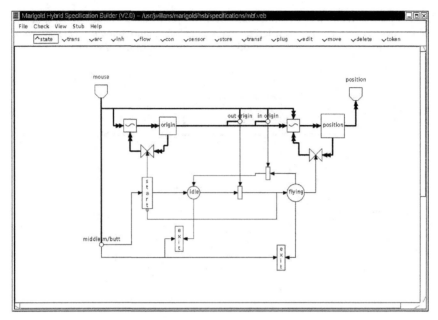

**Fig. 3.** The mouse based flying specification in Marigold HSB

Conditional code. This is placed in some transitions and all sensors. It describes the threshold state of the data for firing the component. Illustrated in figure 4 (b) is the code added to the *middle m/butt* sensor. As can be seen from figure 4 (b), the HSB informs the developer which data flow in and out of the node (i.e. which data they are able to access). The code specifies that when the middle mouse button is pressed, the sensor should fire.

Process code. This is placed in all transformers and denotes how the information flowing into the transformer is transformed when enabled. Illustrated in figure 4 (c) is the code added to the *position* transformer. This describes how *position* should be transformed using the current *mouse* position and the *origin* position.

Once the code has been added, it is necessary to generate a stub of the interaction technique. This is a description of the interaction technique which is independent of an environment. No commitment is made to the inputs and outputs of a technique.

The second stage of implementing a Flownet specification involves integrating the interaction technique stub, generated from the HSB, into an environment (input devices and output devices, for instance). This is performed using the prototype builder (PB). Illustrated in figure 5 is the mouse based flying interaction technique within the PB and connected into an environment. Each node has a set of variables. The variables for the mouse based flying interaction technique (*mbf*) are those that were placed in the plugs within the HSB. The relation between the environment elements are defined by joining these variables enabling the flow of data from one to another. Within the mouse based flying specification, we have linked a *desktop mouse*, as an input to the technique, and a *viewpoint*, as an output from the technique. From this specification, the code for a prototype environment can be generated and compiled. However, for the navigation to be perceived world objects must be present

within the environment. In the next section, we specify and introduce a world object
for the user to observe as they navigate.

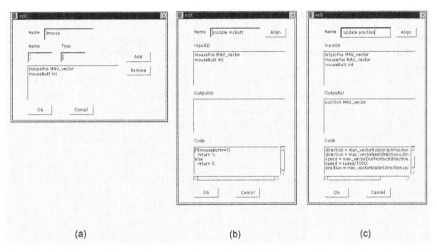

(a)                              (b)                              (c)

**Fig. 4.** a) Adding variables to the mouse input plug b) Adding conditional code to the middle
mouse button sensor c) Adding process code to the position transformer

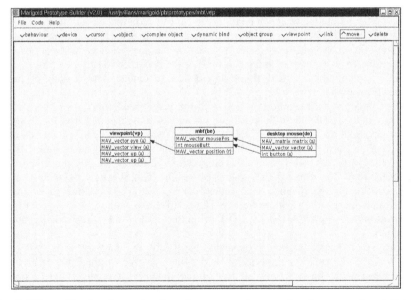

**Fig. 5.** PB specification for a virtual environment prototype using the mouse based flying
interaction technique

## 4  World Objects

Often virtual environments are simply walk-through where the user navigates the environment observing static renderings of world objects. However, more complex environments reflect the real world where the user can interact with and observe world objects. In this section we expand the environment of the previous section to include a virtual gas hob (oven) world object with which the user can interact.

A world object specification is also constructed using the HSB and a stub generated as described in the previous section. Rather than using the PB to integrate the stub of the world object's behaviour into an environment directly, a third tool, the Complex Object Builder (COB), supports an intermediate refinement stage. Within the COB a link is made between a world object's behaviour stub (generated from the HSB) and the visual renderings constructed using a third party 3D-modelling tool such as 3DStudio [2]. Any additional information required from the external environment (the PB) is also made explicit using the COB. From the COB specification, another stub is generated and integrated into the complete environment (interaction techniques and devices, for instance) within the PB specification. This intermediate refinement stage simplifies the PB specification because a single node is an encapsulation of both a world object's behaviour and appearance. In addition, a world object's encapsulated behaviour/appearance may be reused. The COB supports this type of reuse by packaging these components together in a reusable node.

Figures 6 and 7 illustrate the discrete and continuous parts of the behaviour of the hob respectively. Although the HSB supports the construction of a Flownet specification using one view, it can be useful to split the continuous and discrete part of larger specifications to maintain clarity of presentation. As can be seen from these two figures, common constructs (sensors and flow controls) relate the two views. Code is added to some of the nodes constituting the specification in the manner as described for the mouse based flying interaction technique, and a stub of the behaviour is generated.

The next stage is to integrate this stub with the visual renderings using the COB (figure 8). The node labelled *hob* is the stub of the gas hob generated from the HSB. Representations of the renderings of the *gas switch*, *ignition* and *flame* are related to the behaviour because their states are changed according to the Flownet specification. However the *oven body* is not related because, although this is visually perceived, it has no behaviour. The *external link* node specifies that data is required which is not contained within the specification. Consequently, the position of the virtual hand must be linked into the gas hob within the PB. The relative positioning of the rendering representations are also set within the COB. From the COB, a stub of the specification is generated.

Our original PB specification is expanded to include the complex world object. Figure 9 shows the prototype specification that includes the original mouse based flying interaction technique (*mbf*) and a simple manipulation (*sman*) interaction technique. The *sman* technique, also defined using a Flownet, controls the position of a pointer world object by two mappings of the *keyboard* device. The *hob* world object stub can be seen on the left side of the specification. The one variable within this complex world object is linked to the pointer as required by the *external link* within the COB specification.

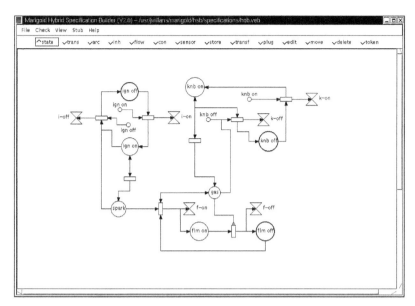

**Fig. 6.** The discrete part of the Flownet specification for the gas hob within the HSB

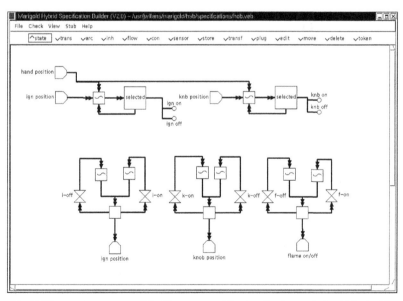

**Fig. 7.** The continuous part of the Flownet specification for the gas hob within the HSB

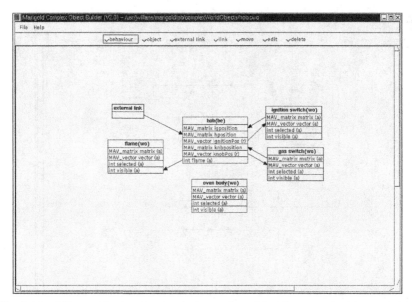

**Fig. 8.** The COB specification showing the integration of the Flownet stub of the gas hob into the external environment

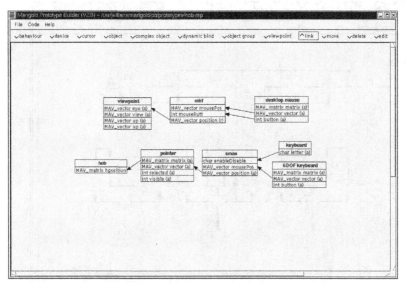

**Fig. 9.** The PB specification extended to include the integration of a complex world object and manipulation interaction technique

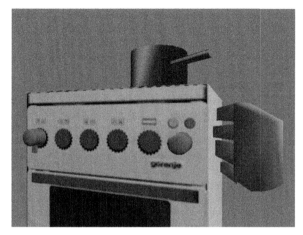

**Fig. 10.** The prototype (as generated from the PB specification) running

The prototype generated from this specification is illustrated in figure 10. It allows the user to navigate around the environment using the mouse and manipulate the virtual pointer using the keyboard. The virtual pointer can be used to interact with the hob which will behave according to its Flownet specification. The gas on/off switch is on the left side of the oven, and the push ignition on the right. Additionally, a pan is situated on top of the hob (this is part of the oven body rendering representation in figure 8).

## 5   Non-static Binding

Flownet specifications are not concerned with the environment external to the behaviour, they abstract from this by interfacing to plugs. The PB and the COB tools provide a means of binding an environment to the plugs of Flownets. The binding style described in the previous sections is static. However, a developer may wish to specify that a selection interaction technique can select one of a number of objects within the kitchen, without explicitly linking every object to the selection technique. Similarly, a developer may wish to state that any object placed within a kitchen drawer is affected by the opening and closing behaviour of the drawer because it is in the drawer. This type of non-static binding is supported by two additional constructs within the Marigold PB namely *world object group* and *dynamic bind*. In this section, we expand the specification we have developed in the previous two sections to include behaviours which make use of these constructs. The expanded PB specification is illustrated in figure 11 with a number of new nodes and links.

Firstly, as illustrated at the top of the specification of figure 11, the *world object group* construct provides a method of grouping world object renderings (*ball one* and *ball two*). In this example, the select interaction technique is using this group to determine which object is selected by the pointer object and to change the state of the ball objects selected variable (from false to true). The *sman* (simple manipulation) technique then controls the position of whichever ball is selected.

Secondly, the *dynamic bind* construct allows conditions such as *in the drawer* to be described so that when an object (rendering) satisfies the statement it binds to the behaviour. In figure 11 such a construct (*db*) can be seen labelled *in drawer* and linked to the position of the *drawer* complex object. When a dynamic bind is inserted into a PB specification, the tool asks the user to specify a visual rendering and set the position of this visual rendering. For the *in drawer* dynamic bind a rendering was constructed to represent the space inside the virtual drawer (this is not displayed in the environment). This rendering is positioned so that it is initially inside the drawer. A number of options can also be set on a *dynamic bind* construct. In this example, it is specified that when an object is fully within the space defined by the rendering (in the drawer) it will bind dynamically to the behaviour of the *drawer* and that the *dynamic bind* rendering itself should also bind to the behaviour. Consequently, when objects, such as *ball one* and/or *ball two* are placed within the *drawer*, they bind to the *drawer's* behaviour and open and close with the drawer. In addition, although the bind rendering cannot be observed in the environment, this always remains inside the virtual *drawer*.

The *dynamic bind* construct also has the potential to specify physical laws within a virtual environment. For instance, a bind could be constructed the size of the environment which is linked to a Flownet specification imposing a gravity behaviour. Consequently, when objects are within the environment, they are linked to this behaviour. Or, a bind could be placed within a swimming pool, such that when an object enters the water, it is linked to the gravity of water. To enable this, it would be necessary for Marigold to support the generation of collision detection.

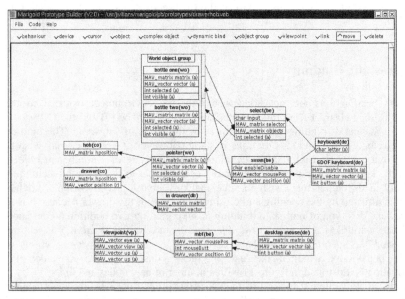

**Fig. 11.** Expanded PB specification with an additional drawer complex world object, selection interaction technique and dynamic binding constructs

The small kitchen example now supports navigation, control of the virtual hand, behaviour of the gas hob, behaviour of the drawer and the selection and manipulation of the two balls. These balls are initially positioned within the drawer and therefore bind to the drawer's opening and closing behaviour. The virtual hand can be used to remove either one or both of the balls from the drawer, whereupon they lose their binding to the drawer's behaviour. A screenshot of this environment is shown in figure 12.

**Fig. 12.** Final prototype with a drawer, hob and three interaction techniques

# 6   Related Work

The distinguishing feature of Marigold is the initial consideration of the virtual environment interface behaviour independent of any specific implementation using Flownets. The approach presented in [13, 17] links higher level abstractions to implementation components, such that when these abstractions are rearranged, the change is propagated into the underlying implementation. This is in the style of user interface management systems (UIMS) which has been researched widely for other styles of interface. However, such an approach can be seen to be implementation driven, rather than requirements driven, since an implementation abstraction must exist in order to satisfy a specification abstraction. We have illustrated that using Marigold the developer is building implementation components that meet the requirements of the specification.

Another approach which has influenced the design of the PB and COB components of Marigold is the data-flow style of specification. In [21] modules are connected together (in a manner similar to the PB) to specify virtual environments. These modules are linked to an underlying implementation which changes according to the configuration of the specification. This style of specification has been extended so that it can be achieved while the developer is immersed in the environment [29]. Like the UIMS approach, the semantics of these modules are static and consequently limit what can be specified . In Marigold, although static components are also used within the PB and COB, these components (devices and rendering representation) are input and output to the behaviour and not the behaviours themselves. The behaviour components are developed specifically for an applications requirements using the HSB.

An approach to designing complete virtual environments is introduced in [15] which contains a component for describing the behaviour of the system using Statecharts. This is an adaptation of an existing real-time system design approach. A number of contrasts can be drawn with Marigold. The behaviour of the systems described are walk-through. It would be difficult using this approach to describe some of the complex user-driven behaviour captured using Flownets, for example. Additionally, Marigold offers a faster and, thus, tighter integration between specification and prototype because of the environment integration method supported by the PB and COB.

The interactive cooperating object formalism (ICO) is a design approach for the development of interactive systems [19]. ICO combines the power of object-oriented structure and Petri-nets for the internal behaviour of objects. A tool is being developed which supports their implementation [6]. The abstractions made within the PB, and the style of PB specification itself, enable a developer to rapidly integrate a specification into an environment. These abstractions are not readily available within ICO and would need to be written using code. In addition, the Flownet formalism itself seems to be a suitably rich formalism for describing virtual environment behaviour compared to Petri-nets alone. Flownets continuous data-flow constructs capture a strong mapping between the user interaction (devices), their state, and the presentation of this state.

# 7  Conclusion

In this paper we have discussed a need for the pre-implementation design of virtual environment interface behaviour and have reviewed the approaches developed to address this concern. We have also motivated a need to integrate this form of design with the building of prototypes so that developers and users can explore the designs at an implementation level. The Marigold toolset has been developed as an approach to providing a rapid transition between such designs and prototypes. We have described this toolset by exemplifying the incremental design and prototyping of a small kitchen prototype. Such a prototype may be used at two levels. Firstly, it is important to determine whether the environment is usable. For instance, in the example, does the behaviour of the environment enable the user to navigate to the drawer, open the drawer, transfer both balls to the pan on the hob from the drawer, and switch the hob on? At another level, does the environment fulfill the broader requirements? If the small kitchen was to aid in training chefs, then the behaviour of the environment is clearly inadequate. At the end of a session the chef maybe able to complete a task within the virtual environment, but this bears no correspondence to the behaviour of the real environment.

It is anticipated that once the developer and users have arrived at a design that satisfies the requirements, through an iterative process of design and prototyping, one of two options may be taken. Firstly, a programmer may construct the final implementation directly from the specifications taking into account issue of performance (paramount to larger environments). Secondly, as suggested in [25], the Flownet specifications may be refined to a more detailed specifications (maybe using HyNets as presented in [25]), from which a programmer would implement the behaviour of the environment.

**Acknowledgements**

We are grateful to Shamus Smith for his comments on this paper.

# References

1.  *Maverik programmer's guide for version 5.1.* Advanced Interface Group, Department of Computer Science, University of Manchester, 1999.
2.  Autodesk-corporation. 3DStudio. 111 McInnis Parkway, San Rafael, California, 94903, USA.
3.  Steve Bryson. *Approaches to the Successful Design and Implementation of VR Applications.* London Academic Press, 1995.
4.  Rik Carey and Gavin Bell. *The Annotated VRML 2.0 Reference Manual.* Developers Press, 1997.
5.  Superscape Corporation. Superscape, 1999. 3945 Freedom Circle, Suite 1050, Santa Clara, CA 95054, USA.
6.  Remi Bastide David Navarre, Philippe Palanque and Ousmane Sy. Structuring interactive systems specifications for executability and prototypeability. In Philippe Palanque and Fabio Paterno, editors, *Design, Specification and Verification of Interactive Systems '00*, pages 97-119. Lecture notes in Computer Science 1946, 2001.
7.  Pierre duPont. Building complex virtual worlds without programming. In Remco C. Veltkamp, editor, *Eurographics'95 STAR report*, pages 61-70. Eurographics, 1995.
8.  J. W. Forrester. *Industrial Dynamics.* MIT Press, 1961.
9.  Mark Green. The design of narrative virtual environments. In *Design, Specification and Verification of Interactive Systems'95*, pages 279-293, 1995.
10. David Harel. Statecharts: A visual formalism for complex systems. *Science of Computer Programming*, 8:231-274, 1987.
11. Michael Harrison and Harold Thimbleby, editors. *Formal Methods in Human-Computer Interaction.* Cambidge University Press, 1990.
12. Robert J. K. Jacob. Specifying non-WIMP interfaces. *In CHI'95 Workshop on the Formal Specification of User Interfaces Position Papers*, 1995.
13. Robert J. K. Jacob. A visual language for non-WIMP user interfaces. In *Proceedings IEEE Symposium on Visual Languages*, pages 231-238. IEEE Computer Science Press, 1996.
14. Kulwinder Kaur, Neil Maiden, and Alistair Sutcliffe. Design practice and usability problems with virtual environments. In *Proceedings of Virtual Reality World '96,1996.*
15. G. Jounghyun Kim, Kyo Chul Kang, Hyejung Kim, and Jiyoung Lee. Software engineering of virtual worlds. In *ACM Virtual Reality Systems and Technology Conference (VRST'98)*, pages 131-138, 1998.
16. Mieke Massink, David Duke, and Shamus Smith. Towards hybrid interface specification for virtual environments. In *Design, Specification and Verification of Interactive Systems '99*, pages 30-51. Springer, 1999.
17. S. A. Morrison and R. J. K. Jacob. A specification paradigm for design and implementation of non-WIMP human-computer interaction. In *ACM CHI'98 Human*

*Factors in Computing Systems Conference*, pages 357{358. Addison-Wesley/ACM Press, 1998.

18. Brad A. Myers. User-interface tools: Introduction and survey. *IEEE Software*, 6(1):15-23, 1989.

19. Philippe A. Palanque, Remi Bastide, Louis Dourte, and Christophe Silbertin-Blane. Design of user-driven interfaces using petri nets and objects. In *Proceedings of CAISE'93 (Conference on advance information system engineering), Lecture Notes in Computer Science*, volume 685, 1993.

20. C. A. Petri. Kommunikation mit automaten. Schriften des iim nr. 2, Institut fur Instrumentelle Mathematic, 1962. English translation: Technical Report RADC-TR-65-377, Griffiths Air Base, New York, Vol. 1, Suppl. 1, 1966.

21. William R. Sherman. Integrating virtual environments into the data flow paradigm. In *4th Eurographics workshop on ViSC*, 1993.

22. Shamus Smith and David Duke. Using CSP to specify interaction in virtual environments. Technical Report YCS 321, University of York - Department of Computer Science, 1999.

23. Shamus Smith and David Duke. Virtual environments as hybrid systems. In *Eurographics UK 17th Annual Conference*, pages 113-128. Eurographics, 1999.

24. Shamus Smith, David Duke, Tim Marsh, Michael Harrison, and Peter Wright. Modelling interaction in virtual environments. In UK-VRSIG'98, 1998.

25. Shamus Smith, David Duke, and Mieke Massink. The hybrid world of virtual environments. *Computer Graphics Forum*, 18(3):C297-C307, 1999.

26. Shamus P. Smith and David J. Duke. Binding virtual environments to toolkit capabilities. *Computer Graphics Forum*, 19(3):C81-C89, 2000.

27. Shamus P. Smith, David J. Duke, and James S. Willans. Designing world objects for usable virtual environments. In *Workshop on design, specification and verification of interactive systems*, 2000.

28. Ian Sommerville. *Software Engineering*. Addison-Wesley, fifth edition, 1996.

29. Anthony J. Steed. *Defining Interaction within Immersive Virtual Environments*. PhD thesis, Queen Mary and Westfield College, UK, 1996.

30. Kari Systa. *A Specification Method for Interactive Systems*. PhD thesis, Tampere University of Technology, 1995.

31. Boris van Schooten, Olaf Donk, and Job Zwiers. Modelling interaction in virtual environments using process algerbra. In *12th Workshop on Language technology: Interaction in virtual worlds*, pages 195-212, 1999.

32. Ralf Wieting. Hybrid high-level nets. In J. M. Charnes, D. J. Morrice, and D. T. Brunner, editors, *Proceedings of the 1996 Winter Simulation Conference, pages 848-855. ACM Press*, 1996.

33. James S. Willans and Michael D. Harrison. A toolset supported approach for designing and testing virtual environment interaction techniques. *Accepted for publication in the International Journal of Human-Computer Studies*, 1999.

34. James S. Willans and Michael D. Harrison. A `plug and play' approach to testing virtual environment interaction techniques. In *6th Eurographics Workshop on Virtual Environments*, pages 33-42. SpringerVerlag, 2000.

35. James S. Willans and Michael D. Harrison. Verifying the behaviour of virtual environment world objects. In *Workshop on design, specification and verification of interactive systems*, pages 65-77. Lecture notes in computer science 1946, 2000.

36. Charles Albert Wuthrich. An analysis and a model of 3D interaction methods and devices for virtual reality. In *Design, Specification and Verification of Interactive Systems '99*, pages 18-29. Springer, 1999.

# Discussion

J. Höhle: Can you model independent behaviour instead of reactive behaviour - such as a time based behaviour?
*J.Willans:* You certainly can add timing constraints in the system but it would be hard to ensure their satisfaction. Satisfaction would depend upon implementation details.

*N. Graham:* How hybrid is your specification? You seem to be modelling dynamic behaviour within discrete framework. It does not seem to involve truly continuous inputs, such as dealing with the effect of momentum when lifting an object, or dealing with sound. Is this natural in flownets?
*J.Willans:* We do not have that kind of example yet. Flownets is hybrid in representation at the design level. There are other lower level representations that capture continutiy in more detail (e.g. HyNet).

# Appendix A

```
#include "maverik.h"
#include "mav_tdm.h"

MAV_vector new_offset;
MAV_vector origin_pos;
int mouseClick;

/* RELATES TO FLOWNET SPECIFICATION   */
/* MIDDLE MOUSE BUTTON + UPDATE ORIGIN */

int     mouseButtonPress(MAV_object    *    o,
MAV_TDMEvent * ev)
{
   int origin_x, origin_y;
   int xx, yy;
   if (ev->tracker == 0 && eb->button ==1) {
      mav_mouseGet(mav_win_all, &origin_x,
            &origin_y, &xx, &yy);
      origin_pos.x = origin_x;
      origin_pos.y = origin_y;
      origin_pos.z = 0;
      mouseClick = !mouseClick;
   }
}

/* RELATES TO FLOWNET SPECIFICATION */
/* UPDATE POSITION              */
void updateViewpoint(void)
{
   mav_win_current->vp->eye =
   mav_vectorAdd(mav_win_current->vp->eye,
         new_offset);
}
```

```
/* RELATES TO PETRI-NET PART OF */
/* FLOWNET SPECIFICATION      */

void interaction(void)
{
   MAV_vector current_mouse;
   MAV_vector direction;
   float speed;
   int curr_mouse_x, curr_mouse_y, xx, yy;
   if (mouseClick) {
      mav_mouseGet(mav_win_all, &curr_mouse_x,
            &curr_mouse_y, &xx, &yy);
      current_mouse.x = curr_mouse_x;
      current_mouse.y = curr_mouse_y;
      current_mouse.z = 0;
      if (outOriginSq(current_mouse)) {
         direction = mav_vectorSub
            (origin_pos,
            current_mouse);
         direction = mav_vectorSet
            (direction.x,
            direction.z,
            direction.y);
         speed = mav_vectorDotProduct
            (direction, direction);
         speed = speed / 1000;
         new_offset = mav_vectorScalar
            (direction, speed);
         origin_pos = mav_vectorAdd
            (new_offset, origin_pos);
         updateViewpoint();
      }
   }
}
```

```
/* RELATES TO FLOWNET SPECIFICATION */
/* OUT OF ORIGIN              */

int outOriginSq(MAV_vector current_mouse)
{
   MAV_vector temp;
   float distance;
   temp        =        mav_vectorSub(current_mouse,
origin_pos);
   distance = mav_vectorDotProduct(temp, temp);
   if (distance > 5) {
     return 1;
   }
   return 0;
}

int main(int argc, char *argv[])
{
   MAV_SMS *sms;
   MAV_composite comp;

   mav_initialise();
```

```
   mav_TDMModuleInit();
   sms = mav_SMSNew
     (mav_SMSClass_objList, mav_objListNew());
   mav_compositeReadAC3D
     ("desk.ac", &comp,MAV_ID_MATRIX);
   comp.matrix = mav_matrixSet(0, 0, 4, 0, 0, 0);
   mav_SMSObjectAdd
     (sms,    mav_objectNew(mav_class_composite,
&comp));
   mav_callbackTDMSet
     (mav_win_all,                    mav_class_world,
mouseButtonPress);
   mav_frameFn0Add(interaction);
   mouseClick = 0;
   while (1) {
     mav_eventsCheck();
     mav_frameBegin();
     mav_SMSDisplay(mav_win_all, sms);
     mav_frameEnd(); }
```

# QTk – A Mixed Declarative/Procedural Approach for Designing Executable User Interfaces

Donatien Grolaux, Peter Van Roy, and Jean Vanderdonckt

Université catholique de Louvain, B-1348 Louvain-la-Neuve, Belgium
{ned, pvr}@info.ucl.ac.be, vanderdonckt@qant.ucl.ac.be

When designing executable user interfaces, it is often advantageous to use declarative and procedural approaches together, each when most appropriate:

- A declarative approach can be used to define widget types, their initial states, their resize behavior, and how they are nested to form each window. All this information can be represented as a data structure. For example, widgets can be records and the window structure is then simply a nested record.
- A procedural approach can be used when its expressive power is needed, i.e., to define most of the UI's dynamic behavior. For example, UI events trigger calls to action procedures and the application can change widget state by invoking handler objects. Both action procedures and handler objects can be embedded in the data structures used by the declarative approach.

The QTk tool uses this mixed approach, tightly integrated with a programming language that has extensive support for records and first-class procedures. This permits *executable model-based UI design*: the UI models are executed at run-time without any compilation. To be precise, each UI model is a record that is transformed at run-time to its QTk specification, which is also a record.

We demonstrate the effectiveness of this approach by writing a context-sensitive clock utility, FlexClock, that changes its view at run-time whenever its window is resized. The utility is written in less than 400 lines. This includes full definitions of a calendar widget, an analog clock widget, and 16 views. Each view is defined as a record with three fields. All 16 views including formatting utilities are written in 80 lines total. The mechanism for creating a running UI from these definitions is written in 60 lines. Here is the definition of one view:

```
view( desc: label(handle:H bg:white glue:nswe)
   update: proc {$ T} {H set(text:{FormatTime T})} end
      area: 40#10)
```

The desc field is declarative; it defines the view's structure as a record. Here it is a label widget with an embedded handler object referenced by H. The handler object is created by QTk when the widget is installed. The update field is procedural; it contains an embedded procedure that will be called once a second with a time argument T to set the displayed time. The area field gives the view's minimum width and height, used to select the best view at run-time.

QTk uses tcl/tk as its underlying graphics subsystem. It is part of the Mozart Programming System, which implements the Oz language. For full information see http://www.info.ucl.ac.be/people/ned/flexclock. This research is supported in Belgium by the PIRATES project, funded by the Walloon Region.

M. Reed Little and L. Nigay (Eds.): EHCI 2001, LNCS 2254, pp. 109–110, 2001.
© Springer-Verlag Berlin Heidelberg 2001

## Discussion

*J. Höhle:* Who is the intended target users of your tookit?
*P. Van Roy:* There is not specialist user group, other than toolkit users.

*J. Höhle:* Have you considered verifying properties interfaces defined using the declarative element of your toolkit?
*P. Van Roy:* This use of the declarative element has not been explored at present.

# Consistency in Augmented Reality Systems

Emmanuel Dubois[1,2], Laurence Nigay[1], and Jocelyne Troccaz[2]

[1] CLIPS-IMAG, IIHM, 385 rue de la bibliothèque, BP 53,
F-38041 Grenoble Cedex 9, France
[2] TIMC-IMAG, GMCAO, Faculty of Medicine (IAB), Domaine de la Merci,
F-38706 La Tronche Cedex, France
{Emmanuel.Dubois, Laurence.Nigay, Jocelyne.Troccaz}@imag.fr

**Abstract.** Systems combining the real and the virtual are becoming more and more prevalent. The Augmented Reality (AR) paradigm illustrates this trend. In comparison with traditional interactive systems, such AR systems involve real entities and virtual ones. And the duality of the two types of entities involved in the interaction has to be studied during the design. We therefore present the ASUR notation: The ASUR description of a system adopts a task-centered point of view and highlights the links between the real world and the virtual world. Based on the characteristics of the ASUR components and relations, predictive usability analysis can be performed by considering the ergonomic property of consistency. We illustrate this analysis on the redesign of a computer assisted surgical application, CASPER.

## 1 Introduction

Integrating virtual information and action in the real world of the user, is becoming a crucial challenge for the designers of interactive systems. The Augmented Reality (AR) paradigm illustrates this trend. The main goal of AR is to add computational capabilities to real objects involved in the interaction. The *Augmented Paper Strip* [15] is an example of such an attempt in the air traffic control domain: the goal is to add computational capabilities to traditional paper strips. Another system called *KARMA* [11] provides information to a user repairing a laser printer, by indicating with 3D graphics which parts of the printer to act on, according to the defined maintenance process. The *Tangible Interface* [14] constitutes another example of the combination of real and virtual entities: everyday life objects are used by the user to interact with the computer. More examples of AR systems are presented in [2]. In [9], we have already illustrated this wide area of interactive systems and highlighted an important classification characteristic:

- some systems (*Augmented Reality systems, AR*), enhance interaction between the user and his/her real environment, by providing additional computer capabilities or data,
- while others (*Augmented Virtuality systems, AV*) make use of real objects to enhance the user's interaction with a computer.

Since our application domain is Computer Assisted Surgery (interaction with the real world, the patient, enhanced by the computer), we particularly focus on the

M. Reed Little and L. Nigay (Eds.): EHCI 2001, LNCS 2254, pp. 111–122, 2001.
© Springer-Verlag Berlin Heidelberg 2001

characterization and description of *Augmented Reality systems*. One of the main design challenges of such *Augmented Reality Systems* (*AR*) is to merge real and virtual entities. Real environment and real entities are prerequisite for the design of such systems. The composition of these two kinds of entities constitutes the originality of *AR* systems. In addition, information or action are defined by the *AR* system to facilitate or to enrich the natural way the user would interact with the real environment. Consequently, the main point of interest during the design should be the outputs of the systems, so that additional information and action are smoothly integrated with the real environment of the user.

In this paper, we first briefly present our application domain: Computer Assisted Surgery. Through the presentation of the taxonomy of the domain, we motivate our approach that aims at studying the two types of entities (real and virtual ones) involved in the output user interfaces of *AR* systems. We then present our notation called ASUR. ASUR is based on the principles of the notation OP-a-S [9], which is enriched by characteristics that describe the user's interaction. The ASUR notation describes a system with a user's task-centered point of view and highlights the links between the real world and the virtual world. Based on our descriptive notation, we then show how predictive evaluation of the consistency ergonomic property can be addressed. We illustrate our approach through the redesign of *CASPER*, a computer assisted surgery system.

## 2 Motivation and Application Domain

There are many application domains of *Augmented Reality*, including construction, architecture [20] and surgery [3], [6], [18]. Our application domain is *Computer Assisted Surgery* (*CAS*). The main objective of *CAS* systems is to help a surgeon in defining and executing an optimal surgical strategy based on a variety of multi-modal data inputs. The objectives aim at improving the quality of the interventions by making them easier, more accurate, and more intimately linked to pre-operative simulations where accurate objectives can be defined. In particular, one basic challenge is to guide a surgical tool according to a pre-planned strategy: to do so robots and 3D localizers (mechanical arms or optical sensors) perform real time tracking of surgical tools such as drills [6]. *AR* plays a central role in this domain because the key point of *CAS* systems is to "augment" the physical world of the surgeon (the operating theater, the patient, the tools etc.), by providing pre-operative information including the pre-planned strategy. Information is transmitted between the real world and the computer world using different devices: computer screen, mouse, pedal, tracking mechanism, robot, etc.

Since 1985, our laboratory is working on designing, developing and evaluating *CAS* systems. Through technological progress and a growing consciousness of the possibilities of real clinical improvements using a computer [18], *Augmented Reality* systems are now entering many surgical specialties. Such systems can take on the most varied forms [3]. Three classes of CAMI systems are identified in [19]:

- The passive systems allow the surgeon to compare the executed strategy with the planned one.

- The active systems perform subtasks of the strategy with the help of an autonomous robotic system.
- The semi-active or synergistic systems help the surgeon in performing the surgical strategy but the surgeon is in charge of its execution. The system and the surgeon are working in a synergistic way.

By comparison with the HCI domain, in which traditional design approaches aim at keeping the user in the loop and at focusing on the task supported by the system, *CAS* design methods are principally driven by technologies. Consequently, instead of assessing the quality of a system in terms of the user's perspective and of the software designer's perspective, *CAS* design methods are constrained and driven by technologies. External properties [1] [13], which establish "how usable a system is" and internal properties that describe the software quality, are replaced by clinical and technical considerations [17]. In this context, our research aims at providing a notation for designers to help them in reasoning about the merging of real and virtual entities.

More generally speaking, *Augmented Reality* systems design is often driven by the latest technology. We place a greater emphasis on interaction between the user and the system as well as the user and the real environment. To do so we propose a notation, namely ASUR: ASUR description of a system is composed of:

- the entities that are involved in the interactive system,
- the relation between these different entities.

In addition, characteristics are identified to describe the entities and the relations involved in the user's interaction with the system. Finally, ASUR provides a common description notation of *AR* systems that enables their comparison and their classification. The next paragraph presents this notation.

# 3   ASUR Interaction Description

In this paragraph, we first briefly present the principles of our ASUR notation. As mentioned above, ASUR is based on the principles of our previously presented OP-a-S notation [9]. ASUR extends OP-a-S by providing characteristics of components and relations involved in the dual interaction of the user with the virtual part of the system and with his/her real environment.

## 3.1  ASUR Principles Overview

The basic idea of ASUR is to describe an interactive system as a set of four kinds of entities, called components. In [12], we have already presented some characteristics of such entities, but the relations among them were not studied. When applying ASUR, a relation between two components describes an exchange of data. ASUR components and relations are described in the two following paragraphs.

### 3.1.1 ASUR Components

The first component is the **User** (component U) of the system. Second, the different parts used to save, retrieve and treat electronic data are referred to as the computer **System** (component S). This includes CPU, hardware and software aspects, storing devices, communication links. To take into consideration the use of real entities, we denote each real entity implicated in the interaction as a component R, **Real** objects. The 'Real object' component is refined into two kinds of components. The first component $R_{tool}$ is a **Real** object used during the interaction as a **tool** that the user needs in order to perform her/his task. The second component $R_{task}$ represents a real object that is the focus of the task, i.e. the **Real** object of the **task**. For example, in a writing task with an electronic board like the *MagicBoard* [7], the white board as well as the real pens constitute examples of components $R_{tool}$ (real tool used to achieve the task), while the words and graphics drawn by the user constitute the component $R_{task}$ (real object of the task). Finally, to bridge the gap between the virtual entities (component S) and the real world entities, composed of the user (component U) and of the real objects relevant to the task (components $R_{task}$ and $R_{tool}$), we consider a last class of components called **Adapters** (component A). **Adapters** for **Input** ($A_{in}$) convey data from the real world to the virtual one (component S) while **Adapters** for **Output** ($A_{out}$) transfer data from the component S to the real world (components U, $R_{tool}$ and $R_{task}$). Screens, projectors and head-mounted displays are examples of output adapters, while mice, keyboards and cameras may play the role of input adapters. The exchange of data between ASUR components is described in the next paragraph.

### 3.1.2 ASUR Relations

A relation is symbolized in an ASUR diagram with a unidirectional oriented arrow. It represents a set of data sent by a component to another one. For example, a relation $A_{out} \rightarrow U$, from a screen (component $A_{out}$) to a user (component U) symbolizes the fact that data are perceivable by the user on the screen. Another relation $U \rightarrow R_{tool}$, from a user (component U) to a pen of the *Magic Board* (component $R_{tool}$) represents the fact that the user handles the pen.

Having defined the ASUR components and relations, we now focus on the user and her/his interaction with the computer system as well as with the real environment.

### 3.2 Focus on the User's Interaction

Due to our definition of *AR* systems, the users' interaction has two facets: (1) interaction between the user and the computerized part, and (2) interaction between the user and the real environment. According to our ASUR notation, the first facet is represented by a relation from an output Adapter ($A_{out}$) to the User (U). Data from the computerized part may only be perceived through an output Adapter. Interaction between the user and the real environment (facet 2) is represented by ASUR relations between the component U (the User) and the components $R_{task}$ (Real object of task) as well as the components $R_{tool}$ (Real tool).

Getting a clear understanding of the interaction between a user and an *AR* system involves analysis of the following relations: $A_{out} \rightarrow U$, $R_{task} \leftrightarrow U$ and $R_{tool} \leftrightarrow U$. In the next two paragraphs, we characterize these components and relations.

### 3.2.1  Three Characteristics of an Adapter and a Real Object

The first characteristic induced by the use of a real object or an adapter is the **human sense** involved in perceiving data from such components. The most common used ones are the haptic, visual and auditory senses. For example, in the *Magic Board*, the visual human sense characterizes the white board, the user looking at the drawings and written texts. The auditory sense may also be involved to perceive alarms indicating a problem for example with the vision-based capturing process.

The second characteristic of an adapter or a real object, is the **location** where the user has to focus with the required sense, in order to perceive the data provided by the adapter or to perceive the real entity. The coupling of the characteristic **human sense** with the **location** defines the **perceptual environment** of an adapter or a real object.

The last characteristic is the ability of the adapter or real object to simultaneously **share** the carried data among several users. For example, displaying data on a head-mounted display (HMD) restricts the perception to the user wearing the HMD. On the other hand, projecting data onto a white board enables $N$ users to perceive the data simultaneously.

### 3.2.2  Two Characteristics of a Relation

An ASUR relation represents a flow of data. The **interaction language** used to express data carried by the relation is the first characteristic of a relation. The data transferred to the user may be expressed in an arbitrary manner [4]; e.g. the user needs to learn the form or syntax of the data. On the other hand, the language may be non-arbitrary; in this case, the data are expressed according to an already known convention. In the task of text selection on the *Magic Board*, the projector displays on the white board a square that follows the user's finger motions and delimitates the area of selection. The visualization of the selected area is thus non-arbitrary, since this is widely used in computer applications.

The second characteristic of a relation denotes the importance, for the user's task, of the data carried by the relation. Defined in [12] as the attention received, the **weight** of a relation is a continuous axis ranging from none to high. We keep three values: none, peripheral and high. During a writing task using the *Magic Board*, the white board receives much attention, while the camera and projector receive none. Weighting a relation enables the designer to identify the number of relevant data that the user must perceive during a given task.

Based on our characteristics, the analysis of the ASUR relations and components linked to the user (U) enables the designer to identify problems in the usability of a system. For example, too many relations heavy weighted and with adapters requiring different locations, may lead to difficulties for the user to perceive all the data useful for performing a task. In case of potential identified usability problems, it is important to notice that the designer could only change the characteristics of the adapters and their relations. Characteristics of the real entities ($R_{task}$ and $R_{tool}$) and their relations to the user are prerequisites of the system. In the next paragraph, we show how predictive analysis of the consistency ergonomic property can be addressed using our characteristics of ASUR components and relations.

## 4  Predictive Analysis Based on ASUR

We base our predictive analysis process on the characteristics of ASUR components and relations as well as on ergonomic properties. As mentioned above, in *AR* systems, information or action are defined by the system to facilitate or to enrich the natural way the user would interact with the real environment. Outputs of the systems consequently constitute the focus of our analysis. Among the existing ergonomic properties, two of them are closely related to outputs of interactive systems, namely observability and honesty:

- **Observability** characterizes "the ability of the user to evaluate the internal state of the system from its perceivable representation" [8][13];
- **Honesty** characterizes "the ability of the system to ensure that the user will correctly interpret perceived information and that the perceived information is correct with regards to the internal state of the system" [13]

Additionally Norman's Theory of Action [16] models part of the users' mental activities in terms of a perception step and of an interpretation step. The above two ergonomic properties are directly related to these two steps: Observability is related to the users' perception while honesty supports users' interpretation.

Observability and honesty are traditionally analyzed in the case of representation of one concept at a given time. Facing the formidable expansion of new technologies in the medical domain for example, the surgeon will be exposed to more and more sources of information: NMR data, Ultra Sound images, needle tracking data, etc. Observability and honesty of multiple concepts at a given time, and of one concept represented in multiple ways (representation multiplicity principle) must therefore be considered and involve consideration of another crucial ergonomic property: **consistency** across the variety of representations available at a given time. In the following table, we refine consistency in terms of **perceptual consistency** (observability level) and **cognitive consistency** (honesty level).

In terms of ASUR characteristics, **perceptual consistency** is ensured if every output adapter and real object that convey data to the user have:

- their corresponding **locations** compatible: their locations must spatially intersect,
- their associate **human senses** compatible: the user must be able to sense the different information without losing some of it.

In other words, perceptual consistency is established if every data conveyed by adapters and real objects, along heavily **weighted** relations, are simultaneously perceivable and do not imply that the user changes her/his focus of attention.

Addressing a problem of perceptual inconsistency requires selecting output adapters having their corresponding **locations** and **human senses** compatible with each other and with the ones associated with the real objects involved in the interaction.

**Table 1**: Four types of consistency.

|  | Perception (**Observability**) | Interpretation (**Honesty**) |
|---|---|---|
| *1* concept, *n* representations | **Four** | **Types** |
| *N* concepts, *1* representation each | **Of** | **Consistency** |

At the cognitive level, consistency extends the notion of honesty. In terms of ASUR, **cognitive consistency** is ensured if every relation from output Adapters (components $A_{out}$) or Real objects (components $R_{tool}$ and $R_{task}$) to the User (component U) are based on the same **interaction language**. (see definition in   ). For example, data displayed by *KARMA* are based on a 3D graphical language (for example, arrows explaining which tray to open) matching the view of the real printer. In this example cognitive consistency is ensured because the view of the printer as well as the view of the 3D graphics match each other. If *KARMA* was displaying textual explanations, cognitive consistency would not be satisfied: indeed two languages are involved, a 3D view of the printer and a textual language.

Addressing a problem of cognitive inconsistency requires changing the **interaction languages** associated with the relations from the output adapters to the user.

During the design phase, if different solutions are designed, each solution envisioned is described with the same notation, allowing thus a precise comparison of the solutions in terms of consistency. The next paragraph illustrates our ASUR based analysis of consistency for one of our computer assisted surgery applications, *CASPER*. For more examples, we describe various *AR* systems using ASUR in [10].

# 5 ASUR Based Analysis: An Example

## 5.1 *CASPER* Application

### 5.1.1 Identity Card
*CASPER* (Computer ASsisted PERicardial puncture) is a system that we developed for computer assistance in pericardial punctures. The clinical problem is to insert a needle percutaneously in order to access the effusion with perfect control of the needle position and trajectory. The danger involves puncturing anatomical structures such as the liver or the heart itself. A detailed medical description of the system can be found in [5]. After having acquired Ultra-Sound images and planned a safe linear trajectory to reach the effusion, guidance is achieved through the use of an optical localizer that tracks the needle position. The left part of Figure 1 shows the application in use during the intervention.

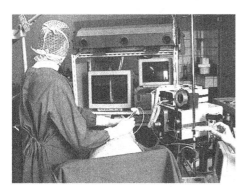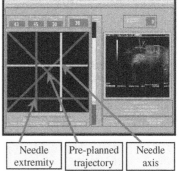

**Fig. 1.** Our *CASPER* application in use (left), *CASPER* guidance information monitored (right).

### 5.1.2    ASUR Description of *CASPER*

During the surgery, the surgeon (U) handles and observes a surgical needle ($R_{tool}$): $U \succ R_{tool}$. The needle is tracked by an optical localizer ($A_{in}$): $R_{tool} \succ A_{in}$. Information captured by the localizer is transmitted to the system (S): $A_{in} \succ S$. The system then displays the current position and the pre-planned trajectory on a screen ($A_{out}$): $S \succ A_{out}$. The surgeon (P) can therefore perceive the information: $A_{out} \prec U$. Finally, the object of the task is the patient ($R_{task}$), who is linked to the needle, and perceived by the surgeon (P): $R_{task} \prec U$. Figure 2 presents the ASUR description of *CASPER*.

## 5.2 Analysis of Cognitive Consistency in *CASPER*

An ASUR based analysis of *CASPER* has lead us to identify inconsistency between the perceived data from the needle and the ones displayed on screen. The arbitrary representation based on three crosses displayed on screen does not match the manipulation of the real needle. Since the puncture is a critical task that involves two relations that both require high attention ($R_{tool} \rightarrow U$ and $A_{out} \rightarrow U$), we deduce that cognitive consistency is not established.

To overcome this problem, we have worked on the way the guidance information is displayed. Instead of using the cross-based graphical representation, we have adopted a cone representation to visualize the trajectory. This design solution has brought up the problem of the point of view for displaying the 3D cone. Three points of view are possible: the needle or the trajectory point of view (the representation of the trajectory -respectively the needle - changes according to the position of the needle) or the user's point of view (the representation depends on the position and orientation of the user's gaze).

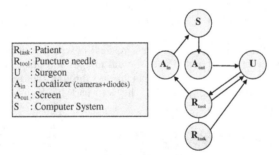

| $R_{task}$ : Patient |
| $R_{tool}$ : Puncture needle |
| U    : Surgeon |
| $A_{in}$ : Localizer (cameras+diodes) |
| $A_{out}$ : Screen |
| S    : Computer System |

**Fig. 2.** ASUR description of *CASPER*.

## 5.3 Analysis of Perceptual Consistency in *CASPER*

ASUR analysis of the output adapter ($A_{out}$) and real objects ($R_{tool}$ and $R_{task}$) involved in the surgery lead us to identify inconsistency between the two locations associated with the screen ($A_{out}$) and the needle in the operating field ($R_{tool}$ and $R_{task}$). While using *CASPER*, the surgeon must always shift between looking at the screen and looking at the patient and the needle.

To address this problem we have designed a new version of CASPER using another output adapter: a see-through head-mounted display (HMD). For cognitive consistency, we display the 3D representation of the trajectory from the user's point of view.

With the new version, the surgeon can see the operating field through the HMD as well as the guidance data in the same location. Additionally, we have used a clipping plane technique to perform a cut-away of a part of the trajectory representation, in order to match the depth of the real and virtual fields of view. Figure 3 presents the HMD we used and a view through the HMD under experimental conditions. We are currently performing acceptance  tests with surgeons.

In order to assess the usability of the new version of *CASPER*, usability experiments are in progress in collaboration with colleagues of the Experimental Psychology laboratory. The goal of the experiment is to evaluate the usability of the new output adapter as well as the representation of the trajectory. We carried out the experiment with 12 participants. Each participant has to reproduce a predefined trajectory in the context of 8 different settings. The settings are defined by the device used to display the guidance information (screen or HMD), the representation of the guidance information (cone or crosses) and the point of view on this information (trajectory point of view or needle point of view). The first global outcome is an overall benefit resulting from the use of the HMD as compared to using the screen. This result is independent of the representation of the guidance information as well as the point of view on the information. Further analysis and results are awaited in the very near future, from data being currently analyzed.

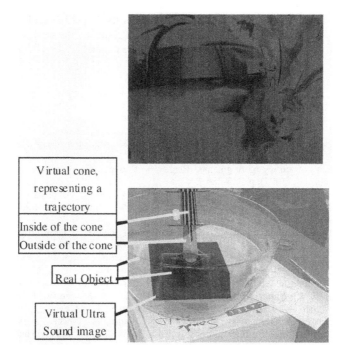

**Fig. 3.** The Sony head-mounted display used to address the perceptual inconsistency (top) and a view through the HMD merging real objects with a virtual conic trajectory (bottom).

# 6  Conclusion and Perspectives

In this paper, we have presented our ASUR notation and the characteristics of its components and relations as a tool to support predictive analysis of the interaction involved using *AR* systems. We showed that an ASUR description of a system could support the analysis of the consistency ergonomic property. The specificity of ASUR relies on the description of both real and virtual entities that are involved in performing a task in the real world. We illustrated our ASUR based analysis using *CASPER*, a *Computer Assisted Surgery (CAS)* system.

Although the inconsistency problems identified in the first version of *CASPER* can be detected without our ASUR analysis, we believe that ASUR provides a tool for systematically studying such usability problems and predicting several usability issues. In addition the simplicity of the ASUR description coupled with ergonomic properties makes it a useful tool for designers of *AR* systems and in particular designers of *CAS* systems who may not be familiar with ergonomic approaches.

In a future work, we plan to study other *AR* systems involving more complex information processes. We would also like to extend the number of ergonomic properties expressed in terms of ASUR in order to be able to cover a wider area of usability requirements.

### Acknowledgements

This work is supported by the IMAG Institute of Grenoble and by the Grenoble University Hospital. We wish to thank the colleagues of the Experimental Psychology Laboratory of the University of Grenoble, who contribute to the new version of CASPER. Many thanks to G. Serghiou for reviewing the paper.

# References

1.  Abowd, D., Coutaz, J., Nigay, L., Structuring the Space of Interactive Properties, in Proceedings of Working Conference IFIP TC2/WG2.7 on Engineering for Human-Computer Interaction, (1992), p. 113-128.
2.  Azuma, R., T., A survey of Augmented Reality, in Presence: Teleoperators and Virtual Environments 6, 4, (1997), p. 355-385.
3.  Bainville, E., Chaffanjon, P., Cinquin, P., Computer generated visual assistance to a surgical operation: the retroperitoneoscopy, in Computers in Biology and Medicine, 25(2), (1995), p. 165-171.
4.  Bernsen, N., O., Taxonomy of HCI Systems: State of the Art, ESPRIT BR GRACE, Deliverable 2.1, (1993).
5.  Chavanon, O., et al., Computer ASsisted PERicardial punctures : animal feasability study, in Conference Proceedings of CVRMed/MRCAS'97, Grenoble, (1997), p. 285-291.
6.  Cinquin, P., Bainville, E., Barbe, C., Bittar, E., Bouchard, V., Bricault, I., Champleboux, G., Chenin, M., Chevalier, L., Delnondedieu, Y., Desbat, L., Dessene, V., Hamadeh, A., Henry, D., Laieb, N., Lavallée, S., Lefebvre, J.M., Leitner, F., Menguy, Y., Padieu, F., Péria, O., Poyet, A., Promayon, M., Rouault, S., Sautot, P., Troccaz, J., Vassal, P., Computer Assisted Medical Interventions, in IEEE Engineering in Medicine and Biology 4, (1995), p. 254-263.

7.  Crowley, J., Coutaz, J., Bérard, F., Things That See, in Communication of the ACM (CACM) Vol. 43 (3), (2000), p. 54-64.
8.  Dix, A., Finlay, A., Abowd, G., Beale, R., Human-Computer Interaction, 2nd Edition, Prentice Hall, (1998).
9.  Dubois, E., Nigay, L., Troccaz, J., Chavanon, O., Carrat, L., Classification Space for Augmented Surgery, an Augmented Reality Case Study, in Conference Proceedings of Interact'99, (1999), p. 353-359.
10. Dubois, E., Nigay, L., Troccaz, J., Un Regard Unificateur sur la Réalité Augmentée : Classification et Principes de Conception, in Revue d'Interaction Homme-Machine, Journal of Human-Machine Interaction, Europia Productions, (2001), 23 pages. To appear.
11. Feiner, S., MacIntyre, B., Seligmann, D., Knowledge-Based Augmented Reality, in Communication of the ACM n•7, (1993), p. 53-61.
12. Graham, T. C. N., Watts, L., Calvary, G., Coutaz, J., Dubois, E., Nigay, L., A Dimension Space for the Design of Interactive Systems within their Physical Environments, in Conference Proceedings of DIS'2000, (2000), p. 406-416.
13. Gram, C., Cockton, G., et al., Design Principles for Interactive Software, Chapman et Hall, (1996), 248 pages.
14. shii, H., Ullmer, B., Tangible Bits: Towards Seamless Interfaces between People, Bits and Atoms, in Conference Proceedings of CHI'97, (1997), p. 234-241.
15. Médini, L. and Mackay, W.E., An augmented stripboard for air traffic control, Technical Report of Centre d'Études de la Navigation Aérienne, France, (1998).
16. Norman, D., A., Cognitive Engineering, User Centered Design, New Perspectives on Computer Interaction, Lawrence Erlbaum Associates, (1986), p. 31-61.
17. Notte, D., Nyssen, A., S., De Keyser, V., Evaluations des techniques de chirurgie minimale invasive par robot: premières constatations, in Conference Proceedings of ErgoIHM'2000, (2000), p. 234-243.
18. Taylor, R., Paul, H., Cutting, C., Mittlestadt, B., Hanson, W., Kazanzides, P., Musits, B., Kim, Y., Kalvin, A., Haddad, B., Khoramabadi, D., Larose, D., Augmentation of human precision in computer-integrated surgery, in Innovation and Technology in Biology and Medicine, Special Issue on Robotic Surgery, 13 (4), (1992), p. 450-468.
19. Troccaz, J., Peshkin, M., Davies, B., The use of localizers, robots and synergistic devices in CAS, in Conference Proceedings of MRCAS'97, (1997), p. 727-736.
20. Webster, A., Feiner, S., MacIntyre B., Massie, W., Krueger, T., Augmented Reality in Architectural Construction, Inspection, and Renovation, in Proceedings of Computing in Civil Engineering, ASCE, (1996), p. 913-919.

## Discussion

*L. Bergman:* The input and output devices may alter the task for the user, for example, see through display altering vision, or sensors on the needle may alter the surgeon's ability to manipulate it. Have you considered these factors?

*L. Nigay:* We are currently working with 2 surgeons to test the important elements of the system such as the head mounted display. To get this right it is important to test it with real tasks with real surgeons. We are iterating with them.

*L. Bergman:* Should these problems be accounted for in the model?
*L. Nigay:* Yes this would be a good idea.

*J.Willans:* There is a tendency to build lots of complexity into new augmented/virtual reality systems. This leads to a general problem in maintaining consistency between

the real world and what the user sees. Satisfying these constraints in real time with complex tracking, computer graphics etc in real time is a generally difficult problem.
*L. Nigay:* Yes there is a need to find out how this system will affect surgeons. This is an important problem that still needs to be solved. Our goal is to be accurate in tracking the needle to the millimetre, however, we are not there yet.

*W. Stuerzlinger:* With head mounted displays tracking the needle to within a millimetre may not be not enough. It is also necessary to track the head position as even small changes could result in dangerous inaccuracy in what can be see.
*L. Nigay:* Yes this is a difficult problem, and needs more research to get it right.

# Heuristic Evaluation of Groupware
# Based on the Mechanics of Collaboration

Kevin Baker[1], Saul Greenberg[1], and Carl Gutwin[2]

[1] Department of Computer Science
University of Calgary
Calgary, Alberta, Canada T2N 1N4
{bakerkev, saul}@cpsc.ucalgary.ca

[2] Department of Computer Science
University of Saskatchewan
Saskatoon, Saskatchewan, Canada S7N 5A9
gutwin@cs.usask.ca

**Abstract.** Despite the increasing availability of groupware, most systems are awkward and not widely used. While there are many reasons for this, a significant problem is that groupware is difficult to evaluate. In particular, there are no discount usability evaluation methodologies that can discover problems specific to teamwork. In this paper, we describe how we adapted Nielsen's heuristic evaluation methodology, designed originally for single user applications, to help inspectors rapidly, cheaply effectively identify usability problems within groupware systems. Specifically, we take the 'mechanics of collaboration' framework and restate it as heuristics for the purposes of discovering problems in shared visual work surfaces for distance-separated groups.

## 1  Introduction

Commercial real-time distributed groupware is now readily available due to improvements in hardware, increased connectivity of the Internet, and demands of increasingly distributed organizations. Yet with the exception of a few systems, groupware is not widely used. One main reason is that these systems have serious usability problems, a situation caused in part by the lack of practical methodologies for evaluating them. As Grudin points out, most groupware systems are complex and introduce almost insurmountable obstacles to meaningful, generalizable and inexpensive evaluation [14]. Without evaluation methods, groupware developers cannot learn from experience. Consequently, even today's collaborative systems contain usability problems that make them awkward to use.

Even though Grudin made this point over a decade ago, we have not yet developed techniques to make groupware evaluation cost-effective within typical software project constraints. One way to address this dearth is to adapt evaluation techniques developed for single-user software usability. Within the field of human computer interaction (HCI), many low-cost evaluation techniques have moved out of the research arena and into accepted practice. Collectively, these become a toolkit. Each

M. Reed Little and L. Nigay (Eds.): EHCI 2001, LNCS 2254, pp. 123–139, 2001.

methodology highlights different usability issues and identifies different types of problems; therefore, evaluators can choose and mix appropriate techniques to fit the situation [23].

While computer supported cooperative work (CSCW) is related to the field of HCI, the standard HCI methodologies have problems when we try to apply them verbatim for the purposes of evaluating groupware, as summarized below.

*User observation* has an evaluator observe a single person use a system to perform tasks within a semi-controlled setting. When applied to groupware, the experimenter now observes how groups use the system. The problem is that observing groups requires many subjects: the logistics of finding them and scheduling these sessions can be difficult. Groups also vary greatly in composition and in how its members interact, which makes observations difficult to analyze and generalize when compared to single user observations. Observation sessions tend to be longer, as group work requires participants to develop a rapport and interaction style. User observations can also be too sterile for testing groupware as it does not account for social, organizational, political and motivational factors that influence how the group accepts and uses the system.

*Field studies* have evaluators study people interacting within their work-a-day world. These can provide the context missing from observational techniques. However, they are complex and expensive in terms of time, logistics and analysis. Evaluators require experience and a considerable amount of time to conduct the evaluation. The number of people that must be observed at each site is high, which can make this task overwhelming. Field studies work best at the beginning and end of a design and are not well suited for iterative design.

*Inspection methods* have evaluators 'inspect' an interface for usability bugs according to a set of criteria, usually related to how individuals see and perform a task. As in single-user applications, groupware must effectively support task work. However, groupware must also support *teamwork*, the work of working together. Inspection methods are thus limited when we to use them 'as-is', for they do not address the teamwork necessary for groupware assessment. For example, Nielsen lists many heuristics to guide inspectors, yet none address 'bugs' particular to groupware usability. Similarly, a cognitive walkthrough used to evaluate groupware gave mixed and somewhat inconclusive results [8]; other researchers are providing a framework for typical groupware scenarios that can form a stronger basis for walkthroughs [4].

In this paper, we adapt Nielsen's popular heuristic evaluation method to groupware assessment. Given heuristics specific to teamwork, we believe this method can be applied to groupware. In particular, we offer a set of heuristics designed to identify usability problems specific to teamwork between distance-separated groups working over a shared visual work surface.

We begin with a brief overview of Nielsen's heuristic evaluation method. We then introduce a previously defined framework called the 'mechanics of collaboration' and use it to derive eight groupware heuristics that we explain in detail. We close by discussing our initial experiences with applying the heuristics and the steps required to validate them as a discount usability methodology.

## 2    Heuristic Evaluation

Heuristic evaluation (HE) is a widely accepted discount evaluation method for diagnosing potential usability problems in user interfaces [22,25,26,27,28,29]. This methodology involves a small set of usability experts visually inspecting an interface and judging its compliance with recognized usability principles (the "heuristics") e.g., 'Provide Feedback' or 'Use the User's Language' [22,25,26,27,28]. Heuristics are general rules used to describe common properties of usable interfaces [27]. They help evaluators focus their attention on aspects of an interface that are often sources of trouble, making detection of usability problems easier.

HE is popular with both researchers and industry. It is low cost in terms of time since it can be completed in a relatively short amount of time (i.e. a few hours). End-users are also not required; therefore, resources are inexpensive. Because heuristics are well documented [e.g., 27,28], they are easy to learn and apply. Finally, HE can be used fairly successfully by non-usability experts. An aggregate of 3-5 evaluators will typically identify 75-80% of all usability problems [27]. We look to capitalize on all these factors as we expand this technique to groupware evaluation.

## 3    The Mechanics of Collaboration

We are adapting HE by developing new groupware-specific heuristics. We started this process in Greenberg et al. [11], where we listed five heuristics based on the Locales Framework. This set was tailored for evaluating comprehensive collaborative environments and how they co-existed with the groups' everyday methods for communicating and collaborating. While likely good for identifying generic groupware problems, these heuristics do not include principles tailored to particular groupware application genres.

In this paper, we specialized our work by developing a more specific but still complementary set of heuristics tailored to the groupware genre of *shared visual workspaces*. These new heuristics should help inspectors evaluate how a shared visual workspace supports (or fails to support) the ability of distance-separated people to communicate and collaborate with artifacts through a visual medium. This application genre is very common e.g., real time systems for sharing views of conventional applications, group-aware versions of generalized and specialized text and graphical editors, and so on.

Specifically, we selected Gutwin's framework for the *mechanics of collaboration* [17] as the basis for this new set of heuristics. This framework is one of the few dealing with teamwork: it was specifically created with groupware in general and shared workspaces specifically in mind. Gutwin's mechanics were developed from an extensive analysis of shared workspace usage and theory [e.g. 1,16,38]. It describes the low level actions and interactions that small groups of people do if they are to complete a task effectively. Basic actions include communication, coordination, planning, monitoring, assistance, and protection. The underlying idea of the framework is that while some usability problems in groupware systems are strongly tied to social or organizational issues in which the system has been deployed, others are a result of poor support for the basic activities of collaborative work in shared spaces. It is these basic activities that the framework articulates.

We believe that the framework can help inspectors identify usability problems of both groupware prototypes and existing systems. While the framework was developed with low-cost evaluation methods in mind, we had to adapt, restructure and rephrase it as heuristics, and augment it with a few other important points that are not included in the framework. Unlike single user heuristics which are somewhat independent—chosen by how well they identified 'standard' usability problems [28]—ours have the advantage that they are linked and interdependent as they collectively describe a partial framework of attributes of how people interact with shared visual workspaces.

The resulting eight heuristics are presented below. For each heuristic, we provide an explanation along with how groupware applications typically realize and support its criteria.

### Heuristic 1: Provide the Means for Intentional and Appropriate Verbal Communication

In face-to-face settings, the prevalent form of communication in most groups is verbal conversations. The mechanism by which we gather information from verbal exchanges has been coined *intentional communication* [e.g. 1,19] and is typically used to establish a common understanding of the task at hand [1]. Gutwin summarizes three ways in which information is picked up from verbal exchanges [15].

1. People may talk explicitly about what they are doing and where they are working within a shared workspace. These direct discussions occur primarily when someone asks a specific question such as "what are you doing?" or when the group is planning or replanning the division of labour.
2. People can gather information by overhearing others' conversations. Although a conversation between two people may not explicitly include a third person, it is understood that the exchange is public information that others can pick up.
3. People can listen to the running commentary that others tend to produce alongside their actions. This "verbal shadowing" can be explicit or highly indirect and provides additional information without requiring people to enter into a conversation. This behaviour has also been called "outlouds" [19].

**Typical groupware support**. Most visual workspace groupware does not support intentional verbal communications directly, as they assume that any communication channel (text, audio, video) is supplied 'out of band'. Depending on the task, an obvious approach to facilitating verbal exchanges is to provide a digital audio link or text chat facility between participants. These can be implemented in many ways, and each has consequences on how well communication activity is supported. Text chat, for example, can be via parcel-post (e.g. type a line and send) [e.g. 41] or real-time (e.g. character by character). While limited, text can be useful for short or sporadic interactions, or where it is impractical to provide an audio connection. Lengthy or highly interactive meetings require an audio channel, typically supplied by a telephone. Digital audio is available in some systems, but currently suffers problems due to poor bandwidth, latency, and quality.

Video can also support verbal conversations. However, there are questions concerning the benefits of adding video to an audio channel. People have found it difficult to engage in conflict negotiation over distributed media since video can introduce distractors that interfere with the accuracy of interpersonal evaluations [7]. Plus, it has been observed that when conversations shifted to a meta-level, the participants turned away from the monitor showing the computerized work surface and would communicate across the video monitor [35]. While video has appeal and utility, the inspector cannot assume that inclusion of video is the panacea for all intentional communication requirements.

## Heuristic 2: Provide the Means for Intentional and Appropriate Gestural Communication

Explicit gestures and other visual actions are also used alongside verbal exchanges to carry out intentional communication. For example, Tang [39] observed that gestures play a prominent role in all work surface activity for design teams collaborating over paper on tabletops and whiteboards (around 35% of all actions). These are *intentional* gestures, where people used them to directly support the conversation and convey task information. Intentional gestural communication takes many forms. *Illustration* occurs when speech is illustrated, acted out, or emphasized. For example, people often illustrate distances by showing a gap between their hands. *Emblems* occur when words are replaced by actions, such as a nod or shake of the head indicating 'yes' or 'no' [34]. *Deictic reference* or *deixis* happens when people reference objects in the workspace with a combination of intentional gestures and communication, e.g., by pointing to an object and saying "this one" [1]. Whatever the type, groupware must provide ways of conveying and supporting gestural communication by making gestures clearly visible, and by maintaining their relation with both objects within the work surface and voice communications [39].

**Typical groupware support.** Because people are distance-separated, gestures are invisible unless they are directly supported by the system. In groupware, this is typically done (if at all) via some form of *embodiment*. Techniques include telepointers [37], avatars, and video images.

*Telepointers* are the simplest means for supporting embodiment in a virtual workspace. A person's cursor, made visible to all, allows one to gesture and point to objects in the workspace. While telepointers are limited 2D caricatures of the rich gestures people do with their hands, they are a huge improvement over nothing at all. Early systems, such as GroupSketch [12], were explicitly designed to facilitate gestural actions: each person had their own large and uniquely identifiable telepointer that they could use simultaneously with the others; telepointers were always visible within the work surface by all participants; they appeared with no apparent delay in order to remain synchronized with verbal exchanges; and they maintained their same relative location to the work surface objects across all displays.

*Avatars* are synthetic bodies representing the people who populate a 3D landscape. While most avatars are extremely crude, some transmit limited hand and body gestures: the idea is to capture real world gestures and have them appear in the simulated space.

*Video* can also recreate real bodies within groupware workspaces. Conventional video windows or monitors are not enough, for gestures are detached from workspace objects. A different approach is to mix and fuse the video of the person, their hands, and the worksurface into a single image [e.g. 20,40]. When done correctly, the final image comprises both the artifacts and the person, where gestures relative to the artifacts are maintained.

### Heuristic 3: Provide Consequential Communication of an Individual's Embodiment

A person's body interacting with a physical workspace is a complex information source with many degrees of freedom. In these settings, bodily actions such as position, posture, and movements of head, arms, hands, and eyes *unintentionally* "give off" information which is picked up by others [33]. This is a source of information since "watching other people work is a primary mechanism for gathering awareness information about what's going on, who is in the workspace, where they are, and what they are doing" [15]. Similarly, visible activity is an essential part of the flow of information fundamental for creating and sustaining teamwork [33]. This form of consequential bodily communication is not intentional in the same manner as explicit gestures (heuristic 2): the producer of the information does not intentionally undertake actions to inform the other person, and the perceiver merely picks up what is available.

Unintentional body language can be divided into two categories. *Actions coupled with the workspace* include such activities as gaze awareness (i.e. knowing where another person is looking), seeing a participant move towards an object or artifact, and hearing characteristic sounds as people go about their activities. *Actions coupled to conversation* are the subtle cues picked up from our conversational partners that help us continually adjust our verbal behaviour [e.g. 2,24,31]. Some of these cues are visual: facial expressions, body language (e.g. head nods), eye contact, or gestures emphasizing talk. Others are verbal: intonation, pauses, or the use of particular words. These visual and verbal cues provide *conversational awareness* that helps people maintain a sense of what is happening in a conversation. This in turn allows us to mediate turn-taking, focus attention, detect and repair conversational breakdown, and build a common ground of joint knowledge and activities [1]. For example, eye contact helps us determine attention: people will start an utterance, wait until the listener begins to make eye contact, and then start the utterance over again [10].

**Typical groupware support.** The goal of supporting consequential communication in real-time groupware is to capture and transmit both the explicit and subtle dynamics that occur between collaborating participants. This is no easy task. While the embodiment techniques previously discussed (heuristic 2) are a start, they are very limited. For example, telepointers allow us to see people moving towards an object. They can also change their shape to reflect a natural action such as pointing or writing [12]. Telepointers may hint at where its owner is looking, although there is no guarantee that the person is really doing so. Avatars can go one step further by linking the 'gaze direction' of the avatar to the point of view, thus signaling its owner's approximate field of view in the environment. While these do help, the impoverished match of these embodiments to a person's actual body movements means that there are many consequential gestures that are not captured and transmitted.

Video systems that mix a person's video embodiment into the workspace are more successful but unfortunately quite limited [e.g. 20]. Special care must be taken with camera placement, otherwise eye contact and gaze awareness will be inaccurate or incorrect—in most desktop systems, we see speakers 'looking' at our navels or hairline simply because cameras are often mounted on top or underneath the monitor. The use of compressed video results in small, jerky and often blurred images that lose many of these subtle body cues. Even with full video, zooming the camera in to capture facial expressions ('talking head' view) means that other body gestures are not visible. Yet zooming out to include the whole body compromises image fidelity and resolution. A notable exception is Ishii's ClearBoard [20], as he goes to great length to keep gaze awareness—whether intentional or consequential—correct.

Audio is also a concern for consequential communication. When the voice channel is of low audio quality, the clarity of a person's speech dynamics are compromised. When the voice is non-directional, people find it difficult to associate a voice with a particular speaker (e.g. multi-point teleconferencing). With half-duplex channels, people cannot speak at the same time, making it harder for listeners to interrupt or to inject back-channel utterances such as 'ok' and 'ums'. Speaker identification and turn-taking is difficult when a teleconference involves four or more people [7].

Advanced systems mimic the spatial relationships between people in a multi-point collaboration by letting individuals turn their heads and speak to one another just as they do in real life. This is accomplished by positioning monitors and cameras within and across sites so that all people are seen and heard in the same relative position on their video and audio surrogates. People's images and voices are projected onto separate video monitors and speakers. One compelling early example was the MIT Media Labs' 'talking heads' project. They fashioned a transparent physical mask of a participant, mounted it on a motorized platform at the remote site, and then projected the video image of the participant into the mask. Through sensors, the mask would move to reflect the person's actual head movement.

## Heuristic 4: Provide Consequential Communication of Shared Artifacts (i.e. Artifact Feedthrough)

In face-to-face settings, consequential communication also involves information *unintentionally* given off by artifacts as they are manipulated by individuals [e.g. 5,9]. This information is called feedback when it informs the person who is manipulating the artifact, and *feedthrough* when it informs others who are watching [5]. Physical artifacts naturally provide visual and acoustic feedback and feedthrough. Visually, artifacts are physical objects that show their state in their physical representation, and that form spatial relationships with one another. In addition, an artifact's appearance sometimes shows traces of its history of how it came to be what it is (e.g., object wear). Acoustically, physical artifacts make characteristic sounds as they are manipulated (e.g. scratch of a pencil on paper) [9]. By seeing and hearing an artifact as it is manipulated, people can easily determine what others are doing to it.

Another resource available in face-to-face interactions is the ability to identify the person manipulating an artifact. Knowing who produced the action provides context for making sense of this action, and helps collaborators mediate their interactions.

Actions within a shared workspace are often used to bid for turn-taking in a conversation; therefore, being able to associate the action with the initiator helps others yield their turn [38].

Due to the spatial separation between artifact and actor, feedthrough tends to be the only vehicle for sharing artifact information amongst groupware participants. However, groupware complicates feedthrough since it limits the expressivity of artifacts. For instance, direct manipulation of artifacts in a virtual workspace (e.g. dragging and dropping a small object) can easily go unnoticed since these actions are not as visible when compared to face-to-face equivalents. They can also happen instantaneously (e.g., a click of a button), leaving little warning of their occurrence and little time to see and interpret them. Similarly, indirect manipulation of a virtual artifact (e.g. menu selections) is difficult to connect to the person controlling the action even when we can determine the action's meaning. Unless feedthrough is properly supported by the system, collaboration will be cumbersome.

**Typical groupware support** At the lowest level, the shared virtual workspace must display the local user's feedback to all remote users. In the event of a direct manipulation of an artifact, the designer must show not only the final position of a moved object but also the selection of the object and the intermediate steps of its move. In groupware, this can be accomplished via *action feedthrough* whereby each individual sees the initial, intermittent, and final state of an artifact as it is manipulated [15]. Early groupware systems imposed "what you see is what I see" (WYSIWIS) view sharing where all participants saw the exact same actions as they occurred in the workspace [e.g. 36]. Similarly, feedthrough must be supported during the indirect manipulation of an artifact. *Process feedthrough* ensures that local feedback of a person selecting an operation or command is also transmitted to all others to help them determine what is about to happen [15]. Intermediate states of indirect manipulation can also be presented via visual techniques such as action indicators and animation. With both types of events, all collaborators must be able to identify the producer of the action. Presenting the interim feedback of an artifact to all participants during an operation ensures changes do not happen instantaneously and that information other people can gather about the activity while it is happening is not reduced. Identifying the producer of the action helps to provide context to it.

Physical objects typically display information regarding how they were created and how they have been manipulated. In contrast, the bland appearance of virtual artifacts reduces what can be learned about their past actions. Objects have a consistent appearance and manipulating them does not automatically leave traces [18]. Their monotony means that people have fewer signs for determining what has happened and what another person has been doing. Techniques for displaying the history of virtual artifacts include 'edit wear' and 'read wear' [18].

In contrast to physical artifacts, virtual ones do not have natural sounds; therefore, many groupware workspaces are silent. If sounds are to be heard, designers must create and add synthetic replacements to groupware [e.g. 9]. Currently, it is difficult to reproduce the subtlety and range of natural workspace sounds. In addition, the directional and proximal components of sounds tend to be weak since workspaces are 2D with limited audio output devices.

**Heuristic 5: Provide Protection**

In face-to-face settings, physical constraints typically prevent participants from concurrently interacting within a shared workspace. Conversely, groupware enables collaborators to act in parallel within the workspace and simultaneously manipulate shared objects. Concurrent access to the shared space is beneficial since collaborators can work in parallel, and because it helps negotiate the use of the space. In addition, it reduces the competition for conversational turn taking since one person can work in the shared space while another is talking and holding the audio floor [38]. On the other hand, concurrent access to objects can introduce the potential for conflict. People can inadvertently interfere with work that others are doing now, or alter or destroy work that others have done. People should be protected from these situations.

Anticipation plays an important role in providing protection. People learn to anticipate each other's actions and take action based on their expectations or predictions of what others will do in the future. Amongst other things, participants can in turn avoid conflicting actions. Therefore, collaborators must be able to keep an eye on their own work, noticing what effects others' actions could have and taking actions to prevent certain kinds of activity.

Social scientists have found that people naturally follow social protocols for mediating their interactions, such as turn-taking in conversations, and the ways shared physical objects are managed [2,38]. Therefore, concurrency conflicts may be rare in many groupware sessions since people mediate themselves—but this can only happen if people have a good sense of what is going on. People are quite capable of repairing the negative effects of conflicts and consider it part of the natural dialog [13]. Of course, there are situations where conflict can occur, such as accidental interference due to one person not noticing what another is doing. In some (but not all) cases, slight inconsistencies resulting from conflicts may not be problematic.

**Typical groupware support.** Many groupware systems give all collaborators equal rights to all objects. To provide protection, they rely on people's natural abilities to anticipate actions, mediate events and resolve conflicting interactions. The system's role is limited to providing awareness of others' actions and feedback of shared objects. For example, remote handles can graphically warn users that someone else is already using an item. Although social protocols will generally work, this approach may not be acceptable under certain situations. For instance, by allowing conflicts to occur systems force the users to resolve these events after they have been detected. This is undesirable if the result is lost work. Users may prefer 'prevention' to 'cure'. The quality of awareness will also not function well with high-latency communications where there is a delay in delivering one user's actions to others.

To assist with social protocols, technical measures such as access control, concurrency control, undo, version control, and turn-taking have been implemented. For example, concurrency control could manage conflicting actions and thus guard against inconsistencies. However, concurrency control in groupware must be handled differently than traditional database methods since the user is an active part of the process. People performing highly interactive activities will not tolerate delays introduced by conservative locking and serialization schemes. Access control can also be used to determine who can access a groupware object and when. Access control may be desirable when people wish to have their own private objects that only they can manipulate and/or view. Within groupware access control must be managed in a light-weight, fine-grained fashion. If not, it will be intrusive: people will fight with the system as they move between available and protected objects.

### Heuristic 6: Management of Tightly and Loosely-Coupled Collaboration

*Coupling* is the degree to which people are working together [32]. In general terms, coupling is the amount of work that one person can do before they require discussion, instruction, information, or consultation with another person. People continually shift back and forth between *loosely- and tightly-coupled collaboration* where they move fluidly between individual and group work. To manage these transitions, people should be able to focus their attention on different parts of the workspace when they are doing individual work in order to maintain awareness of others. Knowing what others are doing allows people to recognize when tighter coupling could be appropriate. This typically occurs when people see an opportunity to collaborate, need to plan their next activity, or have reached a stage in their task that requires another's involvement. For example, assisting others with their task is an integral part of collaboration whereby individuals move from loose to tight coupling. Assistance may be opportunistic and informal, where the situation makes it easy for one person to help another without a prior request. Awareness of others in these situations helps people determine what assistance is required and what is appropriate. Assistance may also be explicitly requested. Gutwin observed one participant making an indirect statement indicating that they wanted assistance, and their partner left their tasks to help out, and then returned to what they were doing [15]. In either case, to assist someone with their tasks, you need to know what they are doing, what their goals are, what stage they are in their task, and the state of their work area.

**Typical groupware support.** The traditional WYSIWIS approach ensures that people stay aware of one another's activities, but is often too restrictive when people regularly move back and forth between individual and shared work [6,9]. More recent systems allow people to move and change their viewports independently, allowing them to view the objects that interest them. This is called relaxed-WYSIWIS view sharing [37]. Unfortunately, when people can look at different areas of the workspace, they are blinded to the actions that go on outside their viewport unless the designer accounts for this. This difficulty of maintaining awareness of others when we are not working in the same area of the workspace is exacerbated because display areas are small and of very low resolution when compared with the normal human field of view. The reduction in size forces people to work through a small viewport, thus only a small part of a large workspace is visible at a time. The reduction in resolution makes artifacts harder to see and differentiate from one another; therefore, visual events can be more difficult to perceive. Groupware must address these issues by providing visual techniques that situate awareness information in the workspace. However, techniques encounter the same visibility problem: the relevant part of the workspace has to be visible for the techniques to be of any use. When the workspace is too large to fit into a single window, the entire area outside the local user's viewport cannot be seen unless special techniques are included that make the relevant parts of the workspace visible. Examples are included here.

*Overviews* provide a birds-eye view of the entire workspace in a small secondary window. A properly designed overview makes embodiments, actions and feedthrough visible, regardless of where they occur in the workspace. They also show where people are working in the workspace and the general structure of their activities. An overview showing additional awareness information like *view rectangles* (an outline showing what another can see) and telepointers is called a *radar view*. With radar views, people can easily pursue individual work by moving their view rectangle in the radar view to a different part of the workspace. Conversely, if they want to work closely together, they can quickly align their view rectangles atop one another. *Detail views* duplicate a selected part of the workspace. This secondary viewport provides a closer look at another person's work area: they show less of the workspace than an overview does, but what they do show is larger and in greater resolution. *Focus+context views* provides both local detail and global context within the same display, usually through information visualization techniques such as fisheye views or special lenses.

## Heuristic 7: Allow People to Coordinate Their Actions

An integral part of face-to-face collaboration is how group members mediate their interactions by taking turns and negotiating the sharing of the common workspace [38]. People organize their actions in a shared workspace to help avoid conflict with others and efficiently complete the task at hand. Coordinating actions involves making some tasks happen in the right order, at the right time while meeting the task's constraints [15]. Symptoms of poor coordination include people bumping into one another, duplication of actions, or multiple individuals attempting to concurrently access shared resources.

Coordination of action is a higher order activity built upon many mechanisms listed in the previous heuristics. Within a shared workspace, coordination can be accomplished via two mechanisms: "one is by explicit communication about how the work is to be performed ... another is less explicit, mediated by the shared material used in the work process" [30]. The first mechanism implies the need to support intentional and appropriate verbal communication (heuristic 1). The second uses workspace awareness to inform participants about the temporal and spatial boundaries of others' actions. In face-to-face interactions, the close physical proximity among the collaborators allows them to mediate actions since they are peripherally aware of others and all actions. Thus collaborators can fit the next action into the correct sequence.

Both mechanisms are beneficial for different levels of coordination. At the fine-grained level, awareness is evident in continuous actions where people are working with shared objects. One example is the way that people manage to avoid making contact with one another while collaborating within a confined space. On a larger scale, groups regularly reorganize the division of labour i.e., what each person will do next as the task progresses. These decisions depend in part on what the other participants are doing and have done, what they are still going to do, and what is left to do in the task. Knowing activities and locations can help determine who should do what task next.

The coordination of activities at both levels is also assisted by anticipation. People take action based on their expectations or predictions of what others will do in the future. Anticipation is integral to fine-grained coordination whereby people predict events by extrapolating forward from the immediate past. If you see someone reaching for a pen, you might predict that they are going to grab it. In turn, you can take action based on this prediction (e.g. pick up the pen and hand it to the other person or alter your own movements to avoid a collision). In this case, anticipation is supported by the up-to-the-moment knowledge of the activity (i.e. where the other person's hand is moving) and the location (i.e. the location of the hand in relation to the pen). In addition, your prediction could have taken into account other knowledge, such as the other person's current activities and if they required a pen. When prediction happens at a larger scale, people learn which elements of situations and tasks are repeated and invariant. People are experts at recognizing patterns in events, and quickly begin to predict what will come next in situations that they have been in before [15].

**Typical groupware support.** People are generally skilled at coordinating their communication and interaction with each other. Consequently, tools used to support collaboration should not impose a structure that attempts to manage the interactions for them. Instead, tools should facilitate the participants' own abilities to coordinate their communication and collaboration. Workspace awareness provides people with information they need to determine whether others' behaviour or current workspace events match the patterns that they have learned. Therefore, groupware must allow individuals to remain aware of others within a shared workspace and the nature of their actions. The visual techniques presented in heuristics 2 through 5 will help to establish this awareness. In addition, collaborators must be able to see all actions within the context of the entire workspace even when people are working in different parts of it. Implementing relaxed WYSIWIS helps to ensure that this is capable within groupware. Finally, collaborators must have the ability to communicate verbally (e.g. via an audio link). It is the inclusion of all this support within groupware systems that enables collaborators to effectively coordinate their activities at both a fine-grain level or on a larger scale.

### Heuristic 8: Facilitate Finding Collaborators and Establishing Contact

One problem with groupware is that it is not clear how people actually begin their groupware meetings. In everyday life, relatively few meetings are *formal* i.e., scheduled in advance with pre-arranged participants. These are usually arranged via e-mail, telephone, formal meeting requests, etc. In reality most meetings are *informal* encounters: unscheduled, spontaneous or one-person initiated meetings. These are facilitated by physical proximity since co-located individuals can maintain awareness of who is around. Under these circumstances, people frequently come in contact with one another through casual interactions (e.g. people bump into each other in hallways) and are able to initiate and conduct conversations with little effort. While conversations may not be lengthy, much can occur: people coordinate actions, exchange information, or offer opportunities. Successful teams rely on regular, informal, and unplanned contact between their members [3,21]. It is more difficult to support informal groupware encounters since the bottleneck to rich spontaneous interactions is distance [21].

In electronic communities, people are distributed. Therefore we need to support how the group determines who is around and their availability if they are to initiate contact in a real-time groupware session. Even when potential collaborators have been identified, many mundane factors now interfere with making contact over computers. People must know electronic addresses and select from many communication channels and applications that are available to the group. People must ready software, equipment, and each other well in advance for real-time remote conferencing. From a technical perspective, sites may not have the same software; workstations may not support the necessary media (e.g., digital audio); specialized equipment may not be available (e.g., video cameras); poor networks may limit interactions; and applications must run across platforms.

**Typical groupware support.** Groupware applications must overcome the distance barrier that inhibits informal encounters in order to be successful. Information on potential collaborators must be provided so that they can be easily found and their availability for group work can be determined. If collaborators are able and willing to engage in a groupware session, you must be able to initiate contact with minimal effort. Instant messaging provides a simple but limited mechanism for seeing who is around. Another more comprehensive approach is used in TeamWave [12]: a room metaphor helps people know who is around and makes it easy to move into conversation and work. Specifically it does the following:

- *Being available.* People can pursue single user activities in a room. Analogous to a physical room used for both individual and group activities, people will be around more often and thus available for real time encounters.

- *Knowing who is around and available for interaction.* Within a spatial setting, we sense who else is around as we walk down hallways, glance into offices, and see others in public spaces. We judge their availability by a variety of cues such as if their door is open and how busy they look. TeamWave's room metaphor provides a similar sense of presence and awareness by displaying the inhabitants of each room within the electronic community (via a user list) as well as status information. To judge availability, four 'door states' indicate a person's desire to be interrupted. As in real life, a wide and partially open door icon indicates a willingness to accept interruptions, while a barred door suggests that the room and its inhabitants are inaccessible.

- *Establishing contact.* There are several ways of establishing contact with individuals. One can just enter a populated room, and they are immediately connected to others: room occupants also see the person enter because their picture appears. A room-specific chat facility allows conversations to occur. If a person would rather initiate a conversation before entering a room, they can page somebody. Phone calls can be quickly established since phone numbers are available in each person's business card. To establish contact with others not currently logged on to TeamWave, a person can leave a note in a room suggesting a meeting time and place.

- *Working together.* The power of the room metaphor is that, once in a room, the working context is immediately available. All tools and room artifacts are at hand and new tools can be easily added.

## 4    Summary and Future Work

Our primary goal in this paper was to describe our groupware heuristics. We are especially concerned about keeping them simple enough to help an inspector look for many of the actions that are crucial to smooth and effortless collaboration within a shared workspace.

Our next step, which we have only just begun, is to validate how inspectors actually use these heuristics for uncovering problems. To gain some preliminary insight into the practicality of these groupware heuristics as a discount usability methodology, we performed an inspection of an object-oriented groupware drawing application. We uncovered many usability problems that we believe we would have otherwise overlooked in a casual inspection of the interface. The heuristics helped us to focus our attention on the critical issues pertaining to effective communication and collaboration among groupware participants. In addition, we were able to perform the inspection within a couple of hours. Consequently, we believe our initial development and application of these mechanics of collaboration heuristics appears promising.

Of course, the above evaluation is not enough to validate our groupware heuristics as a discount usability methodology. The number of usability problems uncovered during our inspection is not truly indicative of a typical practitioner since we (as evaluators) are intimately familiar with the groupware heuristics. Consequently, we are now planning a formal evaluation. Outside evaluators will inspect several real-time shared workspace applications using the groupware heuristics. The aim is to assess the ability of usability specialists to learn and apply the groupware heuristics by analyzing the quantity of problems detected by each evaluator. Ideally, each evaluator will uncover a large relative proportion of all problems found by all evaluators; however, even Nielsen found that the average performance of individual evaluators to be modest [25]. Thus we look to define the average proportion of usability problems found as a function of the number of evaluators performing the heuristic evaluation. As with conventional HE, we hope that only a small number of evaluators (about 3) are required to find a reasonably high proportion of the problems. This would allow the technique to remain low cost in terms of resources.

## References

1.    Clark, H. (1996). Using Language. Cambridge University Press, Cambridge.
2.    Clark, H. and Brennan, S. (1991). Grounding in Communication, in *Readings in Groupware and Computer Supported Cooperative Work*, R. M. Baecker ed., Morgan-Kaufman Publishers, 222-233.
3.    Cockburn, A. and Greenberg, S. (1993). Making contact: Getting the group communicating with groupware. *Proc ACM COCS'93 Conference on Organizational Computing Systems,* 31-41.
4.    Cugini, J., Damianos, L., Hirschman, L., Kozierok, R., Kurtz, J., Laskowski, S. and Scholtz, J. (1997). Methodology for Evaluation of Collaboration Systems. *The evaluation working group of the DARPA intelligent collaboration and visualization program, Rev. 3.0.* http://zing.ncsl.nist.gov/nist-icv/documents/method.html.
5.    Dix, A., Finlay, J., Abowd, G., and Beale, R. (1993). *Human-Computer Interaction,* Prentice Hall.

6.  Dourish, P., and Bellotti, V. (1992). Awareness and Coordination in Shared Workspaces. *Proc. ACM CSCW'92*, 107-114.

7.  Egido, C. (1990). Teleconferencing as a Technology to Support Cooperative Work: Its Possibilities and Limitations, in *Intellectual Teamwork: Social and Technological Foundations of Cooperative Work*, J. Galegher, R. Kraut, and C. Egido ed., Lawrence Erlbaum, 351-372.

8.  Ereback A.L. and Hook, K. (1994). Using Cognitive Walkthrough for Evaluating a CSCW Application. *Proc ACM CHI '94*, 91-92.

9.  Gaver, W. (1991). Sound Support for Collaboration, *Proc. 2nd ECSCW'91*, 293-308.

10. Goodwin, C. (1981). *Conversational Organization: Interaction Between Speakers and Hearers*, Academic Press.

11. Greenberg, S., Fitzpatrick, G., Gutwin, C. and Kaplan, S. (1999). Adapting the Locales Framework for Heuristic Evaluation of Groupware. *Proc. OZCHI'99 Australian Conference on Computer Human Interaction*, 28-30.

12. Greenberg S. and Roseman, M. (1998). Using a Room Metaphor to Ease Transitions in Groupware. Research report 98/611/02, Dept of Computer Science, University of Calgary, Canada.

13. Greenberg, S., Roseman, M., Webster, D. and Bohnet, R. (1992). Human and technical factors of distributed group drawing tools. *Interacting with Computers*, 4(1), 364-392.

14. Grudin, J. (1988) Why CSCW Applications Fail: Problems in the Design and Evaluation of Organizational Interfaces. *Proc ACM CSCW'88*, 85-93.

15. Gutwin, C. (1997). Workspace Awareness in Real-Time Distributed Groupware. Ph.D Thesis, Dept of Computer Science, University of Calgary, Canada.

16. Gutwin, C. and Greenberg, S. (1999). The Effects of Workspace Awareness Support on the Usability of Real-Time Distributed Groupware. *ACM Transactions on Computer-Human Interaction,* 6(3), 243-281.

17. Gutwin, C. and Greenberg, S. (2000). The Mechanics of Collaboration: Developing Low Cost Usability Evaluation Methods for Shared Workspaces. *IEEE 9th Int'l Workshop on Enabling Technologies: Infrastructure for Collaborative Enterprises (WET-ICE'00).*

18. Hill, W., Hollan, J., Wroblewski, D., and McCandless, T. (1991) Edit Wear and Read Wear. *Proc CHI '91*, 3-9.

19. Heath, C., Jirotka, M., Luff, P., and Hindmarsh, J. (1995). Unpacking Collaboration: The Interactional Organisation of Trading in a City Dealing Room. *Computer Supported Cooperative Work*, 3(2), 147-165.

20. Ishii, H., Kobayshi, M., and Grudin, J. (1992). Integration of interpersonal space and shared workspace: ClearBoard design and experiments. *Proc ACM CSCW'92*, 33-42.

21. Kraut, R., Egido, C., and Galegher, J. (1988). Patterns of Contact and Communication in Scientific Research Collaboration Remote Communications. *Proc ACM CSCW'88*, 1-12

22. Mack, R. and Nielsen, J. (1994). Executive Summary. In *Usability Inspection Methods,* Nielsen, J. and Mack, R., Eds. John Wiley and Sons, New York, 1-23.

23. McGrath, J. (1996). Methodology matters: Doing research in the behavioural and social sciences. In R. Baecker, J. Grudin, W. Buxton and S. Greenberg (eds) *Readings in Human Computer Interaction: Towards the Year 2000*, Morgan-Kaufmann, 152-169.

24. McLaughlin, M., (1984). Conversation: How Talk is Organized, Sage, Beverly Hills.

25. Nielsen, J. (1992). Finding usability problems through heuristic evaluation. *Proc ACM CHI'92*, 372-380.

26. Nielsen, J. (1993). *Usability Engineering*. Academic Press.

27. Nielsen, J. (1994). Chapter 2: Heuristic Evaluation. In Usability Inspection Methods, Nielsen, J. and Mack, R., Eds. John Wiley and Sons, New York, 25-62.

28. Nielsen, J. (1994). Enhancing the explanatory power of usability heuristics. *Proc ACM CHI '94*, 152-158.

29. Nielsen, J., and Molich, R. (1990). Heuristic evaluation of user interfaces. *Proc ACM CHI '90*, 249-256.

30. Robinson, M. (1991) Computer-Supported Cooperative Work: Cases and Concepts. *Proc Groupware'91*, 59-75.
31. Sacks, H., Schegloff, E., and Jefferson, G. (1974). A Simplest Semantics for the Organization of Turn-Taking for Conversation, Language, 50, 696-735.
32. Salvador, T., Scholtz, J., and Larson, J. The Denver Model of Groupware Design, SIGCHI Bulletin, 28(1), 52-58
33. Segal, L. (1994). Effects of Checklist Interface on Non-Verbal Crew Communications, NASA Ames Research Center, Contractor Report 177639.
34. Short, J., Williams, E., and Christie, B. (1976). Communication Modes and Task Performance, in *Readings in Groupware and Computer Supported Cooperative Work,* R. M. Baecker Eds., Morgan-Kaufman Publishers, Moutain View, CA, 169-176.
35. Smith, R., O'Shea, T., O'Malley, C., Scanlon, E., and Taylor, J. (1989). Preliminary experiences with a distributed, multi-media, problem environment. *Proc EC-CSCW'89*.
36. Stefik, M., Foster, G., Bobrow, D., Kahn, K., Lanning, S., and Suchman, L. (1987). Beyond the Chalkboard: Computer Supported for Collaboration and Problem Solving in Meetings. *Comm. ACM*, 30(1), 32-47.
37. Stefik, M., Bobrow, D., Foster, G., Lanning, S., and Tatar, D. (1987) WYSIWIS Revised: Early Experiences with Multiuser Interfaces. *ACM Transactions on Office Information Systems*, 5(2), 147-167.
38. Tang, J. (1991). Findings from Observational Studies of Collaborative Work. *Intl J Man-Machine Studies*, 34(2), 143-160.
39. Tang, J. and Leifer, L. (1988). A Framework for Understanding the Workspace Activity of Design Teams. *Proc ACM CSCW'88*, 244-249.
40. Tang, J. and Minneman, S. (1990). VideowDraw : A Video Interface for Collaborative Drawing. *Proc ACM CHI'90*, 313-320.
41. Tatar, D., Foster, G., and Bobrow, D. (1991). Design for Conversation: Lessons from Cognoter. *International Journal of Man-Machine Studies*, 34(2), 185-210.

## Discussion

*D. Salber:* Even though your focus is on shared workspaces your heuristics might apply to email or instant messaging. Did you look at heuristics for a broader range of groupware systems.

*K. Baker:* Not really. Some of these heuristics apply to groupware in general eg finding collaborators, establishing sessions. These are based on a theoretical framework. They describe how people actually work together. They are therefore important in these applications. We don't have (or have not found) a framework for more general groupware such as email yet.

*P. Curzon:* One of the heuristics was about the importance of informal collaboration. However, most real world informal collaboration occurs when you are taking a break from work and the computer. If informal collaboration is some important isn't this always going to be a problem with groupware.

*K. Baker:* The important thing is that these informal conversations are quick to initiate. The heuristic "facilitate finding collaborators and establishing contact" is important here. Setting up communication in groupware has to be lightweight and simple. For most existing groupware systems this is not the case. There are important differences between planned and informal collaboration but this heuristic is important to both.

*P. Van Roy:* More a comment: The book "Understanding comics" by Scott McCloud gives interesting insights on the interaction between words, images, sequence, levels of abstraction for example the concept of closure. Given the two images - the human automatically fills in the gap between them.

*K. Baker:* Interesting, please send us a reference.

*P. Smith:* What do your heuristics have to say about baton passing as a mechanism in collaborative environments.

*K. Baker:* Multiple heuristics may be applicable. For instance, "allow people to coordinate" and "provide protection" are both related to this problem. It is important to apply heuristics with respect to the particular circumstances. For instance, in formal meeting, we need significantly more control/protection. Like Nielsen's heuristics these are anchor points. To take the knowledge of the inspector and make sure that they look for all of the necessary problems. There is training list material that goes with these points. In groupware, training material and experience are more vital as novice evaluators may not have much experience with groupware systems. Even some expert evaluators who looked at early versions of these heuristics were still left hanging, and some asked for checklists. As an example one commercial groupware tool that we looked at initially looked really nice. However, when the authors applied these heuristics they found lots of serious problems. The students didn't report that many problems. However, this was because the system didn't even support fundamental activities. Modern systems can miss really basic things. So such heuristics are necessary. One thing that these heuristics do is help novice evaluators to articulate the problems.

# An Organizational Learning Method
# for Applying Usability Guidelines and Patterns

Scott Henninger

Department of Computer Science & Engineering
University of Nebraska-Lincoln
Lincoln, NE 68588-0115 USA
+1 402 472 8394
scotth@cse.unl.edu

**Abstract.** As usability knowledge and techniques continues to grow, there is an increasing need to provide tools that disseminate the accumulated wisdom of the field. Usability guidelines are one technique that is used to convey usability knowledge. Another is the emerging discipline of usability patterns. This paper presents an approach that combines these techniques in a case-based architecture and utilizes a process to help an organization capture, adapt, and refine usability resources from project experiences. The approach utilizes a rule-based tool to represent the circumstances under which a given usability resource is applicable. Characteristics of the application under development are captured and used to match usability resources to the project where they can be used to drive the design process. Design reviews are used to capture feedback and ensure that the repository remains a vital knowledge source for producing useful and usable software systems.

## 1 Introduction and Motivation

As the body of knowledge on the design of interactive software systems becomes mature, the need for disseminating the accumulated wisdom of the field becomes increasingly important to the design of useful and usable software systems. Design for usability is becoming increasingly important to the success of software systems, but software developers are usually poorly trained in human factors, ergonomics, or usability issues. One solution is to always require a human factors specialist on development teams, but this is often impractical as such specialists continue to be in short supply and budgets do not always allow such specialized personnel. Education and iterative development processes aimed at evaluating and improving the user interfaces are necessary solutions to this problem, but techniques are needed that provide software developers with proactive knowledge and techniques for developing high quality user interfaces.

Usability guidelines have been around in various forms for some time, and have had some impact on design practices for user interface software. Yet the full potential of guidelines have yet to be realized [1, 2]. To date, work in these areas have failed to adequately address concerns facing software designers, developers, and managers, focusing on comprehensive usability issues at the expense of determining which

M. Reed Little and L. Nigay (Eds.): EHCI 2001, LNCS 2254, pp. 141–155, 2001.

guidelines should be used under what circumstances. In addition, usability guidelines often become a static document read only by human factors specialists and used to assess an application's conformance to usability standards. Guideline analyzers, which analyze completed interfaces against guidelines or other usability metrics [3, 4], can assess completed systems, but do little to support the development process.

These methods apply usability knowledge as an assessment, which is often too late in the development process. In one example witnessed by the author, an application was submitted to a human factors group in a large IT department that had a screen with 39 seemingly unordered buttons arranged in an array [5], a poor interface that would cause users to engage in lengthy searches to find desired features. This organization had a well-designed on-line style guide [6], but usability approval was so late in the development process that there was inadequate time and resources to fix the problem before it was shipped. The less mature work on usability patterns take a more proactive view of the design process, but add little to the usability guidelines perspective beyond a different format for documenting the pattern and some concerns for establishing the context of a pattern.

Instead of being relegated to a discretionary reference role and/or an after-the-fact human factors certification process, the knowledge contained in usability resources needs to be delivered as an integral part of the entire development process. Guidelines and patterns can be helpful resources for the developer, but tools for finding applicable resources are lacking. Current approaches are document-based, at best supported with hypertext tools, which relies on individual developers to know of the existence of the resources and understand when they should be applied. Given the potentially copious usability guidelines and patterns, and the lack of training in usability issues, this is not a satisfactory solution.

Tools are needed to turn guidelines and patterns into proactive development resources that can be applied throughout the development process. In this paper, a methodology is presented that represents the context of a given guideline or pattern in the form of applicability rules that formally specify the conditions under which a usability resource is appropriate. We present an exploratory prototype, named GUIDE (Guidelines for Usability through Interface Development Experiences), that we have been using to investigate and demonstrate how this methodology can be used to deliver usability resources to software developers when they are needed. The focus of this work is not the creation or discovery of good guidelines or patterns, but the creation of tools that capture and disseminate knowledge of user interface design principles and experiences at the right time – during the development process. In addition, our organizational learning approach [7] allows the incremental capture of the characteristics of the context of use so the applicability rules and guidelines can evolve as new requirements are encountered, new techniques are used, and new designs are created.

## 2    Usability Guidelines and Patterns

Usability guidelines have become a widely recognized method of bringing the cumulative knowledge of usability issues to bear on the software development process. It is generally accepted that guidelines cannot replace the "golden rules" of interface design - user involvement, user feedback from early prototypes, and iterative development [8]. But guidelines can play a role in improving the quality of the iterative steps, leading to an improvement in quality and reduction (but not elimination) of the number of iterations involved in the design-evaluate-redesign cycle of HCI development.

Guidelines have evolved to take on a number of forms. *Style guides* address how different kinds of windows should look and interact with the user for tasks such as choosing from lists [9-13]. Style guides tend to be platform-specific and focus on interface widgets, such as dialogue boxes, pull-down menus, screen layouts, and naming conventions. Other questions, such as when a particular widget should be used or how the interface elements integrate together, are left unanswered.

While style guides are usually platform-specific, universally applicable *interface guidelines* have also been explored to provide higher-level guidelines on various aspects of human-computer interfaces [14-18]. These guidelines dispense general advice, such as "Allow the user to control the dialogue," "Provide displayed feedback for all user actions during data entry," or "Reduce the user's memory load." At some level, this is sound advice, but this kind of information lacks important contextual information that would allow designers and developers to assess how and when to apply the guidelines to a specific set of circumstances or system requirements.

*Usability standards* also take on the character of guidelines, opting to specify general principles rather than mandating specific techniques, widgets, or tools. International standards have been created [19], and standards have been used within organizations to ensure a degree of consistency across applications [6, 20, 21]. *Domain-specific guidelines* have also been created, the most prominent being guidelines for designing Web pages [22, 23].

All of these efforts have largely focused on the content and structure of the guidelines themselves. On-line versions of guidelines have been created, but have used simple hypertext-based search systems for accessing the guidelines [24-28]. But these systems have done little to address the problems demonstrated in studies using guidelines, such as the time to find relevant guidelines, problems with interpreting guidelines for the task at hand, and generally being too abstract to directly apply [29-31].

While usability guideline have become voluminous to the point that it is difficult to determine which principles are applicable to a given design problem. And the continuing proliferation of technology only exacerbates the problem. Little thought has been given to defining when guidelines are applicable or how guidelines can be refined to meet user task requirements for a specific set of users and a specific type of application. In addition, little research has been done to accumulate knowledge about interface design in a form that can capture relationships between specific contexts and applicable guidelines.

## 2.1    Context and Usability Patterns

A usability patterns community, inspired by the recent work on software patterns, has begun to explore how patterns can be used to provide an intermediate perspective between universally applicable usability guidelines and component-specific style guides [32-35]. The essential idea of a design pattern is to capture recurring problems along with the context and forces that operate on the problem to yield a general solution. Collections of patterns can be organized in a network of higher-level patterns that are resolved or refined by more detailed patterns, resulting in a *pattern language* [36].

Usability patterns explicitly represent context, although approaches vary from a one-sentence description of the design goals [37] to viewing context as the explicit focus of patterns, telling "the designer *when, how* and *why* a solution can be applied." [32]. Usability patterns represent context through text fields such as context and forces that respectively describe how the problem arises and other issues that may impact the outcome. For example, the "Shield" pattern [32] describes the problem of protecting users from accidental selection of a function with irreversible effects. The context states that users need protection against undesired or unsafe system actions, and that the pattern should not be used for easily reversible actions. The forces include severity of the unintended actions and the user's need to work quickly while avoiding mistakes. Patterns can also represent context through a network of linked patterns, the pattern language, from high-level issues to low-level choices, although examples of these networks and tools for traversing the links are currently lacking.


## 2.2    Integrating Usability Guidelines and Patterns

Differences between usability guidelines and usability patterns lie primarily in perspective and representation of the information. In fact, many of the proposed patterns replicate much of the information contained in existing guidelines. The major difference is that pattern languages are intended to be used as a design method. The pattern community is therefore concerned with using patterns to communicate between designers and customers or users [38, 39], a perspective not often seen in guidelines.

Because of this main difference, the perspective of usability patterns tends to be more problem-oriented, focusing on describing a problem and solution, than the more general information or advice perspective of guidelines. Although templates and data structures for describing guidelines and patterns can easily be reconciled, the fields commonly seen in patterns are indicative of the problem-oriented perspective. In addition to the problem-solution (or title-solution) format seen in most guidelines, patterns add fields to describe the context of the problem and the forces that shape the problem and its variants.

The goals of both these approaches are essentially the same: to document and manage experience about usability design issues in a format that is easily disseminated and understood. But much of the work in these fields focus on the development of patterns and guidelines (a notable exception is the recent Tools for Working With Guidelines workshop series [40]). Creating and disseminating this knowledge is important, particularly where empirical validation is present, but little work has been done on creating the computational framework that will supply this information in an effective manner. Our focus is different in that we begin from the perspective of creating resources and tools for software developers. Instead of teaching developers specific usability principles (the proverbial fish), we aim to provide the tools that allow development organizations to create and disseminate usability resources in a manner that helps developers design and develop usable applications (the proverbial teaching them how to fish).

# 3    An Organizational Learning Approach for Usability Guidelines and Patterns

Current research and practice for both guidelines and patterns primarily rely on an educational or information retrieval model. It is entirely up to the designer to either know that usability resource (collectively we will refer to patterns and guidelines as "resources") exists or at least know enough about the repository to realize they should search for applicable resources. One must have an overall understanding of existing guidelines and patterns to recognize when a given resource should be applied. Given the lack of formal usability training, the potential size of a comprehensive pattern or guideline repository, and the frailty of human memory, this assumption does not always hold.

In past research, we have created tools to support organizational memory for usability guidelines. The objective of organizational memory tools is to provide information relevant to organizational practices that "you can't learn in school" [41], such as local terminology, organization and project-specific conventions, lessons learned, policies and guidelines, individuals with expertise, and many others. The Mimir Guidelines system illustrated how organizational memory techniques could be used to collect and disseminate usability guidelines [31]. Project experiences were captured using case-based decision support technology [42], where cases were attached to guidelines as examples of how the guideline had been applied. Dissemination of guidelines was supported through hypertext and searching techniques where users matched project characteristics to appropriate guidelines.

Creating a repository of project experiences, an organizational memory, is valuable in and of itself. But the overarching objective is to learn and improve on past performance. Emphasis must be placed on establishing a continuous improvement process that enhances product quality and developer productivity, while recognizing past experiences as a catalyst for the learning process. We call this an organizational learning approach [7] to emphasize that the knowledge is used as the *basis* for improvement, not just memorizing past experiences [43-45].

## 3.1    The GUIDE Process for Applying Usability Resources

Our approach to supporting organizational learning to usability resources is a combination of tool and process to capture knowledge as it emerges in practice, review and/or otherwise validate that knowledge, and ensure that previous knowledge and known best practices are applied where they exist. The GUIDE (Guidelines for Usability through Interface Development Experiences) methodology supports an organizational learning approach to developing context-specific usability resources through the process shown in Figure 1. A key component of the methodology is a hierarchical structure of usability guidelines delivered in the GUIDE tool shown in

Figure 2. We have chosen to seed our repository with a Web-enhanced Smith and Mosier 944 guidelines corpus [16], although any set of initial guidelines would work equally well.[1]  A rule-based system is then used to match project characteristics (user populations, tasks, GUI tools, etc.) to specific usability resources that project personnel should apply during development.  The result is a set of project activities that are assigned to the project.  For example, if the system is being accessed by users over the Internet and involves access to sensitive data, then guidelines and/or patterns for login interfaces and access to sensitive data will be given to the project as an activity to be considered.

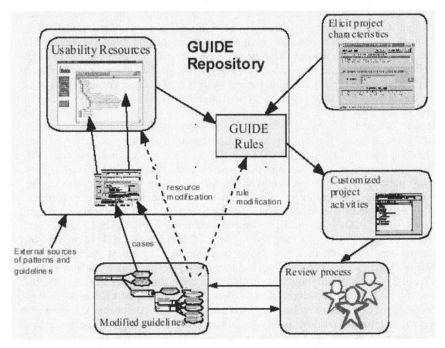

**Fig. 1.** Using and modifying usability resources.

The next step of the process in Figure 1 is the most critical element of the organizational learning approach.  A review process is used to inspect how project personnel answered the options posed by GUIDE and discuss whether the assigned resources are appropriate for the project in question. This review creates an important feedback loop that is used to learn emergent user interface needs.  If there is a mismatch between project needs and the resources assigned by GUIDE, reviewers can recommend that either a different option is chosen (options are described below) or that the knowledge in the repository needs to be updated to meet the needs of this project.  The latter of these two options creates an opportunity for learning.  As shown

---

[1] We are currently investigating the possibility of basing our structure on patterns, adopting one of the usability pattern languages currently under development and augmenting with guidelines from various sources.

in Figure 1, not only are modified project guidelines created, but the conditions for this modification can optionally be fed into the repository, essentially "blazing a trail" for subsequent projects.

**Fig. 2.** Guide interface and a usability case.

For example, suppose a project is the first to have a requirement for both cross-platform and cross-browser Java delivery. A project performs some studies and determines that using the Java Plug-in is the best choice in this instance. While this decision can later be augmented (for example, another project's users may be using 28k modems and could deem downloading the rather sizable Plug-in infeasible), extended, and eventually replaced by subsequent efforts or outside changes in technology, it represents a form of intellectual capital that has considerable research and effort behind it that the organization probably does not want to replicate unnecessarily.

## 3.2 The GUIDE Architecture

GUIDE is an exploratory prototype that has been used to investigate and demonstrate how usability guidelines can be integrated into the software development process. GUIDE borrows from the case-based architecture of the BORE (Building an Organizational Repository of Experiences) project [7], but focuses exclusively on usability issues. Although initial efforts have explored the different issues arising in software engineering processes and usability separately, GUIDE is currently being re-designed to become part of BORE. As with BORE, GUIDE is a Web-based application, using HTML and Java for the user interface, Java for processing, and a database back-end to store information.

Representation of usability resources in GUIDE embodies the intersection of three closely related technologies. Patterns and guidelines, as discussed above, are related by common goals and similar formats. Case-based technology, which uses a problem-solution structure similar to patterns and guidelines, adds an instance-of relationship (cases are context-specific instances of usability resources, which can be patterns or guidelines).

Figure 2 shows a hierarchical view of usability resources in GUIDE and a window showing a specific resource on frame-based Web page navigation. The fields shown on the resource window (the window on the right in Figure 2) include the canonical fields found in usability patterns work, although other formats can easily be integrated into GUIDE's architecture. The system architecture is flexible enough to accommodate many of the different fields suggested in usability guideline and pattern research.

### 3.3    Using GUIDE to Develop Interactive Software Systems

GUIDE uses a case-based structure to associate usability resources to project activities (see the Case Manager windows in Figure 3a and 3b) that document a specific project's use of the resources. At the start of a software development effort, a GUIDE project is created. This will create a number of project initiation cases, some of which will have usability options associated with them, delimited by a '?' inside the icon in the project hierarchy. Clicking on the "Options" tab of the case displays one or more questions about the characteristics or requirements of the project (bottom window, Figure 3a). Selecting a question displays possible answers in the Answers box. Selecting an answer will trigger applicability rules (described in the next section) for resources that are assigned to the project to inform developers of usability principles that need to be followed.

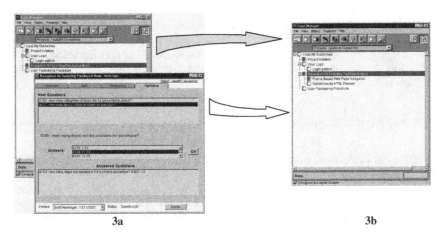

3a                                                          3b

**Fig. 3.** Documenting Contextual factors through project requirements.

For example, the left window in Figure 3a displays a project named "Usability Guidelines." This project has a few initial project cases associated with it that are used to identify tools, techniques, and usability issues that developers should be considering during design. Previous questions have determined that the project involves designing an e-storefront that allows multiple items to be purchased during the same session. This causes a number of cases to be assigned to the project, such as "Navigation for Selecting Purchase Items," that represents recommended project activities. The user has selected the Options tab for that case (bottom window of Figure 3a), revealing a series of questions to disclose further project requirements. One question has already been answered, leading to new questions that explore further project requirements.

### 3.4    Representing Context with Applicability Rules

Usability resources in GUIDE are assigned to a specific project through applicability rules that match project characteristics to appropriate resources. This is accomplished by a forward-chaining inference engine using production rules with an if-then structure. Preconditions are defined as question-answer pairs. Each time a question is answered, the inference engine checks the database to see if any of the rules are satisfied. When this occurs, a set of actions are fired. Actions can cause a variety of events, including placing questions in or taking questions out of the New Question stack, assignment of system variables, and attaching usability resources to the project.

For example, selecting the answer "11-20" in Figure 3a will fire a rule that will place cases in the project that points to frame-based navigation resources, shown as child cases under the "Navigation for Selecting Purchase Items," shown in the selected case in Figure 3b. In essence, the rule base is stating that using frame-based navigation is recommended when the purchasing procedure takes between 1 and 2 steps and there are between 11 and 20 items available for purchase. Rationale for this recommendation is provided in the resource, which states that frames can be used as a solution to the problem of needing to keep a context while accessing multiple pages[2]. Note that these rules encode the context of the resources, a major features of pattern languages [38].

It should also be noted that GUIDE is designed so that questions can be associated with any project case, allowing development teams to incrementally disclose usability issues when they are ready, instead of having to answer all questions at the beginning of a project. In Figure 3b, the "Frame-Based Web Page Navigation" case has further options associated with it, allowing further decomposition to more detailed guidelines or patterns.

Figure 4 depicts a partial decision tree of the kind that can be represented by our forward-chaining inference engine. Through these rules, which are developed by usability professionals (see the following section), the GUIDE system is placing the accumulated wisdom of usability issues at the fingertips of software developers in the context in which they are applicable.

---

[2] This example is intended to demonstrate our approach, not advocate any specific usability principles.

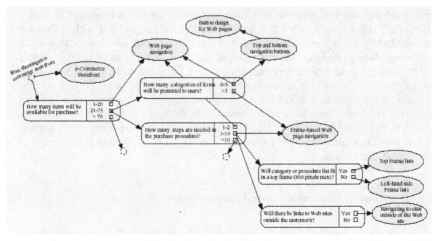

**Fig. 4.** A partial GUIDE decision tree.

### 3.5    Incremental Acquisition of Design Knowledge

Applicability rules in GUIDE are meant to provide a more proactive alternative to the "build a repository and let them search" philosophy that current approaches to usability guidelines and patterns employ. Rules are not meant to automate the design and development of interfaces, but to provide a match to resources that can inform developers of usability issues. They are intended to act as a medium for discussion and debate. Indeed, we view rules as a means to formally state, evolve, and improve the current understanding of usability guidelines and the conditions under which they can provide helpful information to software developers. An important aspect of this philosophy are tools and processes to modify rules to meet the dynamic needs of interactive software systems, as described by process depicted in Figure 1.

The overall objective and philosophy of this research is not to derive universally applicable rules or usability resources. Rather, the aim is to provide the rules and infrastructure that allow an organization or group to accumulate knowledge based on the collective experiences of the organization. This is accomplished by instituting a process that reviews the recommendations given by GUIDE during project design reviews. Suggestions on modifications and improvements to the knowledge base are then forwarded to human factors and GUIDE librarian personnel for consideration. This kind of process has been demonstrated to work in practice [43], provided the repository remains up to date and remains an important corporate asset that provides benefit to the developers.

During periodic design reviews, teams will review and critique project answers to GUIDE questions. Different answers could be negotiated and found to be more appropriate for the project and can easily be changed in GUIDE (the system supports rule backtracking). Review teams could also determine that GUIDE recommendations are either inappropriate or missing. This is seen as an opportunity to improve the knowledge base. The review team provides rationale for why deviations are necessary.  Human factors specialists and/or other GUIDE curators review the requests and either refuses the request, allows a once-only deviation, or turns the rationale into GUIDE rules and modifies the repository with this new knowledge.  New guidelines, examples, and other information could also be created and placed in the repository to document the project's experiences.

For example, suppose that previous projects have noted that users have problems keeping track of where they are in complex procedures and clicking on links to external Web pages that end up in the viewing part of the frame (the second problem is a common HTML frame issue, especially when framed pages are displayed inside of frames).[3]  The review team identify that frames are needed in this project and want to use the review as an opportunity to document and apply the lessons learned from previous efforts.  The external link problem could be handled in a number of ways, including adding sections in the "Guidelines for HTML Frames" document (see Figure 3b) or adding a new guideline and rule stating that if frames are used the new guideline on external page links needs to be followed.  The tracking complex procedures issue could be addressed by creating a new guideline stating, for example, that completion of steps in a procedure are tracked by changing the color or otherwise highlighting completed steps.  Then an action is added to the rule used in Figure 3 so that any project having 3 to 10 steps in the procedure is also given this new guideline.

This process ensures that the repository will evolve to meet the changing needs of the organization while foraging new paths for subsequent projects to follow.  As the repository grows it will accommodate greater portions of projects, minimizing the number of deviations while increasing knowledge reuse.

The rule base can be as specific or vague as needed by the organization.  For example, an organization that services only the medical community will probably have guidelines that are specific to medical terms and procedures, while an organization servicing a broader customer base will probably want guidelines at a higher level of abstraction.  The level of detail is determined by the rules and the amount of effort desired by the organization, not by any limitation or mandate placed in GUIDE itself.

# 4    Conclusions and Future Research

The objective of this research is not an attempt to automate user interface design.  To the contrary, it is recognized that effective user interface design take a degree of talent and careful work with the end users that cannot be captured through rules, patterns or any information system.  Nonetheless, there is recognized knowledge and conventions that can help some designers reach higher levels of competency and help accomplished designers extend their knowledge to areas they have not yet experienced.  This research is an exploration of how resources can be delivered to software developers through a rule-based structure that provides the basis for an organizational memory – capturing the collective intelligence of an organization with respect to usability design issues.  Rules in this context serve as a medium, a formal

---

[3] Empirical studies may have been conducted confirming this, and would be linked to the specific guidelines or patterns addressing these issues.

mechanism for communicating design knowledge and establishing relationships between context and usability resources.

Rule-based systems are often criticized for their inflexibility, which we ameliorate through a process that reviews the relationships established by the rules throughout the development process to ensure that real project experiences are represented. Our integration of rule-based and case-based systems comes closer to the spirit of American Case Law, where statutory law (rules) are contextualized by case law (cases, guidelines, patterns). People then use this structure to argue which cases come closest to the current situation and apply the attached rule, or set new precedents if none of the cases are applicable.

The result is a web of knowledge on usability issues that is continuously updated to meet the evolving needs of the organization. As the repository grows, it will become an important piece of intellectual capital that puts knowledge of proven usability techniques and wisdom at the fingertips of software developers. This approach does not replace the need for iterative software development methodologies and user studies (although knowledge of how to conduct those processes should be contained in the repository), but can reduce the number of iterations by assuring that certain classes of errors are avoided.

This approach requires that some software development staff devote some percentage of their time maintaining the knowledge base. Personnel knowledgeable about usability issues and the structure and content of the usability guidelines and patterns are necessary for this approach to work. In addition, rule-based expertise is necessary to ensure that GUIDE rules are well-structured and operate properly. Such a structure resembles the concept of software factories [46], aligns well with current trends to involve human factors in the design process [6], and can easily be applied to mid-size or small development firms that can only afford to staff usability consultants on a part-time basis.

The intersection of guidelines and patterns needs further investigation. We are currently exploring structures that better integrate the different kinds of knowledge contained in patterns, guidelines, style guides, etc. We are also interested in providing examples as a significant knowledge resource. Given the number and diverse composition and content of existing resources, particularly usability guidelines, finding a proper "seed" [47] for the repository is problematic. A study revealed that there was minimal overlap between 21 different guideline corpuses [48], further underscoring the need for supporting a diverse initial set of usability resources.

Continued research is needed to further understand the issues of using organizational memory repositories as advocated here. Many empirical questions remain, such as whether variance between projects is too great to apply past experiences, whether past experiences stifle creativity or enables it by shifting attention away from re-creating previous solutions, whether the approach is useful for only certain types of organizations, and whether the documentation burden of constructing rues is too great for practical application. The contribution of this work thus far is to provide tool support to turn usability guidelines and patterns into a proactive design tool and design organizational structure and process to capture and disseminate project experiences on usability issues.

The GUIDE and BORE projects will be evaluated through use in the Software Design Studio in the JD Edwards Design Studio at the University of Nebraska-Lincoln. This program, which integrates a combination of business and computer science subjects, has been built around a design studio concept [49] where students are engaged in long-term projects from paying customers external to the University. BORE will be used to deliver and manage the studio's defined software development process. GUIDE will be employed to deliver usability resources as part of scenario-based and other design methodologies. This and other research efforts will further refine the system while further populating the repository with usability knowledge and experiences. We will also seek to apply the tool and technique to pilot studies in industry, as an effort to further study the issues involved with employing an experience-based methodology to the usability design process.

## Acknowledgments

I gratefully acknowledge the efforts a number of graduate students that have helped develop BORE, including Charisse Lu, Kurt Baumgarten, and Peter Hsu. Osama Al Shara has also contributed. This research was funded by the National Science Foundation (CCR-9502461, CCR-9988540, and ITR/SEL-0085788), and through contracts with Union Pacific Railroad.

# References

1. J. D. Gould, S. J. Boies, and C. H. Lewis, "Making Usable, Useful, Productivity-Enhancing Computer Applications," *Communications of the ACM*, vol. 34, 1991, pp. 75-85.
2. F. de Souza and N. Bevan, "The Use of Guidelines in Menu Interface Design: Evaluation of a Draft Standard," *Human-Computer Interaction - INTERACT '90*, 1990, pp. 435-440.
3. M. Y. Ivory, R. R. Sinha, and M. A. Hearst, "Empirically Validated Web Page Design Metrics," *Proc. Human Factors in Computing Systems (CHI 2001)*, Seattle, WA, 2001, pp. 53-60.
4. D. Scapin, C. Leulier, J. Vanderdonckt, C. Mariage, C. Bastien, C. Farenc, P. Palanque, and R. Bastide, "A Framework for Organizing Web Usability Guidelines," *6th Conference on Human Factors and the Web*, Austin, TX, 2000.
5. S. Henninger, "A Methodology and Tools for Applying Context-Specific Usability Guidelines to Interface Design," *Interacting With Computers*, vol. 12, 2000, pp. 225-243.
6. P. A. Billingsley, "Starting from Scratch: Building a Usability Program at Union Pacific Railroad," *interactions*, vol. 2, 1995, pp. 27-30.
7. S. Henninger, "Case-Based Knowledge Management Tools for Software Development," *Journal of Automated Software Engineering*, vol. 4, 1997, pp. 319-340.
8. J. D. Gould and C. H. Lewis, "Designing for Usability - Key Principles and What Designers Think," *Communications of the ACM*, vol. 28, 1985, pp. 300-311.
9. Apple Computer Inc., *Macintosh Human Interface Guidelines*. Reading, MA: Addison-Wesley, 1992.
10. Microsoft Corporation, *The Windows Interface: An Application Design Guide*. Redmond, WA: Microsoft Press, 1992.
11. IBM, *Object-Oriented Interface Design: IBM Common User Access Guidelines*. Carmel, IN: Que, 1992.
12. OSF, *OSF/Motif Style Guide: Revision 1.2*. Englewood Cliffs, NJ: Prentice Hall, 1993.
13. Sun Microsystems, *Open Look Graphical User Interface Application Style Guidelines*. Reading, MA: Addison-Wesley, 1989.
14. C. M. Brown, *Human-Computer Interface Design Guidelines*. New Jersey: Ablex, 1988.
15. P. Heckel, *The Elements of Friendly Software Design*. San Francisco: Sybex, 1991.

16. S. L. Smith and J. N. Mosier, "Guidelines for Designing User Interface Software," Technical Report, The MITRE Corporation ESD-TR-86-278, 1986.
17. J. Vanderdonckt, "Towards a Corpus of Validated Web Design Guidelines," *Proceedings of the 4th ERCIM Workshop on 'User Interfaces for All'*, 1998, pp. 16-31.
18. B. Shneiderman, *Designing the User Interface: Strategies for Effective Human-Computer Interaction, 2nd ed.* Reading, MA: Addison-Wesley, 1992.
19. ISO/WD 9241, "Ergonomic Requirements for Office Work with visual Displays Units," International Standard Organization 1992.
20. E. Rosenweig, "A Common Look and Feel or a Fantasy?," *interactions*, vol. 3, 1996, pp. 21-26.
21. S. Weinschenk and S. C. Yeo, *Guidelines for Enterprise-Wide GUI Design*: Wiley & Sons, 1995.
22. P. J. Lynch and S. Horton, *Web Style Guide : Basic Design Principles for Creating Web Sites*. Princeton, NJ: Yale Univ Press, 1999.
23. J. A. Borges, I. Morales, and N. J. Rodriacuteguez, "Guidelines for Designing Usable World Wide Web Pages," *Proc. Human Factors in Computing Systems (CHI '96) Short Papers*, 1996, pp. 277 - 278.
24. D. Grammenos, D. Akoumianakis, and C. Stephanidis, "Integrated support for working with guidelines: the Sherlock guideline management system," *Interacting with Computers*, vol. 12, 2000, pp. 281-311.
25. J. Vanderdonckt, "Accessing Guidelines Information with SIERRA," *Proceedings Fifth IFIP International Conference on Human-Computer Interaction INTERACT '95*, Lillehammer, 1995, pp. pp. 311-316.
26. R. Iannella, "HyperSAM: A Practical User Interface Guidelines Management System," *Proceedings of the Second Annual CHISIG (Queensland) Symposium - QCHI '94*, Bond Univ., Australia, 1994.
27. L. Alben, J. Faris, and H. Saddler, "*Making it Macintosh:* Designing the Message When the Message is Design," *interactions*, vol. 1, 1994, pp. 10-20.
28. G. Perlman, "Asynchronous Design/Evaluation Methods for Hypertext Development," *Hypertext '89 Proceedings*, 1989, pp. 61-68.
29. L. Tetzlaff and D. R. Schwartz, "The Use of Guidelines in Interface Design," *Proc. Human Factors in Computing Systems (CHI '91)*, 1991, pp. 329-333.
30. H. Thovtrup and J. Nielsen, "Assessing the usability of a user interface standard," *Proc. Human Factors in Computing Systems (CHI '91)*, New Orleans, LA, 1991, pp. 335-341.
31. S. Henninger, K. Haynes, and M. W. Reith, "A Framework for Developing Experience-Based Usability Guidelines," *Proc. Designing Interactive Systems (DIS '95)*, Ann Arbor MI, 1995, pp. 43-53.
32. M. van Welie, G. van der Veer, and A. Eliens, "Patterns as Tools for User Interface Design," *Workshop on Tools for Working With Guidelines*, Biarritz, France, 2000.
33. M. J. Mahemoff and L. J. Johnston, "Principles for a Usability-oriented Pattern Language," *Proc. Australian Computer Human Interaction Conference OZCJI 98*, Adelaide, 1998, pp. 132-139.
34. G. Casaday, "Notes on a Pattern Language for Interactive Usability," *Proc. Human Factors in Computing Systems (CHI '97)*, Atlanta, GA, 1997, pp. 289-290.
35. J. Borchers, "CHI Meets PLoP: An Interaction Patterns Workshop," *SIGCHI Bulletin*, vol. 32, 2000, pp. 9-12.
36. C. Alexander, *The Timeless Way of Building*. New York: Oxford Univ. Press, 1979.
37. A. Granlund and D. Lafreniere, "UPA 99 Workshop Report: A Pattern-Supported Approach to the UI Design Process,", 1999.
38. T. Erickson, "*Lingua Francas* for Design: Sacred Places and Pattern Languages," *Proc. Designing Interactive Systems (DIS 2000)*, New York, 2000, pp. 357-368.
39. J. Tidwell, "The Gang of Four are Guilty,"., 1999.
40. TFWWG, *Tools for Working With Guidelines Workshop (TFWWG2000)*, Biarritz, France, 2000.

41. L. G. Terveen, P. G. Selfridge, and M. D. Long, "From 'Folklore' To 'Living Design Memory'," *Proceedings InterCHI '93*, Amsterdam, 1993, pp. 15-22.
42. J. L. Kolodner, "Improving Human Decision Making through Case-Based Decision Aiding," *AI Magazine*, vol. 12, 1991, pp. 52-68.
43. L. G. Terveen, P. G. Selfridge, and M. D. Long, "Living Design Memory' - Framework, Implementation, Lessons Learned," *Human-Computer Interaction*, vol. 10, 1995, pp. 1-37.
44. J. P. Walsh and G. R. Ungson, "Organizational Memory," *Academy of Management Review*, vol. 16, 1991, pp. 57-91.
45. E. W. Stein and V. Zwass, "Actualizing Organizational Memory with Information Systems," *Information Systems Research*, vol. 6, 1995, pp. 85-117.
46. V. R. Basili, G. Caldiera, and G. Cantone, "A Reference Architecture for the Component Factory," *ACM Transactions on Software Engineering and Methodology*, vol. 1, 1992, pp. 53-80.
47. G. Fischer, R. McCall, J. Ostwald, B. Reeves, and F. Shipman, "Seeding, Evolutionary Growth and Reseeding:    Supporting the Incremental Development of Design Environments," *Proc. Human Factors in Computing Systems (CHI '94)*, Boston, MA, 1994, pp. 292-298.
48. J. Ratner, E. M. Grose, and C. Forsythe, "Characterization and Assessment of HTML Style Guides," *Proc. Human Factors in Computing Systems (CHI '96)*, 1996, pp. 115-116.
49. D. A. Schön,  *The Design Studio:  An Exploration of its Traditions and Potentials*. London: RIBA Publications Limited, 1985.

## Discussion

*P. Smith:* Your approach is to have a set of guidelines and then examples. Commercial systems use the reverse.

*S. Henninger:* Pattern work uses this approach, generalise and then examples. There is not too much difference. The end goal is the same. Communicate a design principle to get a better interface.

# Pervasive Application Development
# and the WYSIWYG Pitfall

Lawrence D. Bergman[1], Tatiana Kichkaylo[2], Guruduth Banavar[1],
and Jeremy Sussman[1]

[1] IBM T.J. Watson Research Center, 30 Saw Mill River Road, Hawthorne, NY 10532
[2] Department of Computer Science, New York University
{bergman1, banavar, jsussman}@us.ibm.com, kichkay@cs.nyu.edu

**Abstract.** Development of application front-ends that are designed for deployment on multiple devices requires facilities for specifying device-independent semantics. This paper focuses on the user-interface requirements for specifying device-independent layout constraints. We describe a device independent application model, and detail a set of high-level constraints that support automated layout on a wide variety of target platforms. We then focus on the problems that are inherent in any single-view direct-manipulation WYSIWYG interface for specifying such constraints. We propose a two-view interface designed to address those problems, and discuss how this interface effectively meets the requirements of abstract specification for pervasive applications.

## 1    Introduction

The challenge of writing applications has broadened in recent years because of a proliferation of portable devices, like personal digital assistants (PDA's) and programmable phones. These devices have varying physical resources, including differing display sizes and special I/O mechanisms. As a consequence, application front-ends, which include the user interface and some control logic such as event handlers, often must be written from scratch for each type of device on which that application is to run. This imposes an immense development and maintenance burden, particularly since the developer may not be aware of all the devices on which the application is to be deployed. Indeed, an application may be run on hardware platforms that were not even in existence when the application was written!

Figure 1 shows an example. The three parts of the figure show a portion of the same application (browsing of information at a job fair) on a PC running java, on a web-browser, and on a mobile phone. Notice the differences in the amount of information contained on a screen, as well as in the layout. The ideal is to have a single specification that produces all of these rendered applications.

There are two common approaches to solving the problem of specifying application front-ends that are intended to run on multiple devices. The first is to use a device- neutral specification language and library, with run-times for that language

M. Reed Little and L. Nigay (Eds.): EHCI 2001, LNCS 2254, pp. 157–172, 2001.

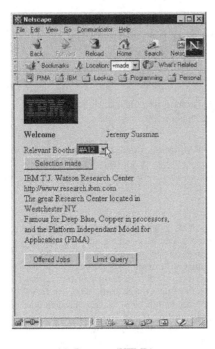
(a) Browser (HTML)

(b) Mobile phone (WML)

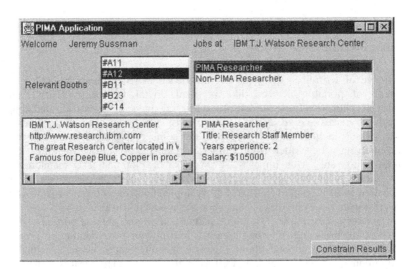
(c) Java

**Fig. 1.** Platform independent application (job fair) deployed on three platforms

deployed on all target devices. This is the approach taken by Java [10]. The problem here is that device characteristics can be radically different, so attempts to run a single user-interface design on multiple platforms is bound to give poor results. Indeed, on some devices, such as cellular phones, the UI assumptions underlying AWT do not hold at all.

The other approach is "style sheet" customization. This allows the designer to craft a device-specific set of layout rules for each individual device, using a style sheet transformation language such as XSLT [18]. The problem with this approach is that a style sheet is not application-specific; the same style sheet is typically applied to all applications to be run on a particular device. The style sheet overrides any application-specific characteristics.

An additional possibility, of course, is to develop a different style sheet for each application-device pair. This leads to the problem of both developing and managing a large number of style sheets – exactly the problem addressed by device-independent application development.

We are developing a framework that attempts to address the problem of creating application front-ends that can be developed once, and that will run reasonably on a wide variety of hardware platforms. Our goal is "write once, right anywhere," application development. Our approach is to provide an application model that allows the designer to create a device-independent specification of the front-end. Part of the specification is a set of device-independent *layout semantics*, which are specified as a set of *constraints*. Although the layout produced by the specification should be reasonable on all devices, the designer may wish to produce interfaces tailored for certain platforms. By providing layout constraints and overrides, the designer can customize the interface to particular sets of devices (e.g., all mobile phones with display screens). We call this process *specialization*. Specializations can be specified broadly, for example, *all* GUI devices, or narrowly, for example, a single device. When specializing for a single device, we allow the designer to use the convenience of a WYSIWYG-style interface. We call this style of specialization *tweaking*. The discussion in this paper will focus on specification of device-independent semantics for specialization, rather than on device-specific specialization or tweaking.

The main contributions of this work are:

- A constraint model for specifying layout semantics that is to be applied to generating user-interfaces for multiple target devices.
- A discussion of problems with single-view direct-manipulation "what you see is what you get" (WYSIWYG) interfaces for specifying device-independent layout constraints.
- A proposed two-view interface design and implementation that addresses these issues.

## 2    Related Work

The work described here builds on previous work in the areas of model-based user interface systems, constraint-based interface specification, and multi-view systems. We do not claim to advance any of these areas in this paper. Instead, we focus on the unique problems that arise when model-based systems with constraints are applied to the area of multi-device application development, and how a multi-view system can potentially solve these problems.

Model-based user interface systems [15], [16] allow interfaces to be specified by constructing a declarative model of how the interface should look and behave. A run-time engine executes the model and constructs the application's displays and interprets input according to the information in the model. Of particular relevance in this area are the various task models of user interaction [4], [11]. Our application development model described in the following section is an example of a model-based user-interface specification. In contrast to previous work, however, the design of our application model was driven specifically by the requirements of multi-device application development.

Model-based systems pose a class of problems having to do with developers' difficulty in maintaining a mental model that connects an abstract interface specification with multiple concrete realizations of it. One problem in this class, called the "mapping problem", was identified by Puerta and Eisenstein [13]. The mapping problem is the difficulty of mapping an abstract interface element to multiple concrete elements, thus reducing the usability of model-based systems. To solve this problem, the authors propose a new model component, called the *design model*, to support the inspection and setting of mappings via the use of intelligent tools. The problem that this paper focuses on – the difficulty of simultaneously viewing device-independent specifications and device-specific layouts derived from those specifications – is another instance in the class of problems mentioned above.

There has been some recent work in the area of "intent-based" markup languages [1], which are textual languages for declaratively specifying multi-device interfaces. However, this work is yet to address the issue of developing tools for creating applications using such languages.

Another related area of research is the creation of user-interfaces via demonstration. The Amulet project is one example of such work [12].

Much work has been done in the area of constraints for user interface development. Constraint-based windowing systems include SCWM [2] and Smalltalk [6]. There are several constraint-based drawing systems such as [14]. Constraints have been applied to a variety of user-interface problems, including database user interfaces [7], programming by demonstration [8], and data visualization [17]. Constraint solvers for user-interface systems have been developed [5], [9]. However, we are not aware of any work that has applied constraints to the problem of multi-device user interface development.

## 3    The Application Development Model

In this section we will give a very brief, high-level view of our application development model. Although the task model described is not particularly novel, we present it in order to lay the groundwork for the central discussion of this paper – mechanisms for specifying layout semantics.

An application developed using our framework consists of one or more *tasks*. A task is a unit of work to be performed by the user. Examples of tasks include registering for a subscription service, placing an online order, or browsing a catalog.

A task may be made up of *subtasks*, with the granularity of the lowest level tasks completely at the discretion of the designer.

Lowest level tasks or *leaf tasks* contain *interaction elements*. Interaction elements provide for user input and/or system output. Interaction elements provide abstract descriptions of entities that may be rendered as widgets on GUI devices, or voice elements in a speech-based interface. Examples of interaction elements include *SelectableList*, which presents the user with a list of data items from which the user is to choose (e.g., a pull-down menu), and *Input* which allows the user to enter information (e.g., a type-in field).

In addition to tasks and interaction elements, the application designer specifies variables and event handlers, navigation between tasks, and sets of layout constraints. The constraints are used to create to page layouts for each particular device, and for determining the number and contents of pages for the rendered application.

# 4    User-Interface Constraint Specification

In this section, we will describe the set of constraints that we wish to specify for our models. It is important to keep in mind that we are not trying to address the problem of laying out user interfaces targeted for particular devices; this ground has been well-covered by previous investigators. We are addressing the problem of devising sets of constraints to be used for generating interfaces for multiple target platforms. For this reason, we focus primarily on semantic constraints.

In this discussion, we distinguish between two types of user-interface constraints that can be specified for a device-independent application. Constraints can either be *generic*, applicable to any type of input or output device, or they can be *graphical*, applicable only to devices with traditional display screens.

## 4.1    Generic Constraints

Generic constraints are completely device-independent. We currently support two types of generic constraints: ordering constraints and grouping constraints.

*Ordering constraints* specify the sequencing of interaction elements in the interface. For example, if an address form has a name field, a street address field, and a city/state field, we want to specify that they be presented in that order. *First* and *last* constraints specify the relative positioning of individual interaction elements within a leaf task. In addition, ordering between pairs of interaction elements can be specified. This type of constraint, which we call *after*, operates on two interaction elements, specifying that one element is to be positioned anywhere after the other[1].

*Grouping constraints* specify that interaction elements are semantically related, and should be kept together (in a GUI, on the same screen and adjacent) in the rendering of the user-interface. An example is that all the elements in the address form discussed above are related, and would constitute a group. The single grouping

---

[1] For simplicity, we limited the interface to this small number of ordering constraints. It may be that others such as *immediately after* are sufficiently valuable to warrant inclusion.

constraint, *group*, can be applied to any number of interaction elements within a leaf task.

For purposes of constraint specification, groups are treated just like interaction elements. In other words, a group can be selected as an argument for any constraint operator, including the grouping operator (i.e., groups can be nested).

*Group* is a high-level semantic construct that can be interpreted in different ways by run-time implementations for different devices. We readily envision more device-specific grouping constraints such as *group by row*, or *group by column* – these particular examples only applying to GUI's.

Note that both generic ordering and grouping constraints are semantic specifications, applicable to any class of device, with device-specific presentation differing from device to device. *First*, for example, on a large-screen visual display would specify the interaction element is to be rendered as a widget positioned in the upper-left corner of the display. On a small-screen display (a mobile phone, for example), rendering of the interaction element might fill an entire screen, thus *first* would indicate the first screen. On a voice interface, on the other hand, *first* would indicate the first event in the voice interactions, either a voice-input or a voice-output item. Similarly *after* clearly specifies temporal ordering for a speech interface, but is subject to interpretation on a GUI – probably producing some sort of text-flow (e.g., down and/or to the left) ordering.

## 4.2    Graphical Constraints

Graphical constraints apply only to devices with visual displays. We currently support two types – sizing constraints and anchoring constraints.

*Sizing constraints* are used to ensure that members of a set of interaction elements are all rendered with the same width (using the *same width* constraint) or the same height (using the *same height* constraint). This facilitates the design of interfaces that conform to standard UI guidelines. Note that we specify sizing and grouping separately. This allows us to specify that all buttons are to be rendered the same size for a particular task, but without requiring that the buttons be adjacent.

*Anchoring constraints* are used to position interaction elements on a screen, giving a designer some control over element placement. The four anchoring constraints, *top*, *bottom*, *left*, and *right* allow a designer to place particular elements at the boundaries of the visual display – permitting a set of buttons to be positioned at the bottom of the screen, for example.

The set of constraints described here is by no means complete. This set does not come close to supporting the degree of control possible with a UI toolkit such as AWT or Motif. This is by intent. Our goal here is provide an easy-to-use, high-level set of constraints that can be used to specify interface characteristics to be applied across a wide range of possible devices. We anticipate that designers who want truly beautiful interfaces for particular target platforms will take the output of our specialization engine, and tweak it by hand, as mentioned in the introduction. Our goal is to produce a usable interface for any device, with provisions for a designer to improve the default interface for particular devices or sets of devices.

# 5    The WYSIWYG Interface Problem

In designing a constraint-specification interface for user-interface layout, it is tempting to develop a single-view, direct-manipulation, "what you see is what you get" (WYSIWYG) interface.    Such an interface would provide the following functionality, required for any design environment of this sort:

1. *Selecting items.* A WYSIWYG view can be used for selecting user-interface components on which operations are to be performed (positioned, sized, etc). This is common practice, allowing the designer to quickly and effectively specify the items in context.

2. *Viewing constraints.* The WYSIWYG view can also be used to display the constraints by visually identifying the interface components and the constraints that apply to them.    This is attractive, because it minimizes the cognitive load on the designer.    S/he can see which interaction elements have constraints applied to them, and what those constraints are, while reviewing the visual appearance of the interface.

3. *Viewing layout.*    It is critical that any interface design system provide visual feedback to the designer.    A WYSIWYG interface shows the effects of changes in the layout specification by presenting a representation of the sizes and positions of the user interface components as they will appear in the final application.

Although it seems desirable to provide all of this functionality in a single view, there are several serious problems in using a single WYSIWYG view as a direct-manipulation interface when specifying constraints for device-independent applications[2].    Some of these problems have to do with the fact that the application will, in general, be laid out across multiple screens, some of them have to do with the fact that the application is to be deployed on multiple platforms. These problems, in order of decreasing importance, are as follows:

1. The most serious drawback to a WYSIWYG interface is that the user is "led down the garden path."    What the user sees is *not* what the user is going to get in general, since only a single device is emulated in the interface.    The user is being tempted to customize the design for *one particular* device, rather than thinking about the general problem of constraints that are appropriate for *all* devices. Even though an interface may provide a capability for toggling between device emulations, thereby allowing a view of multiple layouts, this feature is easily overlooked.    A possible solution is to simultaneously provide multiple views.    This can easily lead to confusion between viewing and control (in the model/view/controller sense), however.

2. A WYSIWYG screen layout changes each time a constraint is specified.    If the WYSIWYG view is also being used as a set of direct-manipulation controls, this has the effect of moving those controls after each operation, a highly undesirable characteristic.    Furthermore, the constraint solver may move some of the interaction elements to different screens (i.e., pages), compounding the sense of dislocation experienced by the user, particularly if only one virtual screen is displayed at a time – interaction elements will disappear from view. This seriously

---

[2] Single-view WYSIWYG interfaces may be effective for customizing a design for a particular device, a process we call *tweaking*, but this is not the device-independent design we are discussing here.

reduces the usability of the interface. It may be possible to partially alleviate this problem by having the user explicitly specify when the interface is to be re-laid out, but this breaks the WYSIWYG model.

3. Constraints cannot always be satisfied. For example, the screen size may not be large enough to contain all interaction elements in a group. A single-view WYSIWYG interface may be unable to display the constraint in that case (e.g., displaying *group* as some form of visual containment is not possible). Switching the emulation to a larger device may allow that constraint to be included, and hence displayed. This is misleading and confusing behavior.

4. The user may wish to specify constraints between interaction elements on separate screens. For example, *after* or *group* constraints may involve interaction elements on more than one screen. This makes the specification a bit awkward, since the user must switch screens while selecting interaction elements. Either multiple screens must be displayed simultaneously, which can strain screen resources, or the designer will need to toggle between screens. Intuitive constraint displays such as arrows for *after* or visual containment for *group* (using a bounding box, for example) becomes problematic. The interface needs to rely more heavily on less obvious visual metaphors and/or user memory.

5. A WYSIWYG interface, because it presents *only* final appearance, lacks information about the structure of the task model, information that might be of value to the interface designer. In specifying layout constraints, it may be important to know with which task or subtask particular interaction elements are associated. Although it would be possible to provide some of this information – by labeling emulated screens with subtask names, for example – it is difficult to envision an interface that will provide all of this structural information in the context provided by a WYSIWYG view.

From this discussion, it should be clear that a better interface paradigm is required. In the next section, we discuss an alternative to the single-view, direct-manipulation WYSIWYG interface, and describe our implementation.

## 6    A Solution: The Two-View Constraint Editor

In this section we will discuss the desired characteristics of a two-view constraint editor, describe our implementation, and explain how the two-view interface solved the "WYSIWYG pitfall."

### 6.1    Desired Characteristics

The fundamental problem with a single-view direct-manipulation WYSIWYG editor is that the single view is not adequate to display both the logical structure of the constraint set within the context of the task model, as well as an indication of the types of layouts produced by that constraint set. What is desired is a single logical view of the task structure and interaction elements that could be used both as an interface to specify constraints, and also as a conceptual view of these constraints. The logical view should be arranged to make it easy to see and think about the entire constraint set.

Additionally, the designer should be able to view the effects of the constraints on interfaces that are generated for various devices, but with minimal opportunity for the user to assume that the preview is a "true" representation of what will be produced for multiple devices. A clear separation of the logical view from the device preview should facilitate this. For these reasons, a two-view system is more likely to embody all the desired characteristics of a constraint-specification interface for pervasive application development than a single-view WYSIWYG editor.

## 6.2    Current Implementation

The constraint-specification interface is one component of an application development (AD) tool for specifying device independent applications. The AD tool is a part of our device-independent application framework, called PIMA (Platform-Independent Model for Applications) [3]. Other components of the AD tool include a task editor for managing task structure and navigation; and a task details editor for specifying interaction elements, variables, and events that comprise individual tasks.

Figure 2 shows the two-view constraint-specification interface with some constraints specified. On the left hand side is a graphical representation of the task structure and interaction elements. We call this the *logical view*. Tasks are represented as labeled, nested rectangles, with gray rectangular icons representing interaction elements contained within them. The logical view serves two purposes. It is used to add and remove constraints, and also to view the set of constraints that are currently specified.

Interaction elements are selected by clicking on them, and then operations are applied to individual interaction elements or groups of interaction elements. The user can select one or more icons using the mouse. Once a set of interaction element icons has been selected, pressing a button corresponding to the desired constraint type specifies a constraint. A visual representation of the constraint is added to the display. In a similar fashion, constraints can be removed.

The right-hand side of the two-view interface is a representation of a screen layout for a single screen on a particular device, with each widget (interaction elements will be presented as widgets in a GUI) represented as a named rectangle, sized and positioned as it would be in the final running application. We call this the *layout view*. In figure 2, the layout view displays an emulation of the first screen of our job fair application, configured for a PC running Java. Each change to the constraint specification triggers an update in the layout view. The user can navigate the screens for a particular device by selecting from a set of radio buttons, or select from a set of different devices by choosing from a selectable list. Each device is emulated using the task/interaction element/constraint description, and a device capabilities file that defines screen size and default widget sizes.

Figure 2 shows the interface after specification of several constraints. Dark circles in the upper left and lower right corners of interaction elements icons in the logical view, indicate first and last constraints, respectively. "Name" has been specified as first, and "URL" as last. Arrows connecting two interaction elements shows ordering. The arrow between "Name" and "Phone" indicates that "Name" is to precede "Phone."

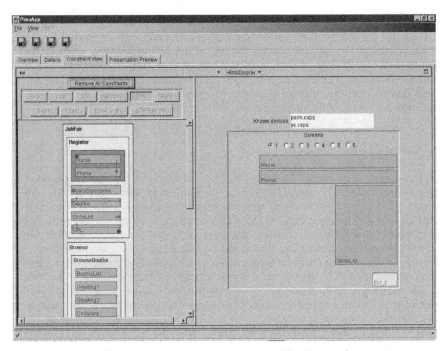

**Fig. 2.** Two-view layout constraint specification interface

To facilitate legibility, we ensure that all arrows point downward, rearranging icon placement if necessary. For obvious reasons, cycles are detected by the interface and disallowed. Small arrow icons point to the side of interaction element icons for which anchoring has been specified. "Skillslist" is anchored on the right, and "Years of experience" on the left. Double-headed arrows are used to indicate same size constraints – vertical arrows for height constraints, and horizontal arrows for width constraints. The arrows have numbers associated with them, the members of a same-size set all displaying the same number. "Degree" and "URL" have the same width in our example.

Groups are represented by repositioning all entities to be grouped so they are contiguous, and drawing a rectangle that encloses them. The figure shows a single group containing "Name" and "Phone."

Figure 3 shows a later version of the same interface. Notice that we have included two device-specific layout views on the right-hand side (layout for a PC on top, for a browser on the bottom). This allows a user to see the effects of constraints on multiple devices simultaneously. We expect this to further reduce the tendency towards "what you see is what you get" design.

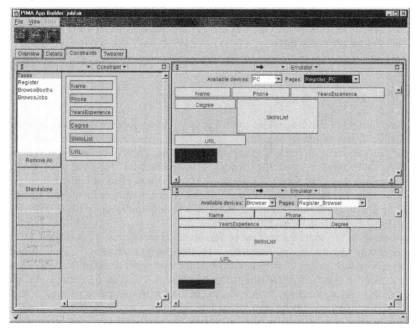

**Fig. 3** A later version of the two-view constraint specification interface showing the constraint view on the left, and the device-specific layouts for two different devices on the right

## 6.3    Solutions to the WYSIWYG Interface Problems

We note that the two-view interface solves the previously described problems of the WYSIWYG interface as follows:

1. Since the logical (interface/constraint display) view is provided as a single, scrollable display, not separate virtual screens; no context switching is required to select multiple interaction elements. Display of ordering is readily achieved on the single panel.
2. Since the logical and layout views are separated, constraints can always be visualized, even if constraints cannot be satisfied for particular devices being emulated.
3. The task model is integrated into the logical view, facilitating high-level, device-independent thinking about the constraint set.
4. Although interface controls can move in this interface (since specification of ordering constraints can lead to re-ordering of the interaction element icons in the logical view), context is much more readily maintained than it would be with a single-view WYSIWYG interface. The task structure helps to retain orientation, and the problem of some of the interaction elements vanishing off-screen for certain operations has been solved. Note that allowing the user to specify when icons are to be reordered would be a viable possibility, and would not pose the problem discussed earlier of breaking a WYSIWYG model.

5. The user is much less tempted to think that the emulation view is "reality" than with a single-view WYSIWYG interface. The separate logical view encourages device-independent thinking.

# 7   Discussion

The main thrust of the two-view interface is to encourage a designer to think generically. When designing a device-independent application intended to run across a variety of pervasive devices, it is critical that design choices be made on general principles, not based on specific display characteristics. The interface should encourage thinking of the form, "This particular interaction element really belongs at the beginning of the task on any device." If the designer is thinking, "I see a blank spot at the top of this screen, I'd like to move that widget to the top to fill it," s/he has missed the point and trouble may ensue.

By separating the logical structure of the application from its presentation view, and integrating the constraint interface with that logical structure, we encourage device-independent design. It is rather difficult for a designer to make the connection between specific constraints and the generated layout. This is by intent. By *not* providing the easy connection, "this widget is at the top of the screen because a 'top' constraint was specified," there will be far less temptation to do "appearance-guided design," in our opinion the major potential pitfall for device-independent UI development.

Even though we have separated the logical view from the UI appearance, presentation of a single concrete representation of the layout could still pose a problem, leading the user to believe it is *the* view. We have recently added multiple device presentations to our interface. Since these presentations are for viewing only, not for interface control, providing multiple views is less problematic than for a WYSIWYG editor. Whether providing multiple views is an effective solution will require future user studies.

# 8   Future Work

Although we have presented a problem with interfaces for device-independent user interface development, and proposed a solution, user studies are clearly necessary.
Two questions need to be addressed. The first is, "does the device-independent application model, with specification of device-independent constraints provide enough control – allowing the designer to create applications that run reasonably well on all platforms, and that are easily tweaked to produce high-quality interfaces on selected platforms?"

The second question broadly is, "what is the 'best' interface for specifying device-independent interface constraints?" We propose a more focused question, namely, "is a two-view interface more usable for specifying device-independent interface constraints than a WYSIWYG interface?" A user study that pits our two-view interface against a single-view WYSIWYG interface, measuring the quality of the interfaces produced and/or the speed of production should provide valuable insight into the nature of the device-independent application development process, and the tools required to support it. Note that to tweak an interface for a specific platform, a single-view direct-manipulation WYSIWYG interface is clearly appropriate.

Another question that we plan to investigate is, "how does one provide a designer with tools for specifying interface characteristics for classes of devices?" We have discussed two points on a continuum – either creating a generic device-independent specification, or customizing an interface for a particular device type. Clearly there are points in between. We can envision classifying device characteristics – graphical vs. voice, large-screen vs. small screen, presence/absence of hardware inputs such as hard buttons, etc., and then providing different sets of constraints or hints for different classes. Essentially design becomes a process of specifying sets of rules – "if the application is running on a small-screened device, place this widget on its own screen," for example. The question then is, does such a design methodology produce usable interfaces with fewer burdens on the designer than producing an interface separately for each device to be supported? We believe that with a carefully developed methodology, the answer will be yes, both in terms of less work being required, and in providing support for devices not in existence when the application is developed.

## 9   Conclusions

In this paper we have explored some issues surrounding design of development tools for creating device-independent applications. In particular, we have pointed out particular pitfalls when designing for multiple devices that do not exist when designing an application for a single device, or a very small set of predetermined devices.

The central problem that we have identified is that a design tool must be carefully structured to not tempt a designer to believe that emulations of the interface represent what will actually be displayed on target devices; the targets cannot be known or adequately represented at design-time. We suggest the need for design environments that facilitate intent-based rather than graphical thinking.

We have proposed one solution to this problem – a two-view system for specifying interface characteristics. Although we have not proven the utility of this approach, we have clearly identified the problems that exist, and suggested how different design interface characteristics might alleviate or exacerbate the problem. We have provided a framework for future studies and interface development in this area.

### Acknowledgments

We would like to thank Noi Sukaviriya and Rachel Bellamy for illuminating conversations, which helped fuel this work. John Richards and Jonathon Brezin identified the "single-view direct-manipulation WYSIWYG" problem, and were instrumental in helping us formulate a viable solution. John Turek provided valuable support and feedback during the course of this work.

# References

1. M. Abrams, C. Phanouriou, A. Batongbacal, S. Williams, and J. Shuster, UIML: An Appliance-Independent XML User Interface Language, in Proceedings of the Eighth International World Wide Web Conference, May 1999, p. 617-630.
2. Greg J. Badros, Jeffrey Nichols, and Alan Borning, SCWM---an Intelligent Constraint-enabled Window Manager, in Proceedings of the AAAI Spring Symposium on Smart Graphics, March 2000.
3. PIMA project home page. http://www.research.ibm.com/PIMA
4. Larry Birnbaum, Ray Bareiss, Tom Hinrichs, and Christopher Johnson, Interface Design Based on Standardized Task Models, in Proceedings of the 1998 International Conference on Intelligent User Interfaces 1998, p.65-72.
   http://www.acm.org/pubs/articles/proceedings/uist/268389/p65-birnbaum/p65-birnbaum.pdf
5. Alan Borning, Kim Marriott, Peter Stuckey, and Yi Xiao, Solving Linear Arithmetic Constraints for User Interface Applications, in Proceedings of the 1997 ACM Symposium on User Interface Software and Technology, October 1997, p. 87-96.
   http://www.acm.org/pubs/articles/proceedings/uist/263407/p87-borning/p87-borning.pdf
6. Danny Epstein and Wilf LaLonde, A Smalltalk Window System Based on Constraints, in Proceedings of the 1988 ACM Conference on Object-Oriented Programming Systems, Languages and Applications, San Diego, September 1988, p. 83-94.
7. Phil Gray, Richard Cooper, Jessie Kennedy, Peter Barclay, and Tony Griffiths, Lightweight Presentation Model for Database User Interfaces, in Proceedings of the 4th ERCIM Workshop on "User Interfaces for All" 1998 n.16, p.14. http://www.ics.forth.gr/proj/at-hci/UI4ALL/UI4ALL-98/gray.pdf
8. Takashi Hattori, Programming Constraint System by Demonstration, in Proceedings of the 1999 International Conference on Intelligent User Interfaces 1999, p.202.
   http://www.acm.org/pubs/articles/proceedings/uist/291080/p202-hattori/p202-hattori.pdf
9. Scott Hudson and Ian Smith, Ultra-Lightweight Constraints, in Proceedings of the ACM Symposium on User Interface Software and Technology 1996, p.147-155.
   http://www.acm.org/pubs/articles/proceedings/uist/237091/p147-hudson/p147-hudson.pdf
10. http://www.javasoft.com/
11. David Maulsby, Inductive Task Modeling for User Interface Customization, in Proceedings of the 1997 International Conference on Intelligent User Interfaces 1997, p. 233-236.
    www.acm.org/pubs/articles/proceedings/uist/238218/p233-maulsby/p233-maulsby.pdf
12. Brad A. Myers, Richard G. McDaniel, Robert C. Miller, Alan S. Ferrency, Andrew Faulring, Bruce D. Kyle, Andrew Mickish, Alex Klimovitski and Patrick Doane. The Amulet Environment: New Models for Effective User Interface Software Development, IEEE Transactions on Software Engineering, Vol. 23, no. 6. June, 1997. pp. 347-365.
13. Angel Puerta and Jacob Eisenstein, Towards a General Computational Framework for Model-Based Interface Development Systems Model-Based Interfaces, Proceedings of the 1999 International Conference on Intelligent User Interfaces 1999, p.171-178.
    http://www.acm.org/pubs/articles/proceedings/uist/291080/p171-puerta/p171-puerta.pdf
14. Kathy Ryall, Joe Marks, and Stuart Shieber, An Interactive Constraint-based System for Drawing Graphs, in Proceedings of UIST 1997, Banff, Alberta Canada, October 1997, p. 97-104. http://www.acm.org/pubs/articles/proceedings/uist/263407/p97-ryall/p97-ryall.pdf
15. Piyawadee "Noi" Sukaviriya, James D. Foley, and Todd Griffith, A Second Generation User Interface Design Environment: The Model and the Runtime Architecture, in Proceedings of ACM INTERCHI'93 Conference on Human Factors in Computing Systems 1993, p.375-382. http://www.acm.org/pubs/articles/proceedings/chi/ 169059 /p375-sukaviriya/p375-sukaviriya.pdf
16. Pedro Szekely, Ping Luo, and Robert Neches, Beyond Interface Builders: Model-Based Interface Tools, in Proceedings of ACM INTERCHI'93 Conference on Human Factors in

Computing Systems 1993, p.383-390. http://www.acm.org/pubs/articles/proceedings/chi/169059/p383-szekely/p383-szekely.pdf

17. Allison Woodruff, James Landay, and Michael Stonebraker, Constant Density Visualizations of Non-Uniform Distributions of Data Visualization, Proceedings of the ACM Symposium on User Interface Software and Technology 1998, p.19-28. http://www.acm.org/pubs/articles/proceedings/uist/288392/p19-woodruff/p19-woodruff.pdf

18. XSL Transformations (XSLT) Version 1.0, W3C Recommendation 16, November 1999. http://www.w3.org/TR/xslt See also, www.xslt.com.

## Discussion

*L. Nigay:* I would like to come back to the specialization mechanism. How do you go from abstract widgets to concrete widgets? Do you have a model to describe devices?
*L. Bergman:* Simple mapping. Only graphical modalities. Thinking of speech input.

K. Schneider: Does the multi-view editor support "tweaking" the user interface? Are the "tweaks" retained when a constraint is added or moved or changed? Can it "tweak" the constraints, such as order, for a particular concrete user interface?
*L. Bergman:* We are just beginning to address those issues. The editor supports "tweaking". The "tweaks" to the properties of an element are retained but the structural "tweaks" are not. And, yes, you can override the constraints when "tweaking" the user interface.

*J. Höhle:* I agree that people want/need a hands-on approach to explore/play with the design. But do you really think that tools like MS FrontPage will continue to be wanted? Couldn't your tool provide hands-on and not produce output that only works with MS-Explorer, only in 640x480, generates incorrect HTML, etc. ?
*L. Bergman:* We have a handful of feedback: these people really want to move around stuff and not learn a new model. We need more feedback, though.

*N. Graham:* What are the limits of this kind of approach? When designing for very different interfaces (eg electronic whiteboard vs palm pilot), the resulting interfaces may be completely different, not just different in choice of widgets.
*L. Bergman:* This approach is very much biased towards form-based approaches. This is a limitation when trying to get cross-platform design.

*C. Roast:* I'm interested in the user's response to a logical view and possible concrete views. There is evidence that the concrete view presists for designers. What plans do you have for your user studies?
*L. Bergman:* Because of the concrete bias we are interested in example driven uses of the editor. As for user studies, we are still planning.

*M. Borup-Harning:* Can your approach cope with situations where e.g. presenting editable information on a GUI vs. a HTML based platform might result in one form-based interface on the GUI platform, whereas the HTML based one will be divided into a presentation page with an edit button and one or more form-based pages for editing.
*L. Bergman:* I am not sure I understand the question? But the system was not meant to deal with arbitrarily long lists of information.

*J. Williams:* Have you considered providing a transition from the concrete interface view to the logical representation? This would allow developers to specify in the concrete, yet you retain reuse across devices. In addition, you can transform existing specifications.
*L. Bergman:* Yes, this is future work.

# A Unifying Reference Framework
# for the Development of Plastic User Interfaces

Gaëlle Calvary, Joëlle Coutaz, and David Thevenin

CLIPS-IMAG,
BP 53, 38041 Grenoble Cedex 9, France
{Joelle.Coutaz, Gaelle.Calvary, David.Thevenin}@imag.fr

**Abstract.** The increasing proliferation of computational devices has introduced the need for applications to run on multiple platforms in different physical environments. Providing a user interface specially crafted for each context of use is extremely costly and may result in inconsistent behavior. User interfaces must now be capable of adapting to multiple sources of variation. This paper presents a unifying framework that structures the development process of plastic user interfaces. A plastic user interface is capable of adapting to variations of the context of use while preserving usability. The reference framework has guided the design of ARTStudio, a model-based tool that supports the plastic development of user interfaces. The framework as well as ARTStudio are illustrated with a common running example: a home heating control system.

## 1 Introduction

Recent years have seen the introduction of many types of computers and devices. In order to perform their tasks, people now have available a wide variety of computational devices ranging over cellular telephones, personal digital assistants (PDA's), Internet enabled televisions (WebTV) and electronic whiteboards powered by high end desktop machines. While the increasing proliferation of fixed and mobile devices fits with the need for ubiquitous access to information processing, this diversity offers new challenges to the HCI software community. These include:

- constructing and maintaining versions of single applications across multiple devices;
- checking consistency between versions for guaranteeing a seamless interaction across multiple devices;
- building into these versions the ability to dynamically respond to changes in the environment such as network connectivity, user's location, ambient sound or lighting conditions.

These requirements induce extra cost in development and maintenance, and complicate the configuration management. In [20], we presented a first attempt at optimising the development process of user interfaces using the notion of plasticity as a foundational property. *Plasticity* refers to the ability of a user interface to mould itself to a range of computational devices and environments, both statically and/or

M. Reed Little and L. Nigay (Eds.): EHCI 2001, LNCS 2254, pp. 173–192, 2001.
© Springer-Verlag Berlin Heidelberg 2001

dynamically, whether it be automatically or with human intervention. In this paper, we go one step further with the presentation of a reference framework for supporting the structured development of plastic user interfaces. In the following section, we present the EDF home heating control system as our running example to illustrate the concepts and principles of the framework. We then recall the definition of our notion of plasticity before describing the framework and ARTStudio, a tool that supports the development of plastic user interfaces.

## 2   An Example: The EDF Home Heating Control System

The heating control system envisioned by EDF (the French Electricity Company) will be controlled by users situated in diverse contexts of use. These include:

- At home, through a dedicated wall-mounted device or through a PDA connected to a wireless home-net;
- In the office, through the Web, using a standard work station;
- Anywhere using a WAP-enabled mobile phone.

A typical user's task consists of consulting and modifying the temperature of a particular room. Figures 1 and 2 show versions of the system for different computational devices:

- In 1 a), the system displays the current temperature for each of the rooms of the house (the bedroom, the bathroom, and the living room). The screen size is comfortable enough to make observable the entire system state;

- In 1 b), the system shows the temperature of a single room at a time. A thumbnail allows users to switch between rooms. In contrast with 1a), the system state is not observable, but browsable [10]: additional navigational tasks, e.g., selecting the appropriate room, must be performed to reach the desired information.

Figure 2 shows the interaction trajectory for setting the temperature of a room with a WAP-enabled mobile phone.

- In 2a), the user selects the room (e.g., "le salon" – the living room).
- In 2b), the system shows the current temperature of the living room.
- By selecting the editing function ("donner ordre"), one can modify the temperature of the selected room (2c).
  When comparing with the situation depicted in Figure 1a), two navigation tasks (i.e., selecting the room, then selecting the edit function) must be performed in order to reach the desired state. In addition, a title has been added to every deck (i.e., a WML page) to recall the user with the current location within the interaction space.
  All of these alternatives have been produced using our reference framework devised for supporting plasticity.

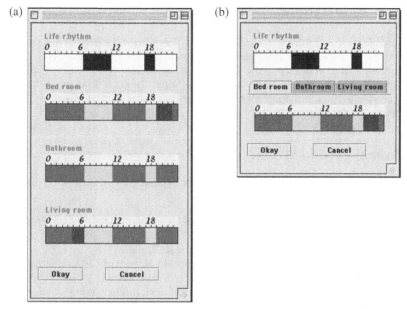

**Fig. 1.** a) Large screen. The temperature of the rooms are available at a glance. b) Small screen. The temperature of a single room is displayed at a time.

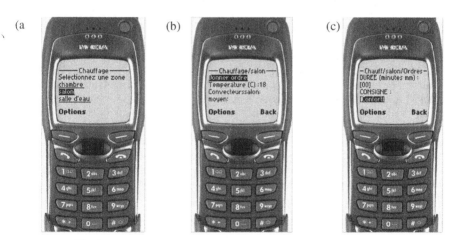

**Fig. 2.** Modifying the temperature using a WAP-enabled mobile phone.

## 3  Plasticity and Related Concepts

The term *plasticity* is inspired from the property of materials that expand and contract under natural constraints without breaking, thus preserving *continuous usage*.

Applied to HCI, plasticity is the "capacity of an interactive system to *withstand variations* of *context of use* while *preserving usability*". In the following subsections, we successively develop the key elements of our definition: context of use, usability, and the process for adapting to changes.

### 3.1  Plasticity and Context of Use

A context of use for a plastic user interface is defined by two classes of physical entities:

- The physical and software platform(s), that is, the computational device(s) used for interacting with the system.
- The physical environment where the interaction takes place.

A *platform* is modeled in terms of resources which, in turn, determine the way information is computed, transmitted, rendered, and manipulated by users. Typically, memory size, network bandwidth and interactional devices motivate the choice for a set of input and output modalities and, for each modality, the amount of information made available. For example, screen size is a determining factor in the design of the EDF Heating Control System.

An *environment* covers "the set of objects, persons and events that are peripheral to the current task(s) but that may have an impact on the system and/or the user's behavior, either now or in the future". According to this definition, an environment may encompass the entire world. In practice, the boundary is set up by domain analysts whose role is to elicit the entities that are relevant to the case at hand. These include observation of users' practice [2, 5] as well as consideration for technical constraints. For example, surrounding noise should be considered in relation to sonic feedback. Lighting condition is an issue when it may influence the robustness of a computer vision-based tracking system. User's location provides context for information relevance: Tasks that are central in the office (e.g., writing a paper) may become secondary, or even irrelevant, in a train. For example, programming the EDF heating system is available on the central wall-mounted device only: according to the domain analysts, this task would not make sense, or would be too complex to be performed with current telephone technology.

In summary,

- a context of use, which consists of the association of a platform with an environment, is definitely anchored in the physical world. Therefore, it does not cover the user's mental models;
- plasticity is not only about condensing and expanding information according to the context of use. It also covers the contraction and expansion of the set of tasks in order to preserve usability.

## 3.2  Plasticity and Usability

The quality of an interactive system is evaluated against a set of properties selected in the early phases of the development process. "A *plastic user interface preserves usability* if the properties elicited at the design stage are kept within a predefined range of values as adaptation occurs to different contexts of use". Although the properties developed so far in HCI [10] provide a sound basis for characterizing usability, they do not cover all aspects of plasticity. We propose additional metrics for evaluating the plasticity of user interfaces.

Figure 3 makes explicit the association of a platform with an environment to define a context of use. We suppose that platforms and environments can be ranked against some criteria computed from their attributes. For example, screen size, computational power and communication bandwidth, are typical attributes of a platform. Using these attributes, a PC would be ranked lower than a PDA since it imposes less constraints on the user interface. Similarly an environment with no noise would be ranked lower than the open street. Then:

– the plasticity of a user interface can be characterised by the sets of contexts it is able to accommodate,
– contexts at the boundaries of a set define the *plasticity threshold* of the user interface for this set,
– the sum of the surfaces covered by each set, or the sum of the cardinality of each set, defines an overall objective quantitative metrics for plasticity. In other word, this sum can be used to compare solutions to plasticity: A user interface U1 is more plastic than a user interface U2 if the cardinality of the set of contexts covered by U1 is greater than that of U2.

We suggest additional metrics to refine the overall measure of plasticity in relation to discontinuity [9]. These include:

– The size of the largest surface: large surfaces denote a wide spectrum of adaptation without technical rupture.
– The number of distinct sets: a large number of sets reveals multiple sources for technical discontinuities. Are these discontinuities compatible with user's expectation? Typically, GSM does not work everywhere. This situation translates as a discontinuity when moving along the environment axis of figure 3. The solution developed for EDF works for the Palm and the mobile phone, but not for the Psion. In this case, there is a discontinuity when moving along the platform axis.
– Surface shapes: a convex surface denotes a comfortable continuous space (cf. Figure 3a). Conversely, concave curvatures may raise important design issues (cf. Figure 3b). Typically, ring shape surfaces indicate that the interior of the ring is not covered by the user interface. It expresses a technical discontinuity for contexts that are contiguous in the ranking scheme. Is this inconsistency, a problem from the user's perspective? A hole within a surface depicts the case where the user interface is nearly plastic over both sets of contexts, but not quite. Is this "tiny" rupture in context coverage expected by the target users?

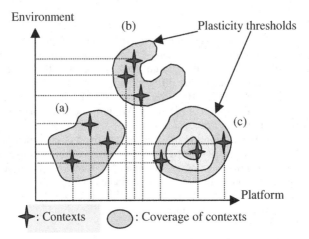

**Fig. 3.** Measuring plasticity from the system's perspective. Greyed areas represent the sets of contexts that a particular technical solution covers. Environments and platforms are ranked against the level of constraints they impose on the user interface.

Intuitively, from a technical point of view, a large unique convex surface characterises a "good" plastic user interface whereas a large number of small concave surfaces denotes a large number of technical discontinuities. Although size, shape, cardinality, and topology of surfaces, are useful indicators for reasoning about the plasticity of a particular technical solution, we need to consider a complementary perspective: that of users. To this end, we suggest two indicators: context frequency and migration cost between contexts.

- *Context frequency* expresses how often users will perform their tasks in a given context. Clearly, if the largest surfaces correspond to the less frequent contexts and/or if a multitude of small surfaces is related to frequent contexts, then designers should revise their technical solution space: the solution offers too much potential for interactional ruptures in the interactional process.
- *Migration cost* measures the physical, cognitive and conative efforts [6] users have to pay when migrating between contexts, whether these contexts belong to the same or different surfaces (cf. Figure 4). Although this metrics is difficult to grasp precisely, the notion is important to consider even in a rough way as informal questions. For example, do users need (or expect) to move between contexts that belong to different surfaces? If so, discontinuity in system usage will be perceived. Designers may revise the solution space or, if they stick to their solution for well-motivated reasons, the observability of the technical boundaries should be the focus of special attention in order to alleviate transitions costs.

As *plasticity threshold* characterises the system capacity of continuous adaptation to multiple contexts, so *migration cost threshold* characterises the user's tolerance to context switching. The analysis of the relationships between the technical and the human thresholds may provide a useful additional perspective to the evaluation of plastic user interfaces.

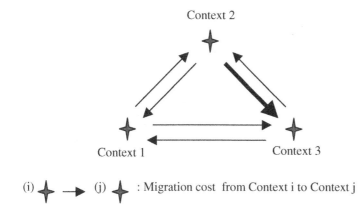

(i) ✦ → (j) ✦ : Migration cost  from Context i to Context j

**Fig. 4.** Measuring plasticity from the human perspective. An arrow expresses the capacity of migrating between two contexts. Its thickness denotes human cost.

Having considered plasticity from the usability perspective, we need now to describe the process through which plastic adaptation may occur.

### 3.3  Plasticity and the Adaptation Process

As any evolutive phenomenon, plastic adaptation is structured as a five step process:

- detection of the conditions for adaptation (here, variations in the context of use),
- identification of candidate user interfaces appropriate to the new context of use,
- selection of a user interface,
- transition from the current user interface to the newly selected solution,
- execution of the new user interface until next conditions for adaptation occur.

Each of the five steps involved in the plasticity process (detection, identification, selection, transition, execution) is performed either by the system, or by the user, or as a mixture of both. At the two extremes,

- The system is able to handle the five steps without human intervention. In this case, the system is capable of *adaptative plasticity*.
- The user performs the five steps manually. The system supports *adaptable plasticity*.
- *Mixed plasticity* covers a combination of both human and system intervention.

Let's now consider the five steps of the adaptation process from a technical perspective.

- *System detection* requires the capacity of sensing the physical environment as well as the capacity of modelling the physical platforms it is supposed to support. Our current implementation of the EDF heating control system includes a platform model. Although environmental conditions define the ontology of the system, these are not exploited yet at the user interface level. Conversely, the Hyperpalette [1], which allows users to "scoop and spread" information between PDA's and a computerised table, is able to detect conditions for compressing or expanding information layout, using an electromagnetic sensing tracker. On the other hand, the HyperPalette does not handle any explicit description of the computational platforms.

- *System identification* of candidate solutions may be either computed on the fly, or selected from a pre-computed set of user interfaces, or from a predefined set of ad-hoc user interfaces. As discussed in Section 6, the EDF heating control system uses a pre-computed set of user interfaces. The Hyperpalette has available a predefined set of two ad-hoc user interfaces dedicated to a single platform (i.e., a PC or a PDA).
- *System selection* of a particular solution relies on a problem solving strategy. The system may be assisted in this task by the user.
- *System transition* between states has been analysed since the early developments of HCI. Norman's evaluation gap, Mackinlay's et al. use of graphical animation for transferring cognitive load to the perceptual level [17], the notion of visual discontinuity [10] etc., have all demonstrated the importance of transitions. Transition between the use of two user interfaces is therefore a crucial point that deserves specific research. The sources of discontinuities identified in Section 3 should provide a sound starting point.
- *System execution* of the newly selected user interface may be launched from scratch or may preserve the system state. For example, the MacOS Location Manager supports switching to a different network protocol without interrupting the running applications. (On the other hand, the detection of the new context of use, as well as the selection step, are performed by the user). Preserving system state, which alleviates discontinuities, requires specific technical mechanisms.

Whether it be adaptative, adaptable or mixed, plasticity covers adaptation to both physical environments and platforms. In current main stream research on adaptation, the focus is either on environmental changes, or on platforms. Context-sensitive systems [18] address environmental conditions whereas resource-sensitive systems [11] exemplified by the development of XML-based standards, address adaptation to computational devices. Although adaptative, these systems and mechanisms demonstrate half-plasticity. The framework presented next is intended to cover all sorts of plasticity: half-full, adaptative-adaptable-mixed (see figure 5 for a simplified classification space).

## 4   A Reference Framework for Plasticity

Our framework is intended to serve as a reference instrument to help designers and developers to structure the development process of plastic interactive systems. We adopt a model-based approach [16] similar to [7]. Modelling techniques support sound design methods (e.g., MAD[8], ADEPT[12,13], MUSE [14]) and pave the way to the development of appropriate tools such as Trident [21] and Mastermind [19].

**Adaptation
performed by**

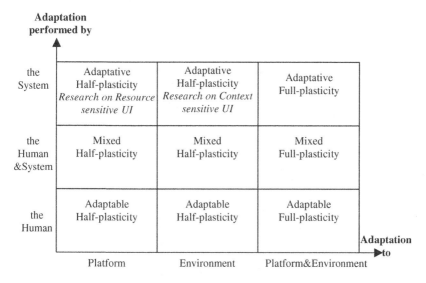

| | Platform | Environment | Platform&Environment |
|---|---|---|---|
| the System | Adaptative Half-plasticity *Research on Resource sensitive UI* | Adaptative Half-plasticity *Research on Context sensitive UI* | Adaptative Full-plasticity |
| the Human &System | Mixed Half-plasticity | Mixed Half-plasticity | Mixed Full-plasticity |
| the Human | Adaptable Half-plasticity | Adaptable Half-plasticity | Adaptable Full-plasticity |

**Fig. 5.** A classification space for plastic user interfaces. Lines indicate the authors of the adaptation (system only, human only, both). Columns correspond to the coverage of the adaptation (platform only, environment only, both). Mixed plasticity may in turn be analysed according to the steps the system covers in the adaptation process.

As shown in Figure 6, the framework:

- builds upon known models such as the domain concepts model and the tasks model, but improves them to accommodate variations of contexts of use;
- explicitly introduces new models and heuristics that have been overlooked or ignored so far to convey contexts of use: the platform, the interactors, the environment and the evolution models.

The *Concepts model* describes the concepts that the user manipulates in any context of use. When considering plasticity issues, the domain model should cover all the contexts of use envisioned for the interactive system. By doing so, designers obtain a global reusable reference model that can be specialised according to the sets of contexts discussed in Section 3. A similar design rationale holds for *task modelling*.

The *Platform Model* and the *Environment Model* define the contexts of use intended by the designers based on the reasoning developed in Section 3. The *Evolution model* specifies the change of state within a context as well as the conditions for entering and leaving a particular context. The *Interactors Model* describes "resource sensitive multimodal widgets" available for producing the concrete interface. These models are specified by the developer. They are said *initial models* in contrast to *transient and final models* that are inferred by the developer and/or the system through the development process. A transient model is an intermediate model (e.g., the abstract and concrete user interfaces), necessary for the production of the final executable user interface. All of the above models are

182    Gaëlle Calvary, Joëlle Coutaz, and David Thevenin

referenced along the development process from the task specification to the running interactive system. The process is a combination of *vertical reification* and *horizontal translation.* Vertical reification covers the derivation process, from top level abstract models to run time implementation. Horizontal derivations, such as those performed between HTML and WML content descriptions, correspond to translations between models at the same level of reification. Reification and translation may be performed automatically from specifications, or manually by human experts. Because automatic generation of user interfaces has not found wide acceptance in the past [15], our framework makes possible manual reifications and translations. Such operations are manual when the tools at hand cannot preserve the usability criteria discussed in Section 3.

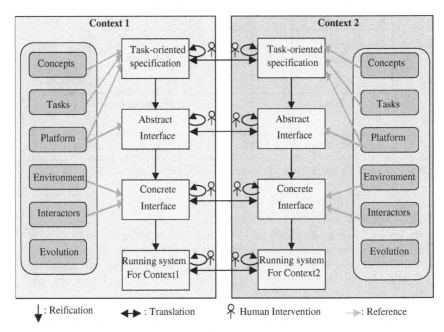

**Fig. 6.** The reference development process for supporting plastic user interfaces. The picture shows the process when applied to two distinct contexts: context1 and context2. Grayed boxes denote initial models.

Automaticity, however, conveys interesting properties that may not be maintained when introducing manual operations within the development process: Let *Reification* $_{ctxt}$ $(M)$ be the reification of model $M$ in context *ctxt*, and *Translation* $_{ctxt, ctxt'}(M)$ be the translation of $M$ from the source context *ctxt* to the target context *ctxt'*, then:

– Identity results from the combination of a translation with its inverse:

$$\forall \text{ ctxt, ctxt' Translation} _{ctxt, ctxt'} \text{ o Translation} _{ctxt', ctxt} = I$$

- Identity results from the combination of a reification and its inverse:

$$\forall \text{ ctxt Reification}_{ctxt} \text{ o Reification}_{ctxt}^{-1} = I$$

- Reification and translation are commutative:

$$\forall \text{ ctxt, ctxt' Translation}_{ctxt, ctxt'} \text{o Reification}_{ctxt'} = \text{Reification}_{ctxt} \text{ o Translation}_{ctxt, ctxt'}$$

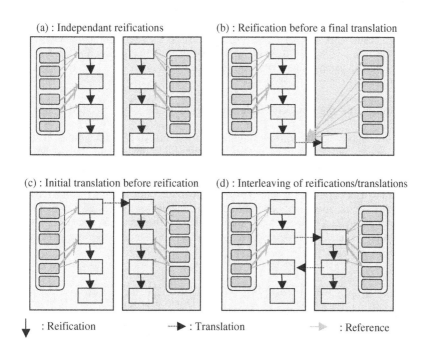

**Fig. 7.** Instantiations of the reference framework.

As shown in Figure 7, our reference framework can be instantiated in many ways:

- In 7a), two running user interfaces are reified in parallel where the initial models are specified for each context of use. This configuration, which depicts current practice, forces to check and maintain consistency manually between the multiple versions.
- 7b) corresponds to the ideal situation: reification is used until the very last step. Consistency maintenance is here minimal. This approach has been used for the Heating Control System shown in Figure 2 for Java-enabled target platforms. All of the interfaces shown in Figure 2 have been derived automatically using ARTStudio presented in Section 6.
- In 7c), the task-oriented specification is translated to fit another context. From there, reifications are performed in parallel. This approach has been adopted for the Heating Control System for WAP mobile phones. Sub trees that correspond to

infrequent tasks have been pruned from the original task tree developed for the Java-enabled platforms. Because ARTStudio does not support Web-based techniques yet, the reification steps have been done manually by a human expert.
- 7d) shows a mix of interleaving between reification and translation.

# 5   ARTStudio

ARTStudio (Adaptation by Reification and Translation) is a tool designed to support the reference framework for plasticity. We first discuss its actual implementation in relation to the framework then describe how a designer should proceed to produce a plastic user interface such as the EDF heating control system.

## 5.1   ARTStudio and the Plasticity Reference Framework

As shown in Figure 8, the current implementation of ARTStudio addresses the reification process only, and does not include the environment and the evolution models. According to the reference framework, the concepts and the task models serve the task-oriented specification which, in turn, leads to the automatic generation of the abstract user interface. Then, the platform model and the interactors model come into play for the automatic generation of the concrete user interface. Each of the transient models (i.e., Task-oriented specification, Abstract and Concrete User Interfaces) may be tuned by the designer to override the "defaults" options used by ARTStudio.

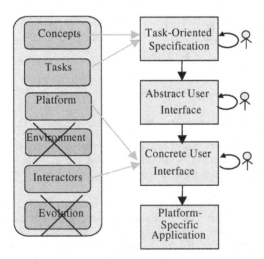

**Fig. 8.** The actual functional coverage of ARTStudio in relation to the plasticity reference framework: translation as well as the environment and evolution models are not currently supported. The "human" symbol denotes the capacity for the designer to tune transient models.

Technically, ARTStudio is implemented in Java and uses the CLIPS rule-based language for generating the concrete user interface. Initial models as well as transient models are saved as XML files. So far, final executable user interfaces are expressed in Java. As a result, the current version of ARTStudio does not support XML-based executables.

## 5.2 ARTStudio and the Development of a Plastic User Interface

We first present an overview of the use of ARTStudio then discuss the models it supports.

**Overview.** Within ARTStudio, the development of a plastic user interface forms a *project* (see Fig. 9). In the current implementation, a project includes the task and the concepts model, the context model and the abstract, the concrete and the executable user interfaces.

By clicking on the "Contexte" thumbnail of Figure 9, the developer has access to the platform and interactors models. The "règles" thumbnail gives access to the generation rules used by ARTStudio during the reification process. These rules, which address presentation issues, can be adapted by the designer to the case at hand. In the current implementation, the rules are not editable. Let's now discuss the process and the associated models for producing a plastic user interface.

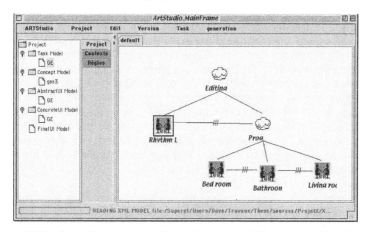

**Fig. 9.** In ARTStudio, a plastic user interface is developed within a project. The frame of the left end-side gives access to the sets of models. The picture shows the task model for a simplified version of the EDF Heating Control System.

**The Task Model.** An ARTStudio task model is a ConcurrTaskTree [3] structure where:

- The leaves correspond to interaction tasks such as reading, selecting, and specifying. As shown in Figure 9, tasks *Bedroom*, *Bathroom*, *Living room*, and

*Rythm*, respectively correspond to the specification of the level of comfort for the bed room, the bath room, the living room, and the rythm of living. The rythm of living expresses the default setting of room temperature based on periods of presence at home.

– The nodes, denoted by a "cloud" symbol, are abstract tasks. They structure the interaction space into sets of logically connected tasks. As such, abstract tasks offer a sound basis for the generation of the abstract user interface. In Figure 9, which shows a simplified version of the EDF Heating Control System, *Prog* groups together the tasks that permit the user to override the default temperature of the rooms. The task *rythm*, which has a specific status in the tasks space, is separated from the "overriding" tasks.

– The vertices express temporal and logical relationships between tasks. For example, in Figure 9, all of the tasks can be performed in an interleaved way (Cf. symbol ///).

The ARTStudio task editor allows the designer to add, cut and paste tasks by direct manipulation. As shown in Figure 10, additional parameters are specified through form filling. These include:

– Specifying the name and the type of a task. For interaction tasks, the designer has to choose among a predefined set of "universal" interaction tasks such as "selection", "specification", "activation". This set may be extended to fit the work domain (e.g., "Switching to out of frost").

– Specifying a prologue and an epilogue, that is, providing function names whose execution will be launched before and after the execution of the task. At run time, these functions serve as gateways between the Dialogue Controller and the Functional Core Adaptor [10] of the interactive system. For example, for the task "setting the temperature of the bedroom", a prologue function is used to get the current value of the room temperature from the functional core. An epilogue function is specified to notify the functional core of the temperature change.

– Referencing the concepts involved in the task and ranking them according to their level of importance in the task. This ordering is useful for the generation of the abstract and concrete user interfaces: typically, first class objects should be observable whereas second class objects may be browsable if observability cannot be guaranteed due to the lack of physical resources.

**The Concepts Model.** Concepts are modeled as UML objects using form fill in. In addition to the standard UML specification, a concept description includes the specification by extension of its domain of values. For example, the value of the attribute *name* of type *string* of the *room* concept may be one among *Living room*, *Bed room*, etc. The type and the domain of values of a concept are useful information for identifying the candidate interactors involved in the concrete user interface. In our case of interest, the *room* concept may be represented as a set of strings (as in Figure 1a), as a thumbnail (as in Figure 1b), or as dedicated icons.

**Fig. 10.** Task specification through form filling.

**The Abstract User Interface.** The abstract user interface is modeled as a structured set of *workspaces* isomorphic to the task model: there is a one-to-one correspondence between a workspace and a task. In addition, a workspace that corresponds to an abstract task includes the workspaces that correspond to the subtasks of the abstract task: it is a compound workspace. Conversely, a leaf workspace is elementary. For example, Figure 11 shows three elementary workspaces (i.e., the *Bedroom*, the *Bathroom* and the *Living Room*) encapsulated in a common compound workspace. This parent workspace results from the *Prog* task of the task model. In turn, this workspace as well as the *Rythm* elementary workspace, are parts of the top level workspace.

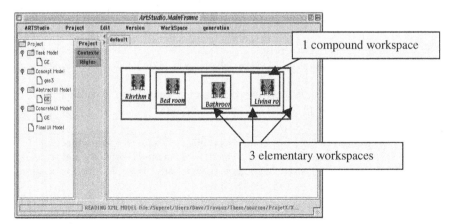

**Fig. 11.** The Abstract User Interface generated by ARTStudio from the Task Model of Figure 9. Thick line rectangles represent compound workspaces whereas thin line rectangles correspond to elementary workspaces.

By direct manipulation, the designer can reconfigure the default arrangements shown in Figure 11. For example, given the task model shown on the top of figure 12, the designer may decide to group the *Rythm* workspace with the *Room* workspaces (Figure 12 a) or, at the other extreme, suppress the intermediate structuring compound workspace (Figure 12 b).

ARTStudio completes the workspace structure with a *navigation scheme* based on the logical and temporal relationships between tasks. The navigation scheme expresses the user's ability to migrate between workspaces at run time.

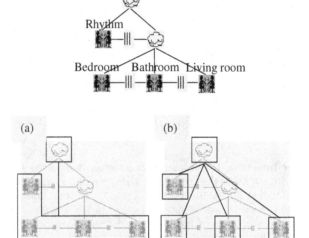

**Fig. 12.** ARTStudio allows the designer to reconfigure the relationships between workspaces.

**The Platform Model.** The platform model is a UML description that captures the following characteristics:

- The size and the depth of the screen,
- The programming language supported by the platform (e.g., Java).

In its present form, the platform model is very primitive but is sufficient for demonstrating the key concepts of our reference framework.

**The Interactor Model.** In [20] interactor modelling is discussed from a theoretical perspective in terms of *representational capacity*, *interactional capacity* and *usage cost*. In ARTStudio, our abstract interactor model boils down to the following simplified technical solution:

- Representational capacity: an interactor can either serve as a mechanism for switching between workspaces (e.g., a button, a thumbnail), or be used to represent domain concepts (e.g., a multi-valued scale as in Figure 1). In the latter case, the interactor model includes the specification of the data type it is able to render.

– Interactional capacity: the tasks that the interactor is able to support (e.g., specification, selection, navigation).
– The usage cost, which measures the system resources as well as the human resources the interactor requires, is expressed as the "x,y" footprint of the interactor on a display screen and its proactivity or reactivity (i.e., whether it avoids users to make mistakes *a priori* or *a posteriori* [4]).

**The Concrete User Interface.** As shown in Figure 8, the generation of the Concrete User Interface uses the Abstract User Interface, the platform and the interactors models, as well as heuristics. It consists of a set of mapping functions:

– between workspaces and display surfaces such as windows and canvases,
– between concepts and interactors,
– between the navigation scheme and navigation interactors.

The root workspace of the Abstract User Interface is mapped as a window. Any other workspace is mapped either as a window or as a canvas depending on the navigation interactor used for entering this workspace. Typically, a button navigation interactor opens a new window whereas a thumbnail leads to a new canvas.

Mapping concepts to interactors is based on a constraint resolution system. For each of the concepts, ARTStudio matches the type and the domain of valuesof the concepts with the interactors representational capacity, the interactors interactional capacity and their usage cost.

Figure 13 shows the Concrete User Interfaces that correspond to the running user interfaces shown in Figure 1. Fig. 13a) targets large screens (e.g., Mac and PC workstations) whereas Fig. 13b) targets small screens (e.g., PDAs).

As for the Abstract User Interface, the Concrete User Interface generated by ARTStudio is editable by the designer. The layout arrangement of the interactors can be modified by direct manipulation. In addition, the designer can override the default navigation scheme.

# 6   Conclusion

Although the prospective development of interactive systems may be fun and valuable in the short run, we consider that the principles and theories developed for the desktop computer should not be put aside. Instead, our reply to the technological push is to use current knowledge as a sound basis, question current results, improve them, and invent new principles if necessary. This is the approach we have adopted for supporting plasticity by considering model-based techniques from the start. These techniques have been revised and extended to comply with a structuring reference framework. ARTStudio, a tool under development, provides a concrete, although incomplete, application of the framework. Plasticity is a complex problem. This article makes explicit our first steps towards a systematic high quality development of plastic user interfaces.

(a)
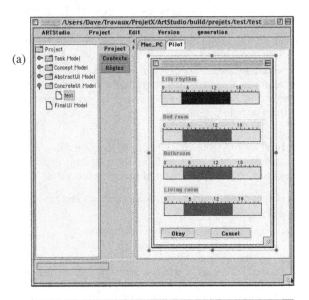

(b)
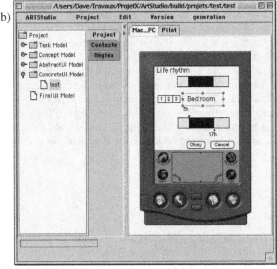

**Fig. 13.** In a), the concrete user interface generated for the system when running on the large screen display of Figure 1a. In b), the concrete user interface that corresponds to Figure 1b for the Palm Pilot.

## Acknowledgments

This work is being supported by EDF, France. We are very grateful to Nicholas Graham for the intensive discussions held in Grenoble about the notion of plasticity.

# References

1.  Ayatsuka, Y., Matsushita, N. Rekimoto, J.: Hyperpalette: a hybrid Computing Environment for Small Computing Devices. In: CHI2000 Extended Abstracts, ACM Publ. (2000) 53–53
2.  Beyer, H., Holtzblatt K.: Contextual Design, Morgan Kaufmann Publ. (1998)
3.  Breedvelt-Schouten, I.M., Paterno, F.D., Severijns, C.A.: Reusable structure in task models. In: Proceedings of DSVIS'97, Design, Specification and Verification of Interactive System, Horrison, M.D., Torres, J.C. (Eds) (1997), 225–240
4.  Calvary, G.: Proactivité et réactivité: de l'Assignation à la Complémentarité en Conception et Evaluation d'Interfaces Homme-Machine, Phd of the University Joseph-Fourier-Grenoble I, Speciality Computer Science, (1998)
5.  Cockton, G., Clarke S., Gray, P., Johnson, C.: Literate Development: Weaving Human Context into Design Specifications. In: Critical Issues in User Interface Engineering, P. Palanque & D. Benyon (Eds), Springer-Verlag: London Publ., ISBN 3-540-19964-0, (1995)
6.  Dowell, J., Long, J.: Toward a conception for an engineering discipline of human factors, Ergonomics, Vol. 32 (11), (1989), 1513–1535
7.  Eisenstein J., Vanderdonckt, J. Puerta, A.: Adapting to Mobile Contexts with User-Interfaec Modeling. In: Proc. of 3rd IEEE Workshop on Mobile Computing Systems and Applications WMCSA 2000 (Monterey, December 7-8, 2000), IEEE Press, Los Alamitos, (2000)
8.  Gamboa,-Rodriguez, F., Scapin, D.: Editing MAD* Task Descriptions for Specifying User Interfaces at both Semantic and Presentation Levels. In: DSV-IS'97, Springer Computer Science, (1997), 193–208
9.  Graham, T.C. N., Watts, L., Calvary, G., Coutaz, J., Dubois, E., Nigay, L.: A Dimension Space for the Design of Interactive Systems within their Physical Environments, DIS2000, ACM Publ. New York, (2000), 406–416
10. Gram, C., Cockton, G. Ed. : Design Principles for Interactive Software. Chapman & Hall, (1996)
11. HUC 2K, First workshop on Resource Sensitive Mobile Human-Computer Interaction, (2000)
12. Johnson, P. Wilson, S., Markopoulos, P., Pycock, Y.: ADEPT-Advanced Design Environment for Prototyping with Task Models. In: InterCHI'93 proceedings, (1993), 66
13. Johnson, P. Johnson, H. Wilson, S.: Rapid Prototyping of User Intefaces Driven by Task Models, Scenario-based design: envisioning work and technology in system development, J. Carroll (ed.), John Wiley&Sons, (1995)
14. Lim, K. Y., Long, J.: The MUSE Method for Usability Engineering, Cambridge Univ. Press, (1994)
15. Myers, B., Hudson, S., Pausch, R.: Past, Present, Future of User Interface Tools. Transactions on Computer-Human Interaction, ACM, 7(1), March (2000), 3–28
16. Paternò, F.: Model-based Design and Evaluation of Interactive Applications, Springer Verlag, (1999)
17. Robertson, G., Mackinlay, J., Card, S.: Cone Trees: Animated 3D Visualizations of Hierarchical Information. In: Proc. CHI90, ACM Publ., (1991), 189–194

18. Salber, D., Dey, A. K., Abowd, G. D.: The Context Toolkit: Aiding the Development of Context-Enabled Applications. In: the Proceedings of the 1999 Conference on Human Factors in Computing Systems (CHI '99), Pittsburgh, PA, May 15–20, (1999), 434-441
19. Szekely P.: Retrospective and Challenges for Model-Based Interface Development, Computer-Aided Design of User Interfaces. In: Proceedings of CADUI'96, J. Vanderdonckt (eds), Presses Universitaires de Namur, (1996)
20. Thevenin, D., Coutaz, J.: Plasticity of User Interfaces: Framework and Research Agenda. In: Proc. Interact99, Edinburgh, A. Sasse & C. Johnson Eds, IFIP IOS Press Publ., (1999), 110–117
21. Vanderdonckt, J.: Knowledge-Based Systems for Automated User Interface Generation; The TRIDENT Experience. RP-95-010, Fac. Univ. de N-D de la Paix, Inst. d'Informatique, Namur, B, (1995)

# Discussion

*H. Stiegler:* Real user interfaces are under constant evolution, maybe done by different people. How much work has to be done, how much knowledge has to be available to accomplish evolutions?
*J. Coutaz:* The same tools have to be available, mainly the same work has to be done as in case of the initial design.

*M. Borup-Harning:* Generating "workspaces" from abstract tasks, will your approach be able to have one canvas/window being used for several workspaces?
*J. Coutaz:* Yes, the designer can decide how it is done.

*K. Schneider:* The framework shows the models repeated in different contexts. Is there a relation between the particular models?
*J. Coutaz:* A good designer would be able to specfy everything in a single set of models, but the reference framework supports mutiple models. There is no relation between the models.

*D. Salber:* You use Fabio's tools for task representation, but you did not mention UML.
*J. Coutaz:* I did. We use UML for domain concepts.

*F. Jambon:* Your workspaces are task oriented. Do you think it is possible for them to be concept-oriented? The example is a database.
*J. Coutaz:* Yes, our tool is today task oriented. In our case study the user is more task-driven. We may improve that. However, the user is mobile and there is a fundamental track in his/her activity.

# Building User-Controlled 3D Models and Animations for Inherently-3D Construction Tasks: Which Tool, Which Representation?

Guy Zimmerman, Julie Barnes, and Laura Leventhal

Department of Computer Science
Bowling Green State University
Bowling Green, OH 43403 USA
+1-419-372-2283
fax +1-419-372-8061
gzimmer@cs.bgsu.edu

**Abstract.** In this paper, we first define a class of problems that we have dubbed *inherently-3D,* which we believe should lend themselves to solutions that include user-controlled 3D models and animations. We next give a comparative discussion of two tools that we used to create presentations: Cosmo™Worlds and Flash. The presentations included text, pictures, and user-controlled 3D models or animations. We evaluated the two tools along two dimensions: 1) how well the tools support presentation development and 2) the effectiveness of the resultant presentations. From the first evaluation, we concluded that Flash in its current form was the more complete development environment. For a developer to integrate VRML into cohesive presentations required a more comprehensive development environment than is currently available with Cosmo™Worlds. From our second evaluation, based on our usability study, we have made two conclusions. First, our users were quite successful in completing the inherently-3D construction task, regardless of which presentation (Shockwave or VRML) they saw. Second, we found that enhancing the VRML models and including multiple perspectives in Shockwave animations were equally effective at reducing errors as compared to a more primitive VRML. Based on our results we believe that for tasks of the 3D-complexity that we used, Flash is the clear choice. Flash was easier to use to develop the presentations and the presentation was as effective as the model that we built with Cosmo™Worlds and Java. Finally, we postulate a relationship between inherently-3D task complexity and the relative effectiveness of the VRML presentation.

## 1 Introduction

In recent years, a wide variety of media representations have become accessible via World Wide Web (WWW) browsers. These newer representations have the potential to greatly expand the types of problems and solutions that are deliverable on the Web. Developers of web presentations would seem to face at least two challenges as a direct result: 1) selecting a best fit of technologies to the problem and solution at hand and 2) achieving a balance between the choice of tool to build the presentation

M. Reed Little and L. Nigay (Eds.): EHCI 2001, LNCS 2254, pp. 193–206, 2001.
© Springer-Verlag Berlin Heidelberg 2001

and the effectiveness of the presentation which can be generated. In general, the designer will likely choose the technologies and tools that have the least cost in terms of resources and training, and at the same time deliver a presentation which is highly effective for the target task and user population.

Making this choice of media representations and tools is often non-trivial; recent advances in computer hardware and software have led to the development of a number of representations including: text, static images, streaming audio/video, computer animation and virtual reality (VR). There are a number of tools that support these formats; some of the tools support only one representation where others may be useful for a variety of representations.

Not all presentations are equally effective for a given task. In our recent work, we have studied the effectiveness of different media combinations for presenting directions for construction tasks. We have found that presentations that include both visual and textual components are more effective than presentations which are either strictly textual or strictly visual. In particular, we have found that textual/visual presentations that incorporate user-controlled 3D models and animations in the forms of simple virtual reality (VR) or three-dimensional (3D) models as visual components can be very effective in the delivery of this type of instruction set. [1]

The development of even the simplest VR models that we have employed is time and resource intensive. In order for such models to have any widespread practical application for the types of tasks that we are studying, it is imperative to have tools that lessen the demands on designer resources. In this paper, we first give some background on the problem domain of current interest: delivery of instructions for an inherently-3D construction task. We argue that presentations that incorporate user-controlled 3D models and/or animations would seem to be a good fit for these types of problems. Secondly, we give a comparative discussion of two tools we used to create presentations for inherently-3D construction tasks: Cosmo™Worlds and Flash. Next we describe the two presentations that we built using these two tools; the presentations included user-controlled 3D models and animations. We evaluate these tools along two dimensions: how well the tool supports presentation development and the effectiveness of the resultant presentations. In our conclusion, we analyze the cost-benefit issues that arise and make recommendations for choosing a tool.

## 2     Problem Domain: Instructions for Construction Tasks

The steady increase of processing power on the local desktop has fostered the introduction of a variety of Web-deliverable media formats including: pictures (JPEG, PNG, GIF), animations (Shockwave, Java applets) and audio/video (MPEG, QuickTime, RealAudio). Many of the information representations now widely used on the Web are simply adaptations of existing media formats that have been in use for many years. For example, animation has been used for many purposes from entertainment to education; the delivery of an animation via a computer display, in and of itself, is unremarkable. There are however some research issues that warrant investigation, such as how these various representations (like animations) can be integrated into a cohesive, interactive presentation.

There are also a small number of new technologies, such as 3D modeling and VR, which have truly expanded the types of information representations that are available via the Web. However, to date most of the widespread use of 3D graphics, particularly on the Web, has been in the "eye-candy" domain, e.g., creating animated logos. There appear to be several reasons for this phenomena. First, there is a general lack of understanding among developers of how to use and integrate 3D graphics effectively for various problem domains. In other words, it is unclear as to what types of tasks and users are best served by presentations that incorporate 3D graphics with other representational forms. [1] Secondly, integrating 3D graphics into presentations is potentially time and resource intensive, due to the limited availability of integrated development tools. [2] Finally, only very recently has the processing power on the local desktop risen to the level necessary to perform real-time 3D processing and thus made this a practical option to Web developers.

It would seem obvious that there is a large class of problems which would benefit from the use of 3D representations of information. That is, there are some problems which seem to be "inherently-3D" in nature and thus a solution to the problem that incorporated 3D graphics would yield an optimal solution. Consider the following problem: *Assemble a toy model car*. This is a non-trivial construction task requiring a series of steps. The car is a real object and it "lives" in 3D space. The pieces of the car are fit together, dynamically, in a multitude of (3D) orientations. While assembling such a car, a person would typically view the car from a variety of (3D) perspectives, choosing vantage points which facilitate the user's understanding of the assembly process. The effectiveness of instructions for assembling a toy car would be enhanced with suitable use of 3D graphics. In particular, given access to a manipulable 3D facsimile of the car, the assembler could compare the facsimile with the actual model car from various perspectives during each step of the assembly.

In this paper, we will refer to problems such as *assemble a model car*, as *inherently-3D construction tasks*. We recognize that the term "inherently-3D" is subjective; in fact, part of our research is to develop metrics to quantify this intuitive notion. We are interested in the delivery of instructions for inherently-3D construction tasks. We observe that the instructions for tasks like assembling a toy car have traditionally been delivered in paper form as combinations of text and still pictures. When such directions are made Web-deliverable they are still typically in the form of text and still pictures. Such instructions are notorious for being difficult for the user to follow. We believe that the effectiveness of directions for inherently-3D tasks, delivered via the Web, could be significantly enhanced using some of the newly available 3D technologies.

To investigate this, we have chosen the construction of origami objects as our inherently-3D construction task. Creating an origami object is similar to assembling a toy car in several ways. Most origami tasks have a number of steps. While assembling an origami object, it is often useful to view it from a variety of perspectives. The paper folding task creates an artifact in the real world. Thus, making an origami object is representative of a very broad class of educational/training problems in which a user is given step-by-step instructions and is expected to build something.

In addition there is a complexity spectrum for origami objects, ranging from simple, basically 2D objects with only a few folds, such as a paper hat, to highly 3D objects such as an origami box. This complexity spectrum provides us with a framework to investigate how the 3D characteristics of the real object are related to the 3D characteristics that are conveyed in the instructions.

Additionally, in terms of a benchmark task for usability research, origami has several useful characteristics. Paper folding is a task which is familiar to most people;

most people have made a simple paper airplane. However, many people do not have explicit experience with traditional origami and thus they would need to follow detailed instructions to actually create an origami (i.e., they need to use our presentation to complete the task). For most people creating an origami object is self-motivating; our users are generally compelled to finish our instruction presentations, because they enjoy the task of building the origami object.

Finally, developing presentations to deliver instructions to construct an origami object is a relevant task for the Web. Traditionally, the instructions for foldings are delivered on paper, using text and still pictures. [3] However, numerous Web sites and commercial multimedia products have recently become available; these sites and products convey the directions for origami objects with text, pictures, video and sound. [4]

# 3    Development Tools

In this paper, we consider the problem of engineering a presentation to deliver instructions for building an origami object, specifically: presentations that include user-controlled 3D models and/or animations. We limited our research to Web-deliverable presentations which integrate the models and/or animations with text and picture versions of the instructions, based on earlier findings that suggest that incorporating visual and verbal information in the presentation aids the user in accomplishing a task.[1] We also required users to be able to control the animations (play, stop, replay). We chose two data formats: VRML and Shockwave. Both VRML and Shockwave permit the creation of user-controlled animations. VRML permits the user to interact directly with the model by rotating and scaling it. Thus the user is able to see the object from any vantage point while a given folding step is being animated. Shockwave animations do not permit the user to manipulate the image, but do permit the display of parallel animations, which allows for a presentation which includes multiple (but static) perspectives. High-end development tools are available for both VRML and Shockwave. Finally, both VRML and Shockwave have Web browser plug-ins, making them Web-deliverable and both can be integrated with text and picture presentations.

In the next two sections, we describe the high-level features of the two tools that we used. In a later section, we will describe how we used the tools.

## 3.1    VRML Development: Cosmo™Worlds

VRML is a *de facto* Web standard and is arguably the best way to provide interactive, 3D environments in that context. To develop our VRML models we used Cosmo™Worlds which is marketed by Silicon Graphics. In addition to the traditional 3D-modeling tools, Cosmo™Worlds provides GUI tools to manipulate virtually every feature of the VRML specification. The developer can manipulate the scene graph directly using an object browser facility or just by clicking on objects in the development window. VRML nodes can be easily grouped together into a Switch node to allow exactly one of the child nodes to be activated at a time. The keyframe animation mechanism allows the developer to easily animate models and it automatically chooses the correct VRML interpolation node for a given keyframe sequence. Users can easily set material properties (color) and/or apply texture maps to a given surface. Script node code and I/O events are handled well; Route statements are illustrated visually. One very useful feature is the PEP tool suite. This set of tools allows individual Points, Edges and Polygons (PEP) to be directly manipulated. For example, a selected polygon can be automatically split into two pieces. Finally, an optimization tool is available to reduce the number of polygons, file size and overall complexity.

With all its functionality, VRML still presents some limitations to the problem at hand. Several of the manipulations we wanted to be able to perform on our VRML objects required the functionality of a full-featured programming language. We used Cosmo™Code to develop Java code to manage a user interface and to allow the user to control aspects of the animation. Other researchers have noted the effectiveness of this combination of VRML and Java. In [2], the authors discuss a CAD tool for the virtual assembly of furniture. Client/server applications using Java and VRML are discussed in [5].

We did consider other formats and tools for our work. Two notable candidates were Java 3D and World Up. [8]  We rejected Java 3D as being too low-level compared to VRML. World Up, a full-featured 3D system, was rejected for several reasons. It is currently in no sense a Web standard. We also felt it unlikely that many users would have the necessary plug-in. Finally, integrating World Up presentations with HTML was problematic. We did feel that the World Up toolkit may be a better overall development tool in the future for problems of this type, in no small part due to its integrated, object-oriented, scripting feature.

### 3.2    Shockwave Development: Flash

One of the most popular tools for creating 2D, Shockwave animations for the Web is Macromedia's Flash. Flash provides a development environment for creating animations for Web pages and optimizes the resultant files for fast delivery over the Internet. [6]

The Flash development environment provides the developer with a vector-graphics editor to create graphical objects to populate scenes in Flash animations. The developer creates multiple layers for each scene in order to more easily manipulate the different objects in a scene. Although a scene is composed of multiple layers, hinting at depth, the objects are 2D. The developer  must create several scenes as the only way to hint at various perspectives. Shockwave models are not directly manipulable by the user.

A major feature of Flash is its ability to create "tweened" animations. With this feature the developer specifies a starting object, position, shape or color and a final object, position, shape or color and Flash generates the frames in-between the two by interpolation. The developer may specify Shape Hints to guide the interpolation in shape tweening. The developer may specify a Path in a Layer Guide to assist the interpolation in a motion tween.

# 4    An Evaluation of Development Tools

In this section we will discuss and compare the specific development tools that we used to develop our VRML models and Shockwave animations. In particular, we will discuss the amount of effort that was involved in creating the models for our work using VRML and Shockwave

In order to compare the development tools for VRML and Shockwave, we first defined a standard interface look-and-feel for the overall presentation. The left side of the presentation included (top to bottom) a label denoting the current step number, a scrollable text window, and a window containing a still picture. The right side of the presentation contained (top to bottom) a window for the 3D model (VRML) or animation (Shockwave) and controls for the user to manipulate the presentation. These controls included buttons to navigate through the different steps of the presentation and controls to manipulate the 3D model or animation. There were some minor differences in the two presentations. (See Figures 1 and 2.)

Next we defined a benchmark task for our user to perform: build an origami whale. The whale consisted of 12 distinct steps and 25 distinct folds. Steps 1 through 5 were essentially 2D in the sense that they were folds or fold-unfold combinations on a flat piece of paper. In Steps 6 through 12, the folds were 3D, in the sense that the paper model became truly 3D. Steps 1-3, 6 and 9 formed the body, Steps 4-6 and 10 formed the fins, Steps 7-8 and 11 formed the mouth, and Step 12 formed the tail. Steps 6 and 12 included "inverted" or "hidden" folds, in which all of the fold could only be truly seen in 3D. We felt that the whale, by virtue of the 3D nature of the final origami object and the inverted folds, was at least marginally an inherently-3D construction task.

## 4.1    VRML

Our VRML whale animation models were integrated into our standard interface using the Cosmo™Player plug-in. The control section contained three spin controls that permitted the user to rotate the model about the three standard axes. There was also a size control that permitted the user to scale the model. A start/stop button allowed the user to control the fold animation at each step. Additionally, users could return the 3D model to its original orientation and size for each step by using a reset button.

We created two different VRML versions for the whale building task. The first version was designated "plain VR" (PVR) and, in some sense, served as a control for us in evaluating Cosmo™Worlds and the effectiveness of the VRML presentation. The construction of the first version used a single VRML Shape node type: indexed face set. There were two vertices corresponding to each fold line (the line along which the paper is to be creased); the coordinates of each of these vertices and the faces to which they belong were manually calculated for keyframes of each animation step. All of these values were stored in the indexed face object. A coordinate Interpolator node and a Script node were used to generate an animation sequence to illustrate each fold. In PVR, the fold line had to be inferred from the lighting/shading of the faces involved in that fold. A Java applet was implemented to generate the user interface which allowed the user to manipulate the model and the animation process. The EAI [7] was used to allow the Java user interface to "communicate" with the VRML model. The VRML model for the first version was created manually with a text editor and required approximately 40 hours to complete. The user interface applet and associated EAI code was developed using Cosmo™Code and required about 10 hours. An additional 5 hours were required to assemble all of the components (text instructions, Java applet user interface and VRML plug-in window) into a coherent set of Web pages; each page corresponded to one step of the construction process.

In the second version, the VRML component was built in Cosmo™Worlds and was designated as "enhanced VR" (EVR). EVR differed from PVR by the addition of explicitly drawn fold lines. In EVR, when the user first initiates the animation of the step, the fold line to be created is "drawn" onto the model and after a brief pause, the animation of the origami fold process begins. It was hoped that the explicit rendering of the fold line would enhance the effectiveness of the VRML model in the representation of each step.

The construction of the second version took more time than the first. This was a direct result of the explicit animated fold line requirement. As in the first version, the basic VRML Shape node used was indexed face set. Because the underlying geometry of the object changed during subsequent steps of the construction process and each new object required additional fold lines, we found it most expedient to create a separate indexed face set object for each step. The addition of the actual fold lines was accomplished using the texture mapping feature in Cosmo™Worlds. However, since there is no integrated facility for creating textures, we created the fold line images using PhotoShop and then imported these into Cosmo™Worlds. This was a non-trivial task since the fold line image geometry had to precisely match the geometry of its corresponding indexed face set. All of the calculations necessary to establish the correct mapping of the image to the corresponding face were done manually. Managing the changing geometry of the indexed face sets was greatly facilitated by the PEP suite of tools in Cosmo™Worlds. For example, creating a fold requires one polygon to become two; the splitting PEP tool made this very easy to do. The additional number of indexed face set objects, combined with the animation of the fold lines, required additional Script node and Java coding to manage the transitions between steps. Fortunately, the Java applet for the control user interface and the Java EAI code from the first version required only modest alterations.

Figure 1 shows a screen from our integrated presentation of instructions for Step 6 of the origami whale. Step 6 is a 3D step in which some of the paper is hidden inside of the object. This figure shows the EVR presentation where the 3D model has been rotated by the user in order to view the model from an alternate perspective.

## 4.2   Shockwave

We also created an integrated presentation of the instructions for folding the whale, using the Flash development tool. We found that in Flash, the animations for the folds were easily created. Motion, shape, and color tweening were all used to create the images of folding paper. In the final Shockwave movie, buttons were included so that the user could interact with the movie. This interaction was limited to 1) starting and stopping the animation, 2) navigating between steps or scenes, and 3) changing viewpoints (selecting an alternate version of some scenes). The user could not directly manipulate the images of the paper model. The user could only replay the Shockwave

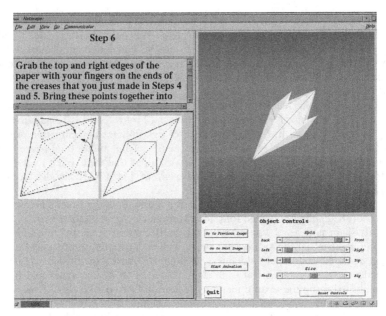

**Fig. 1.** The Enhanced VRML (EVR) Version of the Interface

animations that had been created. The individual images that we used in Flash were either imported GIF images or images that were created with the Flash drawing tool. The fold lines were drawn as part of these images.

The 3D steps of the folding, particularly Steps 6 and 12 with the hidden paper, were a challenge in Flash. We opted to present a second, parallel perspective of these folds. Hence we were able to show the alternate perspectives of some of the scenes. The user was still restricted to the viewpoints determined to be most meaningful by the developer of the movie.

The final interface of text, pictures and Shockwave animation was itself created in Flash and included a scrollable text window, window for a still picture that showed the folds for the current step, the animated folds, controls and directions for the controls. Figure 2 shows a screen from the Flash presentation. It shows the same step as Figure 1. Because the Shockwave model is itself 2D and only shows one perspective for this step with hidden folds we included a second parallel animation from another perspective.

### 4.3   Evaluation

In terms of development, Cosmo™Worlds has a complete set of features for generating objects, color and animations; as a tool for creating stand-alone VRML worlds it is excellent. However, in order to create a presentation incorporating VRML, we found this tool to be inadequate. In particular, the lack of a facility for creating custom texture maps was limiting. We were also hampered in generating our presentation because we had to use Java to create and to manage the integrated text,

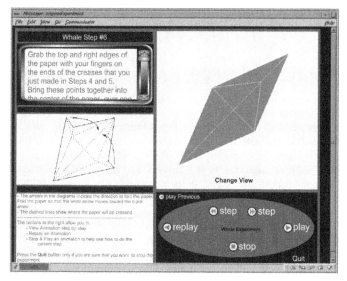

**Fig. 2.** The Flash Version of the Interfaces

picture and VRML components. We note that this is not a deficiency of Cosmo™Worlds *per se*, but rather points to a vacuum in the set of tools that are available for delivery of VRML over the Web.

Flash was quite easy and fast for the application at hand. To develop the animation, we had a keyframe of the initial image of the paper at the current step, several keyframes of intermediate images, and a final frame. Flash tweened these images into our animation. Color interpolation and shape tweening can give the effect of a paper fold. The step buttons permitted the user to step through our intermediate keyframes. The Flash development environment includes a draw tool and the capability of importing GIF images, so it was also quite easy for us to build the keyframes. Finally, we were able to create our entire integrated presentation in Flash, rather than using a second tool, such as Java.

In evaluating the two development tools, Cosmo™Worlds and Flash, we conclude that Flash in its current form is the more complete environment. By contrast, Cosmo™Worlds supports only the development of stand-alone VRML. For a developer to integrate VRML into cohesive presentations requires a more comprehensive development environment than is currently available.


## 5     An Evaluation of the Presentations

We concluded in the previous section that the Flash development tool is the more complete tool for developing integrated Shockwave animations as compared to Cosmo™Worlds for developing integrated VRML presentations. In other words, from the developer's perspective, it is likely to be more work to develop an integrated VRML presentation. In this section, we consider whether that extra work is worthwhile. We compare the effectiveness of the VRML and Shockwave presentations for users who are trying to complete an inherently-3D construction task.

## 5.1    Methodology, Users, and Materials

The task for our users was to construct an origami whale following the standardized presentation and interface look-and-feel that we described in the previous section. Users saw a presentation of instructions for folding a whale, in one of three treatments: (1) PVR: containing text, still-images, and plain VRML (no fold lines), (2) EVR: containing text, still-images, and VRML (enhanced renderings with fold lines), and (3) FLASH: containing text, still-images, and Shockwave animations. User performance was measured and assessed by the number of correct folds and the number of error folds in the whale that each user created.

There were 24 users for this study, with eight per treatment. Users were sophomores and juniors enrolled in computer science classes at Bowling Green State University. All were highly computer literate. Each user viewed only one of the three presentations, PVR, EVR or FLASH. The instructions were presented with a Silicon Graphics O2 computer with a 17 inch monitor. Users viewed the presentations with *Netscape 4.0.*

## 5.2    Procedure

All of the users performed the task in the CHIL lab (Hayes 227, Bowling Green State University). Users arrived at the lab and completed consent materials. Next the users received training. For the users who would see the PVR or EVR presentation, they first completed an interactive training session about VRML and the tool set for the 3D-model presentation in this version of the instructions. Users in the FLASH conditions saw an interactive tutorial that illustrated the animation controls. This training took about 8 minutes to complete, unless the users had questions.

All users then received training in paper folding from computerized instructions. In this phase of the training, users folded a stylized paper airplane which had five steps and eight folds. The computerized instructions were presented in whichever of the three treatments that the user was to receive.

Next the experimenter explained that origami is the ancient art of Japanese paper folding. Users were given a piece of special origami paper and were told to fold the whale by following the presented instructions. There were no time limits for how long subjects were given to fold the whale.

## 5.3    Results

Our dependent variables were the number of correct folds and number of error folds that subjects made. In order to assess the correctness of the whale folds, each existing fold was graded by three criteria:
- Was the fold in the right place?
- Was the direction of the fold correct?
- Was the fold the right size?

In order to be a correct fold, the fold had to be correct on all three of these criteria. Some subjects made a correct fold and then recreased the same fold in a slightly different position. We scored both of these folds as correct.

The subjects in all of the treatments did very well. The average percentage of correct folds was 86.5% across the three treatments and there was no statistical difference by treatment. In other words, on average all groups eventually got more than 86% of the folds right.

We conducted a between-subjects MANOVA of total number of correct folds and total number of error folds, by presentation condition. The MANOVA was significant (Roy's Greatest Root = 0.496 (F(2,21) = 5.2, p < 0.01)). A univariate ANOVA of the total number of correct folds by presentation condition was not significant. A univariate ANOVA of the total number of error folds by presentation condition was significant (F(2,21) =4.6, p < 0.02). The average number of error folds by treatments was PVR, 10.0, EVR, 4.9, and FLASH, 4.3. Post-hoc tests (Fisher's PLSD) indicated that, when considered pairwise, PVR was significantly different from both EVR and FLASH. EVR and FLASH were not significantly different from each other.

In other words, both EVR and FLASH reduced the number of error folds as compared to PVR. Neither EVR or FLASH improved the likelihood that the final whale would be correct, but did reduce the number of errors along the way. The baseline presentation (PVR) was the least effective of the three, although it was still quite effective. Adding explicit fold lines in EVR or having the 2D images with multiple perspectives and fold lines in FLASH were even more effective.

From our usability study we drew two conclusions. First, our users were quite successful in folding the whale, regardless of which presentation they saw. This result is consistent with our previous studies which indicate that presentations that include both visual and textual information are likely to be more useful than either visual or textual information alone. Second, we found that enhancing the VRML models with fold lines and including multiple perspectives in Shockwave animations were equally effective at reducing folding errors as compared to more primitive VRML.

# 6 Summary and Conclusions

In this paper, we identified a class of problems that we termed inherently-3D construction tasks. Our previous research has indicated that when delivering instructions for tasks of this type on the Web, multiple media (visual and textual) are more effective than a single type of media. User-controlled 3D models and animations are examples of visual presentations that can be used effectively with text. We considered two tools for developing 3D models or simulated 3D models for the Web: Cosmo™Worlds to generate VRML models and Flash to generate Shockwave animations. We evaluated these two tools in terms of effectiveness of the development environment and effectiveness of the presentation that could be developed with each tool.

We conclude that in terms of the development environment, Flash can be used to quickly develop a simulated 3D animation, because it provides an integrated environment for developing for the Web. Also, Flash does not require the developer to create an actual 3D model to manipulate. By comparison, development of 3D models in VRML is much more difficult. While Cosmo™Worlds provides a VRML environment, it lacks features for integration as compared to Flash. The VRML models themselves are more complex. For example, in VRML to show fold lines in our model, we were forced to superimpose a texture map on our images. In Flash, lines are simply a part of the image.

In terms of the effectiveness of the resultant presentations, we found that both FLASH and EVR were equally effective as compared to PVR, although all three treatments had a very high success rate for folding the origami whale. We believe that this was because both FLASH and EVR better presented the 3D aspects of the object, via multiple perspectives or fold lines, respectively.

Which tools should a developer select when delivering instructions for inherently-3D tasks? Based on our results we believe that for tasks of the inherent 3D-complexity of our origami whale, Flash is the clear choice. For our whale, Flash was easier to use to develop the presentations and the presentation was as effective as the model that we built with Cosmo™Worlds and Java.

In retrospect, we believe that the construction task did not have enough inherently-3D features to benefit from the additional realism and functionality that the VRML provided. We feel that as 3D-complexity of the task increases, then the extra overhead of VRML would be justified. Figure 3 shows this postulated relationship. We speculate that the whale folding task was at the threshold or cross point. In our future research, we intend to explore this issue further by studying origami objects that have more hidden and inverted folds than the whale did. We are also currently in process of trying to link a cognitive measure with specific characteristics of the origami object as a way to quantify "inherently-3D."

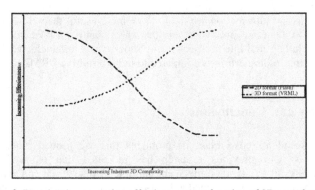

**Fig. 3.** Postulated presentation effectiveness as a function of 3D complexity

We also conclude that future development environments for VRML models could greatly benefit by incorporating some of the features of the Flash development tool. Specifically, we note that future VRML development tools should support an integrated approach to:
- HTML
- VRML 3D modeling
- VRML Script Node programming and ROUTE statements
- Java (for user interface)

- EAI (for Java-VRML interaction)
- Image manipulation: (for textures, etc)

## Acknowledgements

We thank Keith Ebare, Lisa Weihl, Tom Stoltz Brian Ameling, and the conference reviewers for their assistance on this project.

# References

1. Leventhal, L., Barnes, J., Zimmerman, G., Weihl, L.: Multiple Web Representations of Instructions on a Procedural Task: An Empirical Study. Submitted to: *Behaviour and Information Technology*. (2000)
2. Nousch, M., Jung, B.: CAD on the World Wide Web: Virtual Assembly of Furniture with BEAVER. The Fourth International Conference on the Virtual Reality Modeling, Language and Web 3D Technologies (February 1999) http://www.clab.de/vrml99/home.html
3. Honda: The World of Origami, Japan Publications Trading Co., Tokyo (1965)
4. Cassady & Greene: Origami: The Secret Life of Paper (1996) ISBN 1-56482-104-8
5. Jung, B., Milde, J.: An Open Virtual Environment for Autonomous Agents Using VRML and Java. The Fourth International Conference on the Virtual Reality Modeling, Language and Web 3D Technologies (February 1999) http://www.clab.de/vrml99/home.html
6. Macromedia, Inc.: Macromedia Flash 4: Using Flash. Macromedia, Inc., San Francisco (1999)
7. Marrin, C., Couch, J.:   Specification of the VRML/EAI Interface. The web 3D Consortium. Online resource. (January 1999) http://www.web3d.org/WorkingGroups/vrml-eai/
8. Sense8 Corporation: World Up User's Guide.  Sense8 Corporation, Mill Valley, CA (1997)

# Discussion

*J. Willans:* Do you have data on how subjects used each of the representations during construction?
*G. Zimmerman:* Subjects flipped between representations, especially with difficult manipulations. Text was the least favoured. We have video and interaction logs, but we have not analyzed the data yet.

*N. Graham:* Many assembly tasks involve both hands. Do you have ideas about how to adapt a system like yours to these tasks?
*G. Zimmerman:* We have a long wish list of techonologies we'd like to investigate, e.g. immersive virtual reality. We do not see a solution to the problem of requiring a mouse click to move to the next step.

*P. Curzon:* A common problem following construction instructions is that you think you understand what to do, and only realize several steps later that you went wrong. Do you think that giving the user the ability to construct the object virtually on the screen would be a good use of 3D as it would allow the system to give feedback on errors?

*G. Zimmerman:* That's not an area we have looked into. The aim was just to see how a web delivered set of instructions could make things easier.

*J. Willans:* A related question is one that they have been asking at Nottingham University in the UK, which related to what are virtual environment good for?

*G. Zimmerman:* This is one of the questions we are trying to answer also.

# Unconstrained vs. Constrained 3D Scene Manipulation

Tim Salzman, Szymon Stachniak, and Wolfgang Stürzlinger

Dept. of Computer Science, York University
4700 Keele Street, Toronto, ON, M3J 1P3, Canada
{salzman | szymon | wolfgang}@cs.yorku.ca

**Abstract.** Content creation for computer graphics applications is a very time-consuming process that requires skilled personnel. Many people find the manipulation of 3D object with 2D input devices non-intuitive and difficult. We present a system, which restricts the motion of objects in a 3D scene with constraints. In this publication we discuss an experiment that compares two different 3D manipulation interfaces via 2D input devices. The results show clearly that the new constraint-based interface performs significantly better than previous work.

## 1    Introduction

Computer Graphics applications, such as physical simulations, architectural walk-throughs, and animations, require realistic three-dimensional (3D) scenes. A scene usually consists of many different objects. Objects and scenes are usually created with a 3D modeling system. Many different commercial systems are available for this purpose. But in general these products are difficult to use and require significant amounts of training before than can be used productively. Realistic 3D scenes often contain thousands of objects, and complex interfaces can leave the user feeling lost. For example, products such as Maya and 3D Studio Max have dozens of menus, modes and widgets for scene creation and manipulation, which can be very intimidating for an untrained user. Our efforts address these difficulties.

The focus of this work is on the creation of complete 3D scenes, not on the generation of 3D geometric models. We rely on a library of predefined objects from which the user can choose while constructing a scene. The challenge is to provide the user with a technique to easily add objects to the scene and then position them relative to each other.

Interacting with objects in a 3D environment is difficult because six independent variables must be controlled, three for positioning and three for orientation. Even when the task of object manipulation is decomposed into the two separate tasks of positioning and orientation, the user interface is still not intuitive. This is due to the problem of mapping a device with 3 modes (buttons) with 2 degrees of freedom each to 2 different manipulation techniques with 3 degrees of freedom each.

3D input devices, such as a space-ball or tracking devices, make direct interaction with objects in a 3D scene possible, but such devices are uncommon and often expensive, or require training and experience to use. Similarly, 3D output devices

M. Reed Little and L. Nigay (Eds.): EHCI 2001, LNCS 2254, pp. 207–219, 2001.
© Springer-Verlag Berlin Heidelberg 2001

such as head mounted displays or shutter glasses are a possible solution, but again are uncomfortable, uncommon, or expensive.

Our observations of humans rearranging furniture and planning environments indicate that humans do not think about scene manipulation as a problem with 6 degrees of freedom. The rationale is that most real objects cannot be placed arbitrarily in space and are constrained in their placement by physics (e.g. gravity) and/or human conventions (ceiling lamps are almost never placed permanently onto the floor or onto chairs). This lets us to believe that an interface that exposes the full 6 degrees of freedom to the user makes it harder for average persons to interact with virtual environments. Many real objects have a maximum of 3 degrees of freedom in practice – e.g. all objects resting on a plane. Furthermore, many objects are often placed against walls or other objects, thus further reducing the available degrees of freedom. This in turn leads us to believe that a two-dimensional (2D) input device such as a mouse may be sufficient to manipulate objects in a virtual environment. Furthermore, in previous work Poupyrev et al. [13] suggested that all ray-casting techniques can be approximated as 2D techniques as long as objects are relatively close to the viewer. This further supports our argument that a 2D input device is sufficient to manipulate most real objects in a 3D environment.

In our system, information about how objects interact in the physical world is used to assist the user in placing and manipulate objects in virtual environments. Each object in a scene is given a set of rules, called constraints, which must be followed when the object is being manipulated. For example, a photocopier must stand on the floor at all times. When a user interacts with the photocopier by translating or rotating it in the scene, it never leaves the floor. This concept of constraints makes manipulating objects in 3D with 2D devices much simpler.

## 1.1   Previous Work

Previous work can be classified into two categories: those that use 2D input devices and those that use 3D input devices.

For 2D applications Bier introduced Snap-Dragging [1] to simplify the creation of line drawings in a 2D interactive graphics program. The cursor snaps to points and curves using a gravity function. Bier emphasized the importance of predictability in the system; i.e. that the interface should behave as the user expects it to. Hudson extended this idea further to take non-geometric constraints into account and called his technique semantic snapping [9]. Bier subsequently generalized Snap-Dragging to 3D environments [2]. Relationships between scene components were exploited to reduce the size of the interface and the time required to use it Interactive transformations are mapped from the motion of the cursor, which snaps to alignment objects for precision. The main features of this system are a general purpose gravity function, 3D alignment objects, and smooth motion affine transformations of objects. Houde introduced another approach that uses different handles on a box surrounding each object to indicate how it can be manipulated [8].

Bukowski and Sequin [6] employ a combination of pseudo-physical and goal-oriented properties called Object Associations to position objects in a 3D scene using 2D devices (mouse and monitor). They use a two-phase approach. First, a relocation procedure is used to map the 2D mouse motion into vertical or horizontal transformations of an object's position in the scene. Then association procedures align and position the object.

Although fairly intuitive, their approach has a few drawbacks. Firstly, associations apply only to the object that is currently being moved and are not maintained after the current manipulation. Also, when an object is selected for relocation, a local search for associated objects is performed. This can result in lag between the motion of the selected object and the motion of its associated objects. Furthermore, cyclical constraints are not supported.

The fundamental ideas of the Object Associations approach were incorporated into the work of Goesele and Stuerzlinger [10]. Each scene object is given predefined offer and binding areas. These areas are convex polygons, which are used to define constraining surfaces between objects. For example, a lamp might have a binding area at its base and a table might have an offer area defined on its top surface. In this way a lamp can be constrained to a tabletop. To better simulate the way real world objects behave, a constraint hierarchy is used to add semantics to the constraint process. This is a generalization of the Object Association approach. Each offer and binding area is given a label from the hierarchy. A binding area can constrain to an offer area whose label is equal to or is a descendant in the constraint hierarchy to the label of the binding area. In this way a monitor may be defined to constrain to a tabletop, but never to the wall or floor. Also, collision detection is used to add realism to the interface.

Drawbacks are: Once a constraint has been satisfied, there are no means to un-constrain or re-constrain the object to another surface. Further, the constraint satisfaction search is global, in that an object will be moved across the entire scene to satisfy a constraint, an often-undesirable effect for the user, especially because constraints cannot be undone.

A number of authors have investigated the performance of object manipulation with 3D input devices. One of the first was Bolt in 1980 [3]. Most recently Bowman [4], Mine [11], and Pierce [12] proposed different methods that can be applied in a variety of settings. Poupyrev recently also addressed the problem of 3D rotations [14]. For a more complete overview over previous work in this area we refer the reader to [5]. Most relevant to the research presented here is the work by Poupyrev et al [13]. There, different interaction methods for 3D input devices are compared. The item that is of most interest in this context is that the authors suggest that all ray-casting techniques can effectively be approximated as 2D techniques (see also: [5]). This nicely supports our observation that for most situations a user interface that utilizes only a 2D input device may be sufficient to effectively manipulate objects in 3D.

The SmartScene system [16] by Multigen is a Virtual Reality system that employs a 3D user interface based on Pinch-gloves. Object behaviors that define semantic properties can be defined to simplify object manipulation.

To our knowledge, there is no previous work that compares 3D unconstrained and constrained manipulation,

## 2    The MIVE System

The MIVE (Multi-user Intuitive Virtual Environment) system extends the work done in [10] by improving the way existing constraints behave, and adding new useful constraints. This work concerns only the interaction of a single user, therefore we disregard the multi-user aspects of the system here.

The constraint relationships are stored in a directed a-cyclic graph called the scene graph. Figure 1 depicts a simple scene, and it's associated scene graph. When an object is moved in the scene, all of its descendants in the scene graph move with it.

Notice that edges in the scene graph of Figure 1 correspond directly to satisfied constraints in the scene. The user can modify the scene graph structure by interacting with objects in the scene. Constraints can be broken and re-constrained in with ease by simply clicking on the desired object, and pulling away from the existing constraint to break it. This allows us to dynamically change the structure of the scene graph. Figure 2 shows the same scene as Figure 1 after the chair has been pulled away from the large table, and dragged under the smaller table.

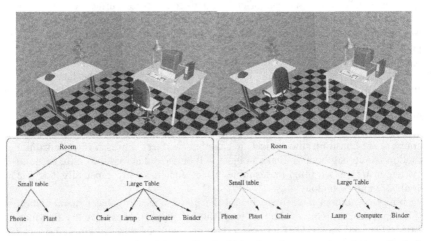

**Fig. 1.** A Scene and its associated Scene Graph. Links describe constraint relations. Both the initial state (left) as well the state after the chair has been moved (right) are shown.

## 2.1  MIVE Constraint Environments

Each object in the MIVE system has a set of constraints associated with it. For example, a table would have a constraint on its base that causes it to stand on the floor, and a constraint on its top that allows other objects (i.e. a phone) to lie on its surface. When the table is added to the scene, it will always lie on the floor. When being moved or rotated the table remains on the floor, and any objects lying on its top surface will move/rotate with it. This is the default constraint mode in MIVE. We call this the Fully Constrained (FC) mode.

Other scene modelers, such as many 3D computer aided design (CAD) programs, use no constraints whatsoever. They provide only an interface to manipulate all six degrees of freedom of an object. We believe that interaction with objects in an unconstrained environment such as this, is much more difficult for the user. To enable an effective comparison we implemented a mode in MIVE, which does not exploit constraints. We call this the Unconstrained (UC) mode.

# 3    User Testing

The main objective of the work presented here is to evaluate if a constraint-based system provides a more effective 3D manipulation interface. We asked participants in our study to design scenes in both the FC as well as the UC mode.

## 3.1    Method

Two representative tasks were chosen to evaluate our objective.

T1) Creation of a scene: See Figure 4 for the target scene image.

T2) Modification of a scene: See Figure 5 and 6 for initial respective final scene.

We choose these two tasks because together they exploit a majority of the operations needed to create and manipulate 3D scenes.

Both scenes consist of 30 objects. For the creation task (T1) the participant can select the objects from a window that displays all 30 objects.

We chose not to attempt a comparison with a user interface that utilizes 3D input devices as these are usually used in a very different setting (standing user and large screen stereo projection or HMD). Although very interesting, this introduces several additional factors into the user tests that may not be easy to account for. Furthermore, even if good 3D output equipment was readily available, in our tests stereo output should not constitute an important factor as most objects are placed clearly in relation to other objects such as a phone on a desk. Also, the test scenes in our environment are limited in range, therefore we believe that stereo viewing will not provide important visual cues.

For task T1 no navigation was necessary to complete the task. For task T2 users had to navigate to move one of the objects that was hidden behind another.

## 3.2    Participants

Fifteen volunteers (3 female, 12 male) participated in this experiment. Half of the participants were computer science students, half had other backgrounds (including 4 persons with artistic talents). Most participants had at least average computer skills, but only 5 had any exposure to 3D scene construction. The ages of the participants ranged from between 20 to 43 years, the average age being 23.

## 3.3    Apparatus

The MIVE interface was designed to be very simple. Figure 2 shows the full user interface of the MIVE program running with its default object list.

The MIVE interface consists of three windows: the scene window, the object selection window, and the button window. The scene window sits on the right hand side of the screen. The participant directly interacts with objects in the scene window by clicking and dragging them.

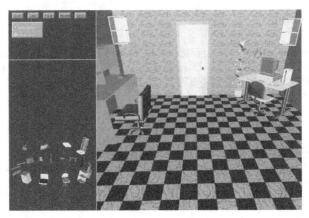

**Fig. 2.** User Interface of the MIVE system.

The lower left-hand corner shows the object selection window. Objects are positioned on an invisible cylinder, which is rotated by clicking any mouse button within the window and dragging left and right. Objects are added to the scene window by simply clicking on the desired object in the object selection window, and clicking on the desired location to add it in the scene window.

The upper left-hand corner of the window contains buttons for performing tasks such as loading or saving the scene, deleting an object, undoing the previous operation, or quitting the program. There is also a radio button which can be used to switch between interaction and navigation mode. This functionality was disabled for the tests in this publication.

The MIVE system is implemented in C++ and runs on an SGI Onyx2 running IRIX 6.5. It is based on the Cosmo3D [7] scene graph API.

### 3.3.1     Interaction

The interface for MIVE was designed to be as simple and uncluttered as possible. All interactions between the participant and the program are done using a 3-button mouse.

The FC mode uses only two of the three buttons. The left mouse button is used to move objects by clicking and dragging them to the desired new location. The middle mouse button is used to rotate the objects. The third mouse button is currently unused in this mode.

The UC mode uses the right and left mouse buttons to move objects in 3D space, and the middle mouse button to perform an Arcball rotation [15] on the object. Using these three buttons it is possible to place any object in any orientation in 3D space. Although constraints are disabled in this mode, collision detection remains active so that objects do not interpenetrate.

### 3.4     Procedure

A five-minute tutorial was given prior to the testing, at which time the experimenter gave the participant instructions on how to use the system and how the mouse worked in each of the two modes. Participants were allowed to familiarize themselves with the interface for 5 minutes.

Each test began with the participant sitting in front of a computer monitor with a scene displayed. A target scene was displayed on an adjacent monitor, and the participant was instructed to make the scene on their screen look like that in the target scene. When the participant deemed that the scene resembled the target closely enough, the participant quitted the program after checking back with the experimenter.

Each participant was asked to perform each the two tasks in each of the two constraint systems for a total of 4 trials for each participant. The starting task for each participant was randomly selected. To counterbalance the design, subsequent tasks were ordered so that a UC task followed a FC task and vice versa. All tests were done in a single session, which took slightly more than one hour on average.

For each of the tests, we recorded the time taken by the participant, the number of actions and the accuracy of the final scene. Accuracy was measured by summing the distances in centimeters between each of the object centers in the user's final result and the target scene.

When the participants had completed all of their tests, they were given a questionnaire that posed questions relating to their preference among the interaction modes in the system.

# 4    Analysis

At the end of each experiment task completion time and the modified scene was stored. The Euclidean distance between the participant's solution and the reference solution was computed later on. Finally, basic statistics and a 2 by 2 repeated measures ANOVA was performed with statistical software.

## 4.1    Adjustments to Data

The data for two participants was excluded from the analysis. One participant tried to visually match the results to the target scene at the level of pixel accuracy by repeatedly navigating back and forth. The experiment was aborted after 1½ hours with 2 tasks completed. Another participant could not understand how Arcball rotations in the 3D unconstrained (UC) mode worked and got too frustrated.

No other adjustments were made to the collected data.

## 4.2    Computed Formulas

Errors are specified as the *sum* of Euclidean distances for all objects, as 3D distances are themselves quadratic measures. We ignore rotation because no ideal measure for rotation differences exists to our knowledge. Moreover it is hard to find a meaningful combination of translation and rotation errors into one number.

# 5    Results

Figure 3 summarizes the result of our user test. The center line of the boxes shows the mean, the box itself indicates the 25th respective 75th percentile, and the 'tails' of the boxes specify the 10th respective 90th percentile. All statistical values in this publication are reported at alpha = 0.05.

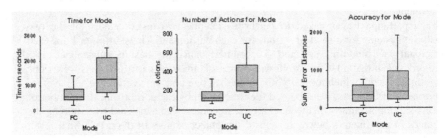

**Fig. 3.** Box-plots of times, actions and accuracy for UC and FC modes.

The analysis of variance showed clear main effects for task completion time ($F_{1,12}$ = 26.36, p < 0.0005), number of actions ($F_{1,12}$ = 16.0, p < 0.005) and accuracy ($F_{2,12}$ = 7.31, p < 0.05) with respect to the two different interaction modes. The statistical power of the first two tests is larger than 0.95, the last test has only a power of 0.699.

The mean completion time (variances are shown in brackets) for FC was 719.7 seconds (543.9) and 1621.4 seconds (1233.1) for UC. This shows that FC is faster by a factor of almost 2.3.

The mean number of actions for FC was 157.96 (110.9) and 395.92 (318.2) for UC. In terms of user actions UC requires more than 2.5 times the number of actions.

The mean sum of error distances for FC was 381.3 cm (242.7) and 762.1 cm for UC (880). This seems to indicate that objects are positioned more accurately in the FC system, although the statistical significance is not as strong as for the other two criteria.

Task T1 is significantly different from T2 in terms of time. (T1 took 30% more time). T1 is not significantly different from T2 in terms of the number of actions (289 vs. 264). Accuracy shows also no significant difference, although we observe that T1 has a slightly larger error on average.

## 5.1    Questionnaire

Questions evaluated participant preferences between each combination of two modes on a 5 point Likert scale. The results show a clear preference for the FC mode. Users found it much easier to manipulate objects when they felt that those objects acted as would their counterparts in the real world. Imposing constraints on objects abstracted difficult transformations that, to the untrained person, were hard to visualize. In particular, most people found rotations hard to grasp in unconstrained mode, since it was much more difficult to orient objects in full 3-space than it was in the more limited yet more intuitive FC mode. Users were noted to find the spinning object selection menu to their liking for it gave quick access to all the objects without cluttering precious desktop space. The only complaint voiced about the menu was the selection of small objects, as users found them hard to select at times.

To aid in the positioning of objects in UC mode, users were quick to point out the benefits of shadows. A visual cue such as a shadow would be necessary to aid in object placement of a fully UC system, but imposes many new problems, i.e. positioning a light source. A FC system eliminates the need to judge height distances from one object to another, and therefore eliminates this problem. The absence of additional (potentially confusing) visual cues makes onscreen manipulation.

# 6   Discussion

The above analysis of the user test data supports the following observations:

First, unconstrained manipulation (UC) is definitely slower than the FC. For scene creation we observe a factor of roughly 2.3. Moreover, we can say that UC manipulation is at least 50% less accurate than the FC mode.

This current test involved complex tasks. In effect this prohibited us from determining precisely the relative performance of the two modes. Another complicating influence is that T2 necessitates navigation.

General observations of users during the tests led to a determination that the majority of inaccuracies and time delay in UC is due to problems in rotation. When constraints are imposed, rotations become much more intuitive, as the user interface restricts rotations to the free axis, which in turn increases speed and accuracy.

Participants seemed to have taken the most time in T1, constructing the scene. We believe that this may be because of the time it takes to 'retrieve' an object. When adding an object the user has to first locate it within the object selection window. For this test we tried to keep the number of objects in the window to a minimum (around 30), but judging from our observations during the test identifying and selecting an object still takes roughly 3 seconds.

A constrained system, unlike an unconstrained system is much more intuitive for users who have little to no experience in 3-dimensional manipulation. Users were able to construct scenes with ease only several minutes after having been introduced to the FC system, yet in UC mode issues relating to 3-dimensional manipulations would often arise.

The main objective of the user test presented here was to determine if constrained scene manipulation is helpful to the user and if it makes a difference. The results indicate that this is true in practice. The work presented here did not look into a comparison with a system that is equivalent to the Object Associations [6] system. This system uses only horizontal and vertical constraints. It is unclear at the moment how powerful these two constraints are compared to the 'semantic' constraints in the MIVE system. An small experiment by the authors with an initial implementation showed that the difference in performance between the two systems is small. Future work will look further into this issue.

# 7    Conclusion

In this publication we presented a system that allows users to easily manipulate a 3D scene with traditional 2D devices. The MIVE system is based on constraints, which enable an intuitive mapping from 2D interactions to 3D manipulations. Phrased differently, the constraints and associated manipulation techniques encapsulate the user's expectations of how objects move in an environment. The user tests showed that the interaction techniques presented here provide clear benefits for 3D object manipulation in a 2D user interface. The unconstrained system was significantly slower and less accurate than the system that uses constraints. Users unanimously preferred the constrained system to the unconstrained one.

Interestingly enough, most systems that use 3D input devices support only unconstrained manipulation. Based on the observation of Poupyrev that many 3D manipulation techniques can be approximated with 2D techniques we are fairly confident that we can speculate that the techniques presented here are directly beneficial to scene manipulation with 3D input devices. To our knowledge, this work is the first to compare 3D unconstrained and constrained manipulation.

The benefits of our interaction techniques become very apparent when one compares the simple MIVE user interface with the complex 3D user interface in commercial packages such as AutoCAD, Maya or 3DStudio Max that are also based on 2D input devices. We can only hypothesize at the outcome of a test comparing our system with e.g. Maya, but are confident that it is clearly easier to learn our user interface due to the reduced complexity. In fairness, we need to point out that these packages are also capable of object creation and the specification of animations, which our system does not currently address.

# 8    Future Work

Our vision is to create a system that makes adding objects to a scene as quick and easy as possible. The user test involved complex tasks that had to be accomplished by the user. In effect this prohibited us from determining the relative performance of individual elements of the user interface precisely. Therefore, future work will focus on determining the relative performance of the individual techniques. As mentioned in the discussion, we will test the developed techniques against an implementation that mimics the interaction methods in the Object Association system [6].

The current set of user tests was designed in such a way that users themselves determine when they have completed the desired tasks. This method was chosen when we realized that it is hard to automatically implement an adequate measure of scene similarity. Factors would need to be taken into account are differences in position, rotation, location in the scene graph and the fact if an object is correctly constrained or not. A viable alternative that we will pursue in the future is that the test supervisor determines when the user has accomplished the task.

Also we noticed that a considerable amount of time was spent selecting new objects in the object menu. This issue warrants further investigation. Furthermore, small objects in the rotating menu are hard to select, which needs to be addressed, too.

Also, we will study how constraint based scene construction performs in immersive virtual environments such as a CAVE.

**Acknowledgements**

Thanks to N. Toth, G. Smith and D. Zikovitz for their help. Furthermore, we would like to thank all persons who volunteered for our user tests.

# References

1.  Bier, E.A., and Stone, M.C. Snap-dragging. ACM SIGGRAPH 1986, pp. 233-240.
2.  Bier, E.A. Snap Dragging in Three Dimentions, ACM SIGGRAPH 1990, pp. 193-204.
3.  Bolt, R., Put-that-there. SIGGRAPH 1980 proceedings, ACM press, pp. 262-270.
4.  Bowman, D., Hodges, L. An Evaluation of Techniques for Grabbing and Manipulating Remote Objects in Immersive Virtual Environments. Proceedings of ACM Symposium on Interactive 3D Graphics, 1997, pp. 35-38.
5.  Bowman, D., Kruijff, E., LaViola, J., Mine, M., Poupyrev, I., 3D User Interface Design, ACM SIGGRAPH 2000 course notes, 2000.
6.  Bukowski, R.W., and Sequin, C.H. Object Associations. Symposium on Interactive 3D Graphics 1995, pp. 131-138.
7.  Eckel, G., Cosmo 3D Programmers Guide. Silicon Graphics Inc. 1998.
8.  Houde, S., Iterative Design of an Interface for Easy 3-D Direct Manipulation, ACM CHI '92, pp. 135-142.
9.  Hudson, S., Adaptive Semantic Snapping – A Technique for Semantic Feedback at the Lexical Level, ACM CHI '90, pp. 65-70.
10. Goesele, M, Stuerzlinger, W. Semantic Constraints for Scene Manipulation. Proceeedings Spring Conference in Computer Graphics (Budmerice, 1999), pp. 140-146.
11. Mine, M., Brooks, F., Sequin, C. Moving Objects in Space: Exploiting Proprioception in Virtual-Environment Interaction. SIGRAPH 1997 proceedings. ACM press, pp. 19-26.
12. Pierce, J., Forsberg, A., Conway, M., Hong, S., Zeleznik, R., et al. Image Plane Interaction Techniques in 3D Immersive Environments. Proceedings of Symposium on Interactive 3D Graphics. 1997. ACM press. pp. 39-43.
13. Poupyrev, I., Weghorst, S., Billinghurst, M., Ichikawa, T., Egocentric object manipulation in virtual environments: empirical evaluation of interaction techniques. Computer Graphics Forum, EUROGRAPHICS '98 issue, 17(3), 1998, pp. 41-52.
14. Poupyrev, I., Weghorst, S., Fels, S. Non-isomorphic 3D Rotational Techniques. ACM CHI'2000, 2000, pp. 546-547.
15. Shoemake, K., ARCBALL: A User Interface for Specifying Three-Dimensional Orientation Using a Mouse, Graphics Interface, 1992, pp. 151-156.
16. SmartScene promotional material, Multigen, 1999.

# Discussion

*K. Schneider:* How does the usability suffer with a large set of objects? Are you able to temporarily override the constraint?

*W. Stuerzlinger:* We are working on a hierarchical approach to selecting objects. The cylinder does allow for larger set of objects than a 2D approach. At any time one is able to move to non-constrained mode. Users in practice did not. One problem is that when going back to constrained mode, all constraints must be satisfied. Since it does not have a full 3D constrain solver, this may be impossible.

*L. Nigay:* Following the previous question, the cylinder view works well, you said. How many objects do you have in the current version?
*W. Stuerzlinger:* About 50 objects. Working on many objects is a real issue.

*D. Damian:* What are the practical applications of your applications besides house decoration?
*W. Stuerzlinger:* Other application domains like plant design or urban planning.

*C. Roast:* The results from the empirical study do not seem surprising, the constraints help with speed and number of actions.
*W. Stuerzlinger:* In fact, despite appearing obvious there are no cases of such results published.

*C. Roast:* Looking at the figures, it is interesting to see that the average time per action is lower for the unconstrained system.
*W. Stuerzlinger:* Interesting point, this is probably because in the constrained system objects are located and positionned accurately in one action, whereas in the other system this involves separate actions.

## Appendix

As more of a joke than anything else we kept a profanity count during the test. The number of unprintable words uttered can be seen as a simple measure of user frustration. In FC mode users uttered 0.33 profanities on average during the test. UC mode featured an average of 3.53 profanities!

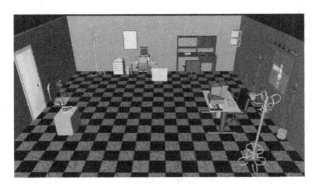

**Fig. 4.** Target Scene for T1) Scene Creation Task. Initial state is an empty room.

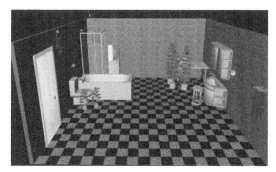

**Fig. 5.** Initial Scene for T2) Scene Modification Task

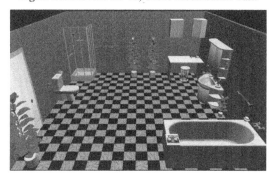

**Fig. 6.** Target Scene for T2) Scene Modification Task

# Toward Natural Gesture/Speech Control
# of a Large Display

Sanshzar Kettebekov and Rajeev Sharma

Department of Computer Science and Engineering
Pennsylvania State University
220 Pond Laboratory, University Park PA 16802
sanshzar@psu.edu, rsharma@cse.psu.edu

**Abstract.** In recent years because of the advances in computer vision research, free hand gestures have been explored as means of human-computer interaction (HCI). Together with improved speech processing technology it is an important step toward natural multimodal HCI. However, inclusion of non-predefined continuous gestures into a multimodal framework is a challenging problem. In this paper, we propose a structured approach for studying patterns of multimodal language in the context of a 2D-display control. We consider systematic analysis of gestures from observable kinematical primitives to their semantics as pertinent to a linguistic structure. Proposed semantic classification of co-verbal gestures distinguishes six categories based on their spatio-temporal deixis. We discuss evolution of a computational framework for gesture and speech integration which was used to develop an interactive testbed (iMAP). The testbed enabled elicitation of adequate, non-sequential, multimodal patterns in a narrative mode of HCI. Conducted user studies illustrate significance of accounting for the temporal alignment of gesture and speech parts in semantic mapping. Furthermore, co-occurrence analysis of gesture/speech production suggests syntactic organization of gestures at the lexical level.

# 1 Introduction

Psycholinguistic studies, e.g., McNeill [1] describes gestures as a critical link between our conceptualizing capacities and our linguistic abilities. People use gestures to convey what cannot always be expressed using speech only. These gestures are sometimes considered as medium for expressing thoughts and ways of understanding events of the world. Kendon [2] distinguishes *autonomous gestures* from *gesticulation*, gestures that occur in association with speech. In fact when combined with speech they were found to be efficient means of communication for coping with the complexity of the visual space, e.g., pen-voice studies [3]. Motivated by this, there has been a considerable interest in integrating gestures and speech for natural human-computer interaction .

To date, speech and gesture recognition have been studied extensively but most of the attempts at combining speech and gesture in an interface were in the form of a predefined controlled syntax such as "*put <point> that <point> there*", e.g., [4]. In a number of recent applications user was confined to the predefined gestures (signs) for

M. Reed Little and L. Nigay (Eds.): EHCI 2001, LNCS 2254, pp. 221–234, 2001.
© Springer-Verlag Berlin Heidelberg 2001

spatial browsing and information querying, e.g. [5]. In both cases the intent of making interaction natural was defeated.

Part of the reason for the slow progress in multimodal HCI is the lack of available sensing technology that would allow sensing of natural behavior on the part of the user. Most of the researchers have used some device (such as an instrumented glove or a pen) for incorporating gestures into the interface, e.g., [6]. This leads to constrained interaction that result in unnatural multimodal patterns. For instance, in the pen-voice studies [5] the order and structure of incoming linguistic information significantly deviated from the ones found in canonical spoken English. However, the availability of abundant processing power has contributed to making computer vision based continuous gesture recognition robust enough to allow the inclusion of free hand gestures in a multimodal interface [7, 8, 9].

An important feature of a natural interface would be the absence of predefined speech and gesture commands. The resulting multimodal "language" thus would have to be interpreted by a computer. While some progress has been made in the natural language processing of speech, there has been very little progress in the understanding of multimodal human-computer interaction [10].

Although, most of gestures are closely linked to speech, they still present meaning in a fundamentally different form from speech . Studies in human-to-human communication, psycholinguistics, and etc. have already generated significant body of research. However, they usually consider a different granularity of the problem outside the system design rationale. When designing a computer interface even when it incorporates reasonably natural means of interaction, we have to consider artificially imposed paradigms and constraints of the created environment. Therefore exclusively human studies cannot be automatically transferred over to HCI. Hence, a major problem in multimodal HCI research is the availability of valid data that would allow relevant studies of different issues in multimodality. This elaborates into a "chicken-and-egg" problem.

To address this issue we develop a computational methodology for the acquisition of non-predefined gestures. Building such a system involves challenges that range from low-level signal processing to high-level interpretation. In this paper we take the following approach to address this problem. We use studies in an analogous domain; that of TV weather broadcast [7] for bootstrapping an experimental testbed, *i*MAP. At this level, it utilizes an interactive campus map on a large display that allows a variety of spatial planning tasks using spoken words and free hand gestures. Then we address critical issues of introducing non-predefined gesticulation into HCI by conducting user studies.

The rest of the paper is organized as follows. In Section 2 we review suitable gestures and their features for the display control problem. Subsequently (in Section 3) we present structural components of gestures as pertinent to a linguistic structure. A semantic classification of co-verbal gestures is also presented in this section. Section 4 introduces a computational framework of the gesture/speech testbed and describes results of continuous gesture recognition. Section 5 presents results of *i*MAP user studies.

## 2  Introducing Gesticulation into HCI

Kendon [2] distinguishes *autonomous gestures* from *gesticulation*, gestures that occur in association with speech. McNeill [1] classifies major types of gestures by their relationship to the semantic content of speech, in particular a person's point of view towards it. He defines *deictic* gestures that indicate objects and events in the concrete world when their actual presence might be substituted for some metaphoric picture in one's mind. *Iconic* gestures were those that refer to the same event as speech but present a somewhat different aspect of it. For example: "and he bends <*extending bent arm in the elbow*> ...". The other types are metaphoric and beat gestures, which usually cover abstract ideas and events. See [1] for a detailed review. The last two gesture classes have currently a limited use for HCI.

As for HCI, deictic gestures seem to be the most relevant. This term is used in reference to the gestures that draw attention to a physical point or area in course of a conversation [1]. These gestures, mostly limited to the pointing by definition, were found to be co-verbal [1, 11]. Another type of gestures that may be useful is sometimes referred to as manipulative or act gestures. These serve to manipulate objects in virtual environment e.g., [12]. They usually characterized by their iconicity to a desired movement, e.g., rotation. Later in the section 3.3 we will present classes of co-verbal gestures based on their deixis that were found useful during narrative display control.

To date, very limited research has been done on how and what we actually perceive in gestures. In face-to-face communication some of the gestures may be perceived on the periphery of the listener's visual field [2, 13]. In that case, the gestures were not visually attended and their meaning may be mainly conveyed by their kinematical patterns. Supporting arguments exist in recent eye tracking studies of anthropomorphic agents while gesturing [14]. They revealed that attention of the listener shifts to the highly informative one-handed gesture when it had changes in a motion pattern, i.e., a slower stroke.

In addition, a shape of the hand while gesturing can be affected by an uncountable number of cultural/personal factors. Therefore, considering manipulation on a large display our approach concentrates on the kinematical features of the hand motion rather then hand signs. Section 4.2 will present review of continuous recognition of gesture primitives[1] based on their kinematical characteristics.

Even though gestures are closely linked to speech, McNeill [1] showed that natural gestures do not have a one-to-one mapping of form to meaning and their meaning is context-sensitive. In the next section we will attempt to clarify nature of co-verbal gestures as pertinent to a linguistic structure. Our approach is dictated by a design rational that begins from acquisition of  kinematical gesture primitives to extraction of their meaning.

---

[1] In the gesture literature it is also referred as a gesture form or a phoneme (sign language).

# 3 Understanding of Gestures

Armstrong [15] views gestures as a neuromuscular activity that ranges from spontaneously communicative to more conventional gestures that poses linguistic properties when presented as conventualized signs. The formalist approach assumes that speech can be segmented into a linear stream of phonemes. Semantic phonology introduced by [16] is a reasonable approach for representing structural organization of gestures and similar to Langacker's notion of cognitive grammars; it reunites phonology with the cognitive, perceptual, and motoric capacities of human.

Since gestures realized in spatial medium through optical signal we would expect that gesture phonemes defined in the spatio-temporal domain progress to form morphemes (morphology), the words in turn have different classes that are used to form phrase structures (syntax), and finally those yield meaning (semantics).   A semantic phonology ties the last step to the first, making seamless connection throughout the structure. Cognitive grammar [17] claims that a linguistic system comprises just three kinds of structures: phonological, semantic, and symbolic (residing in the relationship between a semantic and a phonological structure). In the following subsections we will address those structures from a computational rationale.

## 3.1 Multidimensionality of Meaning

At the phonological level, a continuous hand gesture consists of a series of qualitatively different kinematical phases such as movement to a position, hold, and transitional movement [2]. Kendon [18] organized these into a hierarchical structure. He proposed a notion of gestural unit (*phrase*) that starts at the moment when a limb is lifted away from the body and ends when the limb moves back to the resting position. This unit consists of a series of gesture *morphemes* [2], which in turn consists of a sequence of *phonemes: preparation—stroke—hold—retraction* (Fig. 1). The *stroke* is distinguished by a peaking effort. When multiple gesture morphemes are present, the *retraction* can be either very short or completely eliminated. Kendon concluded, a morpheme is the manifestation of a so-called idea unit. [18] found that different levels of movement hierarchy are functionally distinct in that they synchronize with different levels of prosodic structuring of the discourse in speech. These results will be elaborated in the next subsection.

At the morphological level, McNeill [1] argued that gestures occurring in face-to-face communication have global–synthetic structure of meaning.  The meaning of a part of a gesture is determined by the whole resulting gesture (e.g., global), where different segments are synthesized into a single gesture (e.g., synthetic). Then if a gesture consists of several strokes, the overall meaning of each stroke is determined by the meaning of the compounded gesture.

---

[2] Kendon refers to this unit as gesture phrase (G-phrase)

Stokoe [19] proposed to treat sublexical components of visual gestures (signs) at as motion, irrespective of what is moving; the hand configuration as it was at rest; and the location where the activity occurs. Figure 1 shows compositional semantics of gesture strokes applicable in HCI; we define it as being distributed among, a spoken clause, and a physical (spatial) context. In addition the world-level units of the language of a visual gesture can be viewed as a marriage of a noun and a verb; an action and something acting or acted upon. For simplicity (not incompleteness): such a system encapsulates both words and sentences, for the words can be symbolically, SVO (subject-verb-object) [19]. We will consider these notions in classification of the co-verbal gestures in Section 3.3.

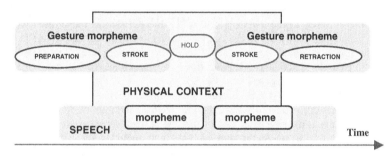

**Fig. 1.** Structural semantics of co-verbal gestures. Strokes are meaningful parts of gesture morphemes as defined in the context. Strokes and holds are temporally aligned with the relevant spoken words.

## 3.2 When Meaning Is Formed

The issue of how gestures and speech relate in time is critical for understanding the system that includes gesture and speech as part of a multimodal expression. McNeill [1] defines a semantic synchrony rule which states that speech and gestures cover the same idea unit supplying complementary information when they occur synchronously.

The current state of HCI research provides partial evidence to this proposition. Our previous studies of weather narration [7] revealed that approximately 85% of the time when any meaningful gestures are made, it is accompanied by a spoken keyword mostly temporally aligned during and after the gesture. Similar findings were shown by Oviatt et al. in the pen-voice studies [11].

At the phonological level, Kita et al. [20] assert that a pre-stroke hold is more likely to co-occur with discourse connectors such as pronouns, relative pronouns, subordinating temporal adverbials (e.g., when, while, as) compared to a post-stroke hold. In an utterance, the discourse cohesion is established in the initial part, and new information is presented in the final part. It is thought that a pre-stroke hold is a period in which gesture waits for speech to establish cohesion so that the stroke co-occurs with the co-expressive portion of the speech. [20] found that a stroke with repetitive motion is less likely to follow post-stroke hold. It was suggested that a post-stroke hold was a way to temporally extend a single movement stroke so that the stroke and post-stroke hold together will synchronize with the co-expressive portion of the speech. Despite the fundamental structural differences of gestures and speech the co-occurrence patterns imply syntactic regularities during message production in face-to-face communication.

### 3.3 Semantic Classification of Gestures

After extensive studies of the weather channel narration [7, 8] and pilot studies on the computerized map we separated two main semantic categories. *Transitive* deixis embraces all the gesture acts that reference concrete objects (e.g., buildings) in the context where no physical change of location was specified as. In the literature these gestures are also referred as the *deictic* gestures [1]. In HCI studies those often were found to be supplemented by a spoken deictic marker (this, those, here, etc.) [11]. Unlike traditionally accepted gesture classification [1] where only the pointing gesture is referred as deictic, we define three subcategories of deictic strokes independently of their form (Figure 2).

Categorization of *transitive* deixis is inferred from the complementary verbal information. They are defined as follows: *nominal, spatial,* and *iconic. Nominal* refers to object selection made through reference of the object itself by an assisting noun or pronoun. *Spatial* deixis is made through an area that includes the reference and usually complemented by adverbials "here", "there", "below", and etc. The *iconic* category is a hybrid of *nominal* and *spatial* deixis. The gesture in this case not only refers to the object but also specifies its attributes in a spatial continuum, e.g., shape. For example, "**this** road <*contour* stroke along>".

*Intransitive* deixis is defined as a gestural act causing spatial displacement. This category is analogous to the spoken motion verb (e.g., move, enter). Unlike *transitive* deixis instead of an object, it requires a spatio-temporal continuum, i.e. direction, as a reference to complete the meaning. Analogous to Talmy's classification [21] of intransitive motion verbs, we distinguish three classes: *initial, medial,* and *final.* These are characterized by a restructuring of space, i.e. spatial displacement, induced by reference-location. These gestures are expected to co-occur with the dynamic spatial propositions (from, through, toward, etc.). For example, in "go **through** <stroke *through* > the building" the stroke corresponds to the *medial* motion gesture.

## 4  Addressing the "Chicken -and -Egg" Problem

Realization of the issues discussed so far in a computation framework, derived from audio and video signals, is a difficult and challenging task. The integration of gesture and speech can occur in three distinct levels - data, feature, or decision level [22]. Data fusion is the lowest level of fusion. It involves integration of raw observations and can occur only in the case when the observations  are of the same type which is not common for multiple modalities. Feature level fusion techniques are concerned with integration of features from individual modalities into more complex multimodal features and decisions. These frequently employ structures known as probabilistic networks,  such as artificial neural networks and hidden Markov models (HMMs).

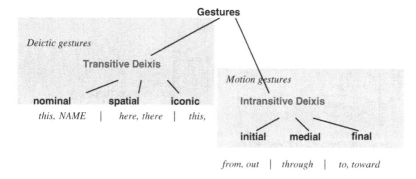

**Fig. 2.** Semantic classification of gestures and examples of the associated keywords

Decision fusion is the most robust and resistant to individual sensor failure. It has a low data bandwidth and is generally less computationally expensive than feature fusion. However, a disadvantage of decision level fusion is that it potentially cannot recover from loss of information that occurs at lower levels of data analysis, and thus does not exploit the correlation between the modality streams at the lower integration levels. Frame-based decision level fusion have been commonly used ever since Bolt's early "Put- that-there" [4].

Both decision and feature level require *natural* multimodal data, e.g., for utilizing statistical techniques. This controversy leads to a "chicken-and-egg" problem. One of the solutions is to use Wizard-of-Oz style of experiments [23]. Where, the role of a hypothetical multimodal computer is played by a human "behind the scene". However, this method does not guarantee a timely and accurate system response which is desirable for eliciting adequate user interaction. Therefore, in the next subsection we present a gesture/speech testbed enabled with gesture recognition (primitives) and timely feedback.

### 4.1 Computational Framework: *i*MAP

Figure 3 presents an overview of the conceptual design of the multimodal HCI system composed of three basic levels. At the topmost level, the inputs to the *i*MAP are processed by low-level vision based segmentation, feature extraction and tracking algorithms [24], and speech processing algorithms. The output of this level goes to two independent recognition processes for speech and gesture. The gesture recognition framework has been inspired and influenced by a previous work for continuous gesture recognition in the context of a weather map [7].

In order to achieve an adequate perceptual user interface *i*MAP utilized both audio and visual feedback. The position of the cursor is controlled by the hand by means of a 2D vision-based tracking process, for details see [24]. It enabled rather steady and near real time positioning of the cursor with up to $110^{\circ}$ (visual angle) sec$^{-1}$. The visual (building name) and audio (click) feedback was enabled by the recognition of a basic *point* and consecutive *contour* gesture. When the pointing gesture was recognized the cursor shape is changed from "✎" to "✂" (Fig. 4.b).

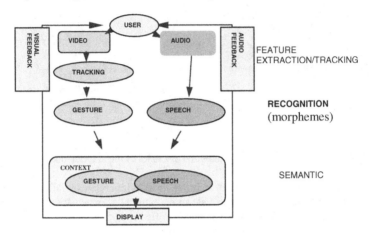

**Fig. 3.** Framework diagram for gesture/speech testbed.

### 4.2 Continuous Gesture Recognition

We sought a solution to bootstrapping continuous gesture recognition (phonemes) through the use of an analogous domain, which does not require predefinition of gestures. We refer to it as the *weather domain*. The weather domain is derived from the weather channel on TV that shows a person gesticulating in front of the weather map while narrating the weather conditions (Fig. 4.a).

Here, the gestures embedded in a narration are used to draw the attention of the viewer to specific portions of the map. Similar set of gestures can also be used for the display control problem. The natural gestures in the weather domain were analyzed with the goal of applying the same recognition techniques to the design of a gesture recognition system for *i*MAP.

a)                                        b)

**Fig. 4.** Two analogous domains for the study of natural gestures: a) The weather narration and b) The computerized map interface (*i*MAP)

To model the gestures both spatial and temporal characteristics of the hand gesture need to be considered. In [7] the natural gestures were considered to be adequately characterized by 2D positional and time differential parameters. The meaningful gestures are mostly made with the hand further away from the body. Little or no movement characterizes other gestures that do not convey any information with the hands placed close to the body e.g., beat gestures.

Analysis of the weather narration [7] made it possible to discern three main strokes of gestures that are meaningful in spite of a large degree of variability. Those were termed as pointing, circle, and contour primitives (Fig. 5).

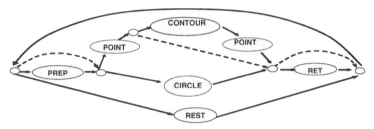

**Fig. 5.** Morphological network of gesture primitives. Dashed lines imply an alternative transition graph to a next phoneme. Meaningful strokes of gestures are identified as gesture primitives: *contour*, *circle*, and *point*.

Since compound gesture can be formed by a combination of the strokes of one or more of the aforementioned three primitive categories. The preliminary experiments led to the choice the following HMMs which were constrained as shown in the morphological network (Fig. 5) [7]. Left-to-right causal models were defined, with three states for each of the preparation, retraction, and point HMMs and four states for each of the contour and rest HMMs.

### 4.3  Results in Natural Gesture Recognition

The results from extensive experiments conducted in both the weather domain and the *i*MAP framework indicated that it is possible to recognize natural gestures continuously with reasonably good rates, 78.1 and 79.6% respectfully [7, 8]. Method of the preliminary bootstrapping in the analogous domain proved to be an effective solution.

Consequent application of speech/gesture co-occurrence model in the weather domain through the associated keyword spotting (noun, pronoun, and spatial adverbials) significantly improved gesture recognition. A correct rate of 84.5% was attained [7]. The next section presents a series of user studies that contributes to formalization of a framework for meaningful gesture processing at the higher levels.

## 5  Toward Eliciting Syntactic Patterns

The goal of the following study is to elicit empiric clarification on the underlying semantics of natural gestures and syntactic patterns for the display control problem. We adopt classes of gestures from [14, 5]: point, contour, and circle. We consider a series of spatial tasks in the context of a campus map. Those included combination of object selection/localization tasks (e.g., "What is the nearest parking lot?") and their narrative displacement to other locations (e.g., "How would you take your car to another location?"). The presented a set of spatial tasks did not require map manipulation, e.g., zooming, rotation, scrolling. See [25] for detailed review.

### 5.1 User Study Results: Multimodal Patterns

This section presents a summary *i*MAP user study results, c.f. [25], that was conducted with 7 native English speakers. Analysis of a total of 332 gesture utterances revealed that 93.7% of time the adopted gesture primitives were temporally aligned with the semantically associated nouns, pronouns, spatial adverbials, and spatial prepositions. Unlike previous attempts e.g., [11, 8] it permitted us to take a first step in the investigation of the nature of *natural* gesture-speech interaction in HCI.

*Transitive* deixis was primarily used for the selection/localization of the objects on the map. *Intransitive* deixis (gestures that were meant to cause spatial displacement) were mostly found to express the direction of motion.

The original classification of gesture primitives was supported by stronger intrinsic separation for *intransitive* deixis compared to the *transitive* category. This is due to the kinematical properties of the *point* and *contour* gestures, which are more likely to express the direction of the motion. The utterance analysis revealed presence of motion complexes that are compounded from the classes of *intransitive* deixis. *Point* was found to be the only primitive to express *initial movement* deixis which is followed by either *point* or *contour* strokes depending on the type of motion presented, i.e. implicit or explicit respectively. The explicit type was mostly encountered if precise path need to be specified.

Due to associative bias of the *contour* primitive with the motion, the temporal co-occurrence analysis was found effective in establishing pattern for the following primitive. If an *initial* spatial preposition co-occurred with the pre-stroke hold of the *contour* it was followed by *point* primitive to establish destination point of motion. In, general this finding agrees with Kita [20] that pre-stroke hold is indicative of the discourse function.

The results indicate that the proposed method of semantic classification together with co-occurrence analysis significantly improve disambiguation of the gesture form to meaning mapping. This study, however, did not consider perceptual factors in the visual context. Studies by De Angeli et al. [3] showed that salience and topological properties of the visual scene, i.e., continuity and proximity affect the form of gesture, similar to the *iconic* deixis in our classification. Inclusion of this notion should contribute to the disambiguation in the *transitive* deixis category.

## 5.2 Syntactic Patterns

In human-to-human studies McNeill argues that gestures do not combine to form larger, hierarchical structures [1]. He asserts that most gestures are one to a clause, but when there are successive gestures within a clause, each corresponds to an idea unit in and of itself. In contrast to those findings the current results suggest that the gesture primitives (phonemes), not gestures as defined by McNeill, could be combined into a syntactic binding. The primitives were treated as self-contained parts with the meaning partially supplemented through the spatial context and the spoken counterpart. From the established patterns we can combine gesture phonemes that were mapped into the *intransitive* deixis, into the motion complexes (Fig. 6).

Since *transitive* deixis is inherently a part of the motion complex, which is likely to have a compounded structure, we distinguish two syntactic structures: *transitive* and *intransitive* motion complexes. In case of *intransitive* complex a gesture stroke conveying movement, e.g., *contour*, classified as *medial* is not usually followed by another stroke if the stroke precedes or occurs synchronously with a spoken keyword. E.g., "take this <point> car <contour> **out** of the lot". *Transitive* motion complex is characterized by a *medial* pre-stroke hold often followed by the final *pointing*. This formulation of spatio-temporal semantics of gesture links individual stroke and sequence of strokes used in "go form here to here" constructions. The emergence of the above patter is similar to SVO (subject-verb-object) structure proposed by Stokoe [16]. This system encapsulates both word-level and sentence-level structures.

This formulation became possible when our method considered semantic categorization of spatio-temporal properties of gestures (Fig. 2). Unlike McNeill's classification of gestures [1] it distinguishes between morphological (primitives) and functional characteristics (spatio-temporal context).

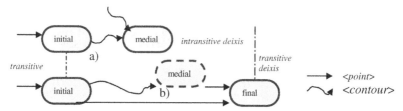

**Fig. 6.** Motion complexes of gestures and their relationships to the deixis categories: a) *Intransitive* complex: *initial* or *medial* gestures that precede or occurs synchronously with the speech do not result in the *final* stroke; b) *Transitive* complex: *medial* class that has a pre-stroke hold and eventually is followed by the *transitive* gesture (*point*)

Our method of combining phonological level and spatio-temporal context (semantic classification and co-occurrence models) enabled us to distinguish two levels of language structure in the co-verbal gestures: morphological and lexical. These findings are applicable in the context of spatial manipulation on a large display and may not have analogies in the face-to-face communication domain.

# 6  Conclusions

There is a great potential for exploiting the interactions between speech and gesture to improve HCI. A key problem in carrying out experiments for gesture speech recognition is the absence of truly natural data in the context of HCI. Our approach was to use the study of gesture and speech in the analogous domain of weather narration to bootstrap an experimental testbed (*i*MAP) achieving reasonably good recognition rates. This in turn provided the testbed for relevant user studies for understanding the gestures in the context of HCI. *i*MAP user studies indicated highly multimodal interaction. Results of the current study allow us to sketch mechanics of interpretation process when dealing with non-predefined gestures. Proposed semantic classification together with gesture speech co-occurrence allowed us to draw causal links of mapping gesture forms to their meaning. Current methodology of combining phonological level and spatio-temporal context (semantic classification and co-occurrence models) enabled us to distinguish morphological and lexical syntactically driven levels of gestural linguistic structure. In general, such systematic approach may provide a foundation for effective framework for multimodal integration in HCI.

### Acknowledgment

The financial support of the National Science Foundation (CAREER Grant IIS-00-81935 and Grant IRI-96-34618) and U. S. Army Research Laboratory (under Cooperative Agreement No. DAAL01-96-2-0003) is gratefully acknowledged.

# References

1. D. McNeill, *Hand and Mind.* The University of Chicago Press, Chicago, (1992).
2. A. Kendon. *Current issues in the study of gesture.* In J. L. Nespoulous, P. Peron, and A. R. Lecours, editors, The Biological Foundations of Gestures: Motor and Semiotic Aspects, Lawrence Erlbaum Assoc., (1986) 23—47.
3. A. De Angeli, L. Romary, F.Wolff. *Ecological interfaces: Extending the pointing paradigm by visual context.* P. Bouquet et al. (Eds.).:CONTEXT'99, LNAI 16888, (1999) 91—104.
4. R.A. Bolt. Put-that-there: Voice and gesture at the graphic interface. *In SIGGRAPH-Computer Graphics,* New York, ACM Press (1980)
5. S. Oviatt. *Multimodal interfaces for dynamic interactive maps.* In Proceedings of the Conference on Human Factors in Computing Systems (CHI'96), (1996) 95—102.
6. D. J. Sturman and D. Zeltzer. *A survey of glove-based input.* IEEE Computer Graphics and Applications, (1994) 14(1):30—39.
7. R. Sharma, J. Cai, S. Chakravarthy, I. Poddar and Y. Sethi. *Exploiting Speech/Gesture Co-occurrence for Improving Continuous Gesture Recognition in Weather Narration.* In Proc. International Conference on Faceand Gesture Recognition, Grenoble, France, (2000).
8. I. Poddar. *Continuous Recognition of Deictic Gestures for Multimodal Interfaces.* Master Thesis. The Pennsylvania State University, University Park, (1999).
9. V. I. Pavlovic, R. Sharma, and T. S. Huang. *Visual interpretation of hand gestures for human- computer interaction: A review.* IEEE Trans on Pattern Analysis and Machine Intelligence, (1997) 19(7):677—695.

10. R. Sharma, V. I. Pavlovic, and T. S. Huang. *Toward multimodal human-computer interface.* In Proceedings of the IEEE , (1998) 86(5):853—869.
11. S. Oviatt, A. De Angeli, and K. Kuhn. *Integration and synchronization of input modes during multimodal human-computer interaction.* In Proceedings of the Conference on Human Factors in Computing Systems (CHI'97), 95—102, ACM Press, New York, (1997) 415-422.
12. P. R. Cohen. *Synergic use of direct manipulation and natural language.* In Proc. Conference on Human Factors in Computing (CHI), (1989) 227—233.
13. M. Argule and M. Cook. *Gaze and Mutual Gaze.* Cambridge: Cambridge University Press, (1976).
14. S. Nobe, S. Hayamizu, O. Hasegawa, H. Takahashi. *Are listeners paying attention to the hand gestures of an anthropomorphic agent: An evaluation using a gaze tracking method.* Lecture Notes in Computer Science. Springer Verlag Heidelberg. (1998) 1371: 0049.
15. D.F. Armstrong, W.C. Stokoe, and S.E. Wilcox. Gesture and the nature of language. *Cambridge University Press* (1995).
16. Stokoe, W.C. Sign language structure: an outline of the communication systems of the American deaf. *Studies in Linguistics, Ocasional Papers,* 8 (1960)
17. R.W. Langacker. Foundation of cognitive grammar. *Stanford University Press,* V.2 (1991).
18. A. Kendon. *Conducting Interaction.* Cambridge: Cambridge University Press (1990).
19. Stokoe, W.C. Semantic phonology. *Sign Language Studies* (1991) 71:107-114.
20. S. Kita, I.V. Gijn, and H.V. Hulst. *Movement phases in signs and co-speech gestures, and their transcription by human coders.* In Proceedings of Intl. Gesture Workshop, (1997) 23—35.
21. L. Talmy. *How language structures space.* Spatial Orientation: Theory, Research and Application. In Pick and Acredolo (Eds.), Plenum Publishing Corp., NY, (1983).
22. B. Dasarathy. Sensor fusion potential exploitation - innovative architectures and illustrative approaches. Proc. IEEE, 85(1):24;38, January 1997.
23. D. Salber and J. Contaz. Applying the wizard-of-oz techniques to the study of multimodal systems. In *EWHCI'93,* Moscow, Russia, 1993.
24. Y. Azoz, L. Devi, and R. Sharma. *Reliable tracking of human arm dynamics by multiple cue integration and constraint fusion.* In  Proc. IEEE Conference on Computer Vision and Pattern Recognition, (1998).
25. S. Kettebekov and R. Sharma. Understanding gestures in multimodal human computer interaction. *International Journal on Artificial Intelligence Tools,* (2000) 9:2:205-224.

## Discussion

*L. Bass:* You derived all this data from looking at the weather channel. Did you validate this from a different source of gestures, since it seems that different channels have different styles? The styles are actually taught to the weather readers by the networks.

*S. Kettebekov:* Yes, we did look at different stations and found that the presentation on the Weather Channel included more diectic gestures and fewer "noise" gestures. Since we were just using the weather channel to find the gestural primitives this didn't seem like a significant problem.

*W. Stuerzlinger:* In the weather channel the presenter is not interacting directly with the screen, which is quite different from your application domain. Did you have a problem with 2D versus 3D gestures?

*S. Kettebekov:* In our system there is a cursor on the screen which changes shape as the user gestures. The users rapidly learned to use this, which made the results more consistent. Of course we're trying to minimize the learning.

*J. Williams:* You use key words from the speech to aid in interpretation of the gestures. Have you considered just using the tone of the voice rather than the actual speech?'
*S. Kettebekov:* In order to really correllate keywords with the gestures you have to go all the way to the semantic level. However for the prosodic features you can just use tone. We are just beginning to look at this.

*S. Greenberg:* Can you explain how you came up with the numbers in your studies?
*S. Kettebekov:* We structured the study according to the categories listed in the paper. We gave them tasks involving fairly lengthy narrations and studied the gestures. Each narration was between 20 and 30 minutes which is why we only needed seven subjects to gather a significant data set.

*J. Coutaz:* In Grenoble we have done a finger tracking task on a whiteboard and found an ambiguity between pointing for the system and pointing to explain something to another person. Was this a factor in your study?
*S. Kettebekov:* In a narration task it's easier to observe the feedback. The on-screen cursor probably helped too.

*B. Miners:* Problems in continuous gesture recognition include segmentation of consecutive morphemes. You mentioned use of continuous gesture. Were any coarticulation or segmentation issues encountered?
*S. Kettebekov:* Not in our observations of the weather domain.

*B. Miners:* Does this mean that segmentation did not occur at the gesture level but relied on voice keywords.
*S. Kettebekov:* Right.

# An Evaluation of Two Input Devices
# for Remote Pointing

I. Scott MacKenzie[1] and Shaidah Jusoh[2]

[1] Department of Mathematics & Statistics
York University, Toronto, Ontario, Canada M3J 1P3
mack@yorku.ca
[2] Department of Computer Science
University of Victoria, Victoria, British Columbia, Canada V8W 3P6
shaidah@csc.uvic.ca

**Abstract.** Remote pointing is an interaction style for presentation systems, interactive TV, and other systems where the user is positioned an appreciable distance from the display. A variety of technologies and interaction techniques exist for remote pointing. This paper presents an empirical evaluation and comparison of two remote pointing devices. A standard mouse is used as a base-line condition. Using the ISO metric throughput (calculated from users' speed and accuracy in completing tasks) as the criterion, the two remote pointing devices performed poorly, demonstrating 32% and 65% worse performance than the mouse. Qualitatively, users indicated a strong preference for the mouse over the remote pointing devices. Implications for the design of present and future systems for remote pointing are discussed.

## 1 Introduction

Pointing operations are fundamental to human interaction with graphical user interfaces (GUI). The most common pointing device is the mouse, but other devices are also used, such as trackballs, joysticks, or touchpads. Pointing is the act of moving an on-screen tracking symbol, such as a cursor, by manipulating the input device. The tracker is placed over text, icons, or menu items, and actions are selected by pressing and releasing a button on the pointing device.

Remote pointing is an emerging variation of mouse pointing whereby the user is positioned an appreciable distance from the display. Common applications are presentation systems and interactive TV.

This paper presents an empirical evaluation of two remote pointing devices. We begin by describing applications and two representative devices chosen for evaluation. Following this, the method and results are presented.

### 1.1 Presentation Systems

For *presentation systems*, a typical setup includes a notebook computer and a projection system driven by the computer's video output signal. The user interacts

M. Reed Little and L. Nigay (Eds.): EHCI 2001, LNCS 2254, pp. 235–250, 2001.

with the system using the built-in pointing device of the notebook computer, but this displaces the presenter's focus of attention and can diminish the impact of the presentation. Alternatively, a remote pointing device is substituted, allowing the presenter to engage the audience more directly. Pointing operations, when necessary, are performed without physically moving to the notebook computer to acquire the system's built-in pointing device. A variety of technologies are available for remote pointing, and we will present some of these later. The general idea of remote pointing for presentation systems is illustrated in Fig. 1.

**Fig. 1.** Remote pointing for presentation systems

## 1.2    Interactive TV

We use the term *interactive TV* with reference to a variety of current and anticipated developments for home computing and entertainment. There is considerable debate on technology convergence in telephone, cable, and internet services, and it is not our intent to enter into this forum here. A reasonable assumption, however, is that home entertainment systems will change substantially in the near future. New applications are anticipated, such as web browsing, email, home banking, travel reservations, and so on. All of these require an input mechanism more sophisticated than a typical remote control. Today's remote control units are optimized for *selecting* (via buttons), but are less capable of *choosing*, *pointing*, or *entering*. Because of the rich task space, a new interaction paradigm is likely to emerge — the *TV-GUI* perhaps. Regardless of the form, it is likely that pointing tasks will be part of the interaction. Indeed, remote controls are now available with built-in trackballs or isometric joysticks, although a standard on-screen interface has not emerged. For interactive TV, therefore, we are interested in investigating a variety of interaction issues, and, in particular, the mechanism of pointing. This is illustrated in Fig. 2.

**Fig. 2**. Remote pointing for interactive TV

Related to the anticipated insurgence of devices for remote pointing, is the need to empirically test the devices and interaction techniques that users will engage. Will they afford facile interaction or will they encumber the user with an awkward, slow, and error prone interface? Before presenting our empirical study, we describe the technology and devices used.

## 2    Devices for Remote Pointing

Two devices were chosen for this study. They use different technologies and, therefore, are good choices since they represent different points in the design space. Both function as mouse replacement devices and do not require special interface hardware. They interact with the system's installed mouse driver. Although these are intended for presentations, the interaction styles also apply to interactive TV.

### 2.1    GyroPoint

The *GyroPoint* is a product of Gyration, Inc. (Saratoga, CA). The device has two distinctly different modes of operation. First, it can function as a regular mouse. It includes a ball mechanism and can operate on a mousepad in the traditional manner. Second, it includes a solid-state gyroscope, permitting operation in the air. For the latter, the angular movement of the hand/device side-to-side or up-and-down maps to *x-y* tracker motion on the system's display.

According to the manufacturer, "a unique electromagnetic transducer design and a single metal stamping utilize the Coriolis effect to sense rotation. Analog voltages proportional to angular rates around the two sensed axes are provided relative to a voltage reference output" [5]. The technology has several applications, including computer pointers, TV remote controllers, robotics, factory automation, antenna stabilization, and auto navigation.

There are corded and cordless versions of the *GyroPoint*. The corded version sells for less than one hundred dollars. The cordless version sells for several hundred dollars. In this study we used the corded version.

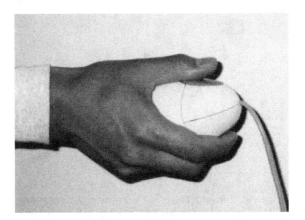

**Fig. 3.** A user holding the *GyroPoint*

When operated on the desktop, there are two buttons in the usual position. Two additional buttons for operation in the air are located on the sides near the front of the device. The usual grip for air operation is illustrated in Fig. 3. The grey button under the thumb is the primary button for selection. Another button on the opposite side is operated by the index finger and acts as a clutch. When depressed, rotary motion of the device by the wrist, forearm, and arm maps to $x$-$y$ motion of the tracker. Motion is relative, so the clutching action of the index finger is the same as lifting and repositioning a mouse on a mousepad.

## 2.2    RemotePoint

The *RemotePoint* is a product of Interlink Electronics (Camarillo, CA) [6]. It is a cordless pointing device with a built-in isometric joystick, similar to the "eraser tip" joystick on a notebook computer. The device includes an infrared transmitter that communicates with a base receiver at distances up to 40 feet. The base receiver plugs into the system's mouse port. It is priced under one hundred dollars.

The device is held with the thumb positioned over the joystick which has a large, round rubber casing. The primary button is pressed by the index finger in a trigger-like fashion (see Fig. 4).

Users may position the *RemotePoint* and *GyroPoint* to point at the system's display, but this is not a requirement of the devices per se.

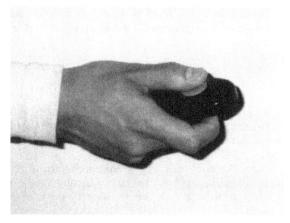

**Fig. 4**. A user holding the *RemotePoint*

# 3    ISO Testing of Pointing Devices

All pointing devices are not created equal. Nor will they perform equally. The evaluation of a device's performance is tricky at best since it involves human subjects. (This is in contrast to, say, the performance evaluation of system hardware using standardized benchmarks.) Although there is an abundance of published evaluations of pointing devices, the methodologies are ad hoc. Experimental procedures are inconsistent from one study to the next, and this greatly diminishes our ability to understand or generalize results, or to undertake between-study comparisons. As a consequence, we have much to examine, but we are in a quandary on what it means.

Fortunately, there is a recent standard from the International Standards Organization that addresses this particular problem. The full standard is ISO 9241, *Ergonomic design for office work with visual display terminals (VDTs)*. The standard is in seventeen parts. Part 9 of the standard is called *Requirements for non-keyboard input devices* [8].

ISO 9241-9 describes a battery of tests to evaluate computer pointing devices. The procedures are well laid out and, if followed, will result in a strong and valid performance evaluation of one or more pointing devices.

The basic quantitative test is the serial point-select task (see [13] for an example). The user manipulates the on-screen tracker (viz., cursor) using the pointing device and moves it back-and-forth between two targets, and selects the targets by pressing and releasing a button on the device. A serial task is used because it is easy to implement and affords rapid collection a large quantity of empirical data. The selections are blocked within, say, 20 back-and-forth selections per task condition. As the task is carried out, the test software gathers low-level data on the speed and accuracy of user actions. Three dependent measures form the basis of the subsequent quantitative evaluation: movement time, error rate, and throughput.

*Movement time* (*MT*), or task completion time, is the mean time in milliseconds for each trial in a block of trials. Since the end of one trial is the beginning of the next, the mean is simply the total time for a block of trials divided by the number of trials. *Error rate (ER)* is the percentage of targets selected while the tracker is outside the target.

## 3.1    Throughput

*Throughput (TP)* is a composite measure of both the speed and accuracy of performance. The measure was introduced by Fitts in 1954 [4], and it has been widely used in human factors and experimental psychology ever since.[1] See [9, 18] for extensive reviews.

Throughput is a very powerful measure. Unlike movement time or error rate, throughput is relatively *independent* of the task difficulty or the range of task difficulties in the experiment. (In fact, this is the very thesis upon which Fitts' original work was based.)

Throughput is calculated as follows:

$$Troughput = \frac{ID_e}{MT} \tag{1}$$

where

$$ID_e = \log_2\left(\frac{D}{W_e} + 1\right). \tag{2}$$

The term $ID_e$ is the effective index of difficulty, and carries the unit "bits". It is calculated from $D$, the distance to the target, and $W_e$, the effective width of the target. Since $MT$, or movement time, carries the units "seconds", throughput carries the units "bits per second", or just "bps".

The use of the *effective* width $(W_e)$ is important. $W_e$ is the width of the distribution of selection coordinates computed over a block of trials. Specifically,

$$W_e = 4.133 \times SD_x \tag{3}$$

where $SD_x$ is the standard deviation in the selection coordinates measured along the axis of approach to the target. Thus, $W_e$ captures the spatial variability or accuracy in a block of trials. As a result, throughput is a measure of both the speed and accuracy of the user. In a sense, throughput reflects the overall efficiency with which the user was able to accomplish the task given the constraints of the device or other aspects of the interface.

It is important to test the device on difficult tasks as well as easy tasks; so, multiple blocks of trials are used, each with a different target distance and/or target width.

---

[1] Fitts used the term *index of performance* instead of throughput. The term *bandwidth* is also used to imply throughput.

# 4    Method

## 4.1    Participants

Our study included twelve paid volunteer participants, composed of nine males and three females. Eleven of the participants were students at a local university; one was a member of the staff. All participants used computers with a GUI and mouse on a daily basis. None had prior experience with remote pointing devices.

## 4.2    Apparatus

The experiment was conducted on a PC-class desktop computer with a 15" VGA display. The experiment used the *Generalized Fitts' Law Model Builder* software [17] to present the tasks and capture the data from four pointing devices. The following devices were used:

    *1 GyroPoint*-air by Gyration
    *2 GyroPoint*-desk by Gyration
    *3 RemotePoint* by Interlink Electronics
    *4 Mouse 2.0* by Microsoft

The *GyroPoint* was treated as two devices, since it was tested in each of its two modes: in the air and on the desktop. By having the same device operated in two ways, differences between the "air" and "desk" conditions can be more narrowly attributed to the interaction technique (air vs. desk) rather than to the ergonomics of the device. We also included a Microsoft *Mouse 2.0* in the study, for a similar reason. By including a standard desktop pointing device, we have a "baseline" condition. This is important, because we want not only to evaluate and compare the performance of remote pointing devices, but to provide context for our results in the highly tested world of desktop pointing.

So, of the four device conditions, two are examples of remote pointing (*GyroPoint*-air and *RemotePoint*) and two are examples of desktop pointing (*GyroPoint*-desk and *Mouse 2.0*).

## 4.3    Procedure

A simple point-select task was used, conforming to the pointing test in Section B.6.1.1 of ISO 9241-9 [6]. The task was explained and demonstrated to participants and a warm-up block of trials was given. In each trial, two targets of width *W* separated by distance *D* appeared on the display. A cross-hair tracker appeared in the left target and a purple X appeared in the right target. Participants moved the cross-hair tracker by manipulating the pointing device. The goal was to move the tracker back and forth between the targets and alternately select the targets by pressing and releasing the primary device button.

An error occurred if the centre of the tracker was outside the target when the button was pressed. An error was accompanied by an audible beep. As each selection occurred the purple X moved to the opposite target to help guide the participant through a block of trials. For each distance/width condition, 20 back-and-forth selections were performed. Participants were instructed to proceed as quickly as possible while trying to minimize errors. Approximately one error per block of 20 trials (5%) was considered acceptable. Participants could rest between blocks of trials at their discretion.

For the two desktop device conditions (*GyroPoint*-desk and *Mouse 2.0*), participants were seated in front of the system as customary with desktop computers. For the two remote pointing conditions (*GyroPoint*-air and *RemotePoint*), participants stood in front of the system at a distance of 1.5 meters (see Fig. 5).

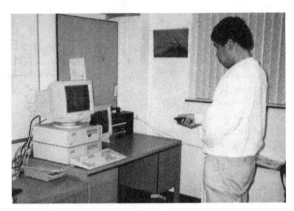

**Fig. 5.** Operating the *RemotePoint* while standing in front of the display

The operation of the remote pointing devices was explained and demonstrated. Participants held the device in their hand and manipulated it as noted earlier. For the *RemotePoint*, the base station was positioned beside the display; thus the device was pointing roughly at the display during the trials. For the *GyroPoint*-air, the device operates in relative mode, so it need not point at the display for normal operation. However, as participants also had to visually monitor the display, they held the device with it pointing at the display.

### 4.4    Design

The experiment was a $4 \times 3 \times 3 \times 4 \times 20$ repeated measures factorial design. The factors and levels were as follows:

Device      *GyroPoint*-air, *GyroPoint*-desk, *RemotePoint*, *Mouse 2.0*
Distance    40 mm, 80 mm, 160 mm
Width       10 mm, 20 mm, 40 mm
Block       1, 2, 3, 4
Trial       1, 2, 3 ... 20

To balance for learning effects, participants were randomly assigned to one of four groups. Each group received the device conditions in a different order using a Latin square.

For each device condition, participants performed 4 blocks of trials. Each block consisted of the 9 target distance/width conditions presented in random order. For each target distance/width condition, 20 trials were performed. Thus, a block consisted of $9 \times 20 = 180$ trials. Participants were able to perform all four device conditions in about one hour. The conditions above combined with 12 subjects resulted in 34,560 total trials in the experiment.

The nine target distance/width conditions were chosen to cover a range of task difficulties. The easiest task combined the shortest distance (40 mm) with the widest target (40 mm). The index of difficulty was

$$ID = \log_2\left(\frac{D}{W}+1\right) = \log_2\left(\frac{40}{40}+1\right) = 1.0 \text{ bits}. \tag{4}$$

The hardest task combined the largest distance (160 mm) with the narrowest target (10 mm):

$$ID = \log_2\left(\frac{D}{W}+1\right) = \log_2\left(\frac{160}{10}+1\right) = 4.1 \text{ bits}. \tag{5}$$

The dependent measures were movement time (ms) error rate (%), and throughput (bps).

At the end of the experiment, participants were interviewed and asked to complete a questionnaire. Comments were sought on their subjective impressions of the four device conditions.

# 5    Results and Discussion

The grand means on the three dependent measures were 957 ms for movement time, 2.6% for error rate, and 3.0 bps for throughput.

It is possible that even with counter balancing asymmetrical skill transfer effects may occur across device conditions [14]. This would surface as a significant main effect for *order of presentation*, that being the group to which participants were assigned. This was tested for and found not to have occurred as the group effect was not statistically significant on each dependent measure.

The main effects and interactions on each dependent measure are presented in the following sections.

## 5.1    Speed

The *GyroPoint*-desk was the fastest device condition with a mean movement time of 598 ms. The other device conditions were slower: by 11% for the *Mouse 2.0* (666 ms), by 56% for the *GyroPoint*-air (930 ms), and by 173% for the *RemotePoint* (1633 ms). Thus, the two remote pointing conditions were substantially slower than the two desktop conditions. The differences were statistically significant ($F_{3,24} = 199.4$, $p < .0001$). The main effect for block was also significant ($F_{3,24} = 20.7$, $p < .0001$), as was the device × block interaction ($F_{9,72} = 3.1$, $p < .05$). The main effects and interaction are illustrated in Fig. 6.

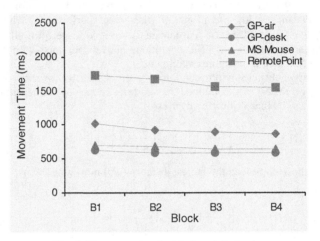

**Fig. 6.** Movement time by block and device

The very long movement time for the *RemotePoint* is an indication that participants had a difficulty in controlling the movement of the cursor with this device. Problems with finger tremor in controlling isometric joysticks have been noted before [15], so the poor showing here is not surprising.

## 5.2    Accuracy

The most accurate device was the *RemotePoint* with an error rate of 1.6%. It was followed by the *Mouse 2.0* at 2.4% errors, the *GyroPoint*-desk at 2.7% errors, and the *GyroPoint*-air at 3.5% errors. On the whole, these error rates are quite low. The main effect of device on error rate was significant ($F_{3,24} = 8.6$, $p < .0005$), but the main effect of block was not ($F_{3,24} < 1$, ns). The device × block interaction was significant ($F_{9,72} = 3.0$, $p < .05$). The main effects and interaction are illustrated in Fig. 7.

Standard deviation bars are omitted from Fig. 7 to avoid clutter. However, the lack of statistical significance in the block effect suggests that an interpretation of the trends illustrated over the four blocks in Fig. 7 is not warranted.

The primary observation for accuracy is on the significant main effect of device and the very good showing of the *RemotePoint*. The latter is surprising given the dismal performance on movement time. It is possible that participants proceeded with great caution, given their inability to move expeditiously with this device. Since movement time was much longer, perhaps the very slow positioning tended to make selection — once the cursor was finally positioned inside the target — less error prone.

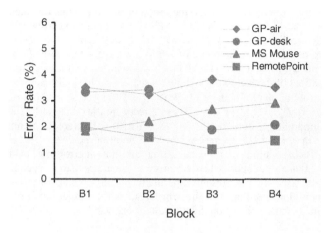

**Fig. 7.** Error rate by block and device

## 5.3    Throughput

Throughput is an important dependent measure because it combines speed and accuracy in a single metric, as noted earlier. Thus, tendencies to trade speed for accuracy or vice versa will not necessarily yield differences in throughput. The highest throughput was observed for the *GyroPoint*-desk at 4.1 bps. The other devices exhibited lower throughputs by 10.4% for the *Mouse 2.0* (3.7 bps), by 32% for the *GyroPoint*-air (2.8 bps), and by 65% for the *RemotePoint* (1.4 bps).

These measures are in line with others we have observed. Although there are numerous figures for throughput in the literature, very few were calculated as per the ISO 9241-9 standard, and, therefore, comparisons are difficult. For example, a 1991 study [13] reported throughput for the mouse; however, it was obtained from a regression model of the following form:

$$MT = a + b \times ID_e .\tag{6}$$

The model includes coefficients for the intercept, $a$ in ms, and the slope, $b$ in ms/bit, reported as $a = -107$ ms and $b = 223$ ms/bit. Throughput was reported as the slope reciprocal, namely $1 / 223 = 4.5$ bps. Importantly, the calculation of $ID_e$ included the adjustment noted earlier for the effective target width (see equations 2 and 3). However, the presence of a non-zero intercept tends to weaken the comparison. According to ISO 9241-9, throughput is obtained from the division of means (see equation 1), not from the slope reciprocal in a regression model. All else being equal, the two calculations should yield reasonably similar results provided the intercept is zero, or close to zero.

In our empirical studies, we have recently standardized our calculation of throughput to conform to the ISO standard. As a result, a payoff is now appearing. The payoff is the ability to compare results across studies with confidence that the comparison is "apples with apples".

We consistently obtain measures in the range of 3.0 to 5.0 bps for mice [e.g., 10]. Obtaining a similar figure herein is like a "self check" on the experimental apparatus, procedures, data collection, analysis, etc. We also consistently find lower figures for other pointing devices — in the range of 2.6 to 3.1 bps for trackballs [10], 1.0 to 2.0 bps for touchpads [3, 11], and 1.6 to 2.6 bps for isometric joysticks [3, 16].

The main effect of device on throughput was statistically significant ($F_{3,24} = 91.2$, $p < .0001$), as was the main effect of block ($F_{3,24} = 18.7$, $p < .0001$) and the device × block interaction ($F_{9,72} = 2.4$, $p < .05$). These effects are illustrated in Fig. 8.

The most important observation in Fig. 8 is that the two remote pointing devices faired poorly in comparison to the two desktop pointing devices. These trends are consistent with observations on movement time, but not on error rate. Although, this might suggest that throughput is less sensitive to accuracy than to speed, this is not necessarily the case. In the present experiment the differences in movement time were much more dramatic that those in error rates; so, the similar rank ordering of the movement time results to the throughput results is expected.[2]

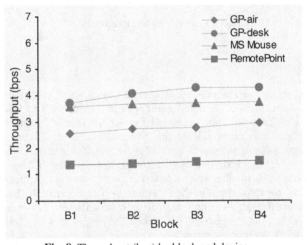

**Fig. 8.** Throughput (bps) by block and device

## 5.4    Qualitative Results

Participants were surveyed for their impressions and their perceived performance. They completed a questionnaire on ease of use, giving each device a rating from 1 (very difficult) to 5 (easy). The results are shown in Table 1. All participants gave the *Mouse 2.0* and the *GyroPoint*-desk the top rating: "easy to use". Nine of twelve participants rated the *RemotePoint* either a 1 "very difficult" or 2 "difficult". The *GyroPoint*-air faired reasonably well with nine participants rating its ease of use as either "neutral" or "slightly easy".

---

[2] Note that throughput is calculated from movement time and "spatial variability". Error rates per se, although highly correlated with spatial variability, are not used in the calculation of throughput.

**Table 1.** Results of Questionnaire on Ease of Use

| | | Device | | |
| Participant | Mouse 2.0 | GyroPoint desk | GyroPoint air | RemotePoint |
|---|---|---|---|---|
| 1 | 5 | 5 | 4 | 2 |
| 2 | 5 | 5 | 3 | 2 |
| 3 | 5 | 5 | 3 | 4 |
| 4 | 5 | 5 | 4 | 3 |
| 5 | 5 | 5 | 3 | 3 |
| 6 | 5 | 5 | 2 | 1 |
| 7 | 5 | 5 | 2 | 1 |
| 8 | 5 | 5 | 4 | 2 |
| 9 | 5 | 5 | 2 | 1 |
| 10 | 5 | 5 | 3 | 2 |
| 11 | 5 | 5 | 4 | 1 |
| 12 | 5 | 5 | 3 | 1 |
| Mean | 5.0 | 5.0 | 3.1 | 1.9 |

Note: 1 = easy, 2 = slightly easy, 3 = neutral, 4 = difficult, 5 = very difficult

Although the learning trends over the four blocks given in the preceding sections are not dramatic (due to the simplicity of the task), all participants indicated it took effort to get accustomed to the *GyroPoint*-air and the *RemotePoint*. Given that participants had substantial mouse experience but no previous remote pointing experience, it is possible that an extended study would reveal less difference between the devices after prolonged practice.

Participants were also asked to rate their performance with each device, and they all recognized that they took the longest with the *RemotePoint* and were quickest with the *Mouse 2.0* or the *GyroPoint*-desk.

## 5.5 Experiment and Interaction Considerations

Ideally, in experimental research one tries to hold all variables constant except those under investigation. In this experiment, we had difficulty in achieving this because remote pointing and desktop pointing, by their very nature and by the representative devices chosen, involve a complex shift in interaction styles and psychomotor issues. For example, two devices were operated while seated, two while standing. For three devices, the primary button was activated by the index finger, whereas for one device (*GyroPoint*-air) the button was activated by the thumb. Three devices used position control in which tracker motion was effected by motion of the wrist, forearm, and arm, whereas one device (*RemotePoint*) used velocity control in which motion was effected by force of the thumb.

The use of the corded version of the *GyroPoint* should be considered. Although this may have compromised performance for the air condition, we received no comments on this during the interview.

While we do not feel the issues above impact our overall results or conclusions, it is important to acknowledge the inherit limitations in the experimental design. As remote pointing evolves, further experimentation is warranted in which issues are investigated in isolation (e.g., standing vs. sitting).

Finally, the control-display relationship is altered by the distance of the user from the display. For the seated conditions the eye-to-display distance was about 0.75 meters, for the standing position the distance was 1.5 meters. Thus the "same" targets appeared smaller for the standing position. There is an abundance of inconclusive research on the human performance impact of control-display relationships (e.g., [1, 2, 12]). In one study, viewing distance was actually controlled over a range of 1.5 to 8 meters [7]; however, no difference was found in the operating characteristics of three remote pointing devices. So, we do not feel this is an issue in the present study. For distances greater than a few meters however, display size and the size and form of the tracker are certainly issues to be explored further.

## 6   Conclusions

This study has shown that two representative devices for remote pointing performed poorly in comparison to a standard mouse. Although the *RemotePoint* had the lowest error rate, it was by far the slowest of the four devices tested and had the lowest throughput, measured as per the ISO standard on pointing devices. Furthermore, subjects gave this device the lowest subjective rating. The *GyroPoint*, while operated in the air, faired slightly better than the *RemotePoint*, but was favoured much less than a standard mouse. These results suggest that remote pointing, while a requirement of some systems, needs further development to support facile interaction.

### Acknowledgements

We thank Shawn Zhang for creating the two concept sketches, and Aleks Oniszczak for helpful discussions while we were designing the experiment. This research is funded by the Natural Sciences and Engineering Research Council (NSERC) of Canada.

## References

1.  Arnault, L. Y., and Greenstein, J. S. Is display/control gain a useful metric for optimizing an interface? *Human Factors 32* (1990) 651-663
2.  Buck, L. Motor performance in relation to control-display gain and target width, *Ergonomics 23* (1980) 579-589
3.  Douglas, S. A., Kirkpatrick, A. E., and MacKenzie, I. S. Testing pointing device performance and user assessment with the ISO 9241, Part 9 standard, *Proceedings of the ACM Conference on Human Factors in Computing Systems - CHI '99*. New York: ACM (1999) 215-222
4.  Fitts, P. M. The information capacity of the human motor system in controlling the amplitude of movement, *Journal of Experimental Psychology 47* (1954) 381-391
5.  Gyration, Inc., Saratoga, California. http://www.gyration.com/.
6.  Interlink Electronics, Inc., Camarillo, California. http://www.interlinkelec.com/.

7.  Ishiyama, K., and Yano, S. A study of characteristics of pointing devices for television operation, *IEEE Conference on Systems, Man, and Cybernetics - SMC 2000*. New York: IEEE (2000) 1307-1312

8.  ISO. Report Number ISO/TC 159/SC4/WG3 N147: *Ergonomic requirements for office work with visual display terminals (VDTs) - Part 9 - Requirements for non-keyboard input devices* (ISO 9241-9), International Organisation for Standardisation (May 25, 1998)

9.  MacKenzie, I. S. Fitts' law as a research and design tool in human-computer interaction, *Human-Computer Interaction 7* (1992) 91-139

10.  MacKenzie, I. S., Kauppinen, T., and Silfverberg, M. Accuracy measures for evaluating computer pointing devices, *Proceedings of the ACM Conference on Human Factors in Computing Systems - CHI 2001*. New York: ACM (in press)

11.  MacKenzie, I. S., and Oniszczak, A. A comparison of three selection techniques for touchpads, *Proceedings of the ACM Conference on Human Factors in Computing Systems - CHI '98*. New York: ACM (1998) 336-343

12.  MacKenzie, I. S., and Riddersma, S. Effects of output display and control-display gain on human performance in interactive systems, *Behaviour & Information Technology 13* (1994) 328-337

13.  MacKenzie, I. S., Sellen, A., and Buxton, W. A comparison of input devices in elemental pointing and dragging tasks, *Proceedings of the ACM Conference on Human Factors in Computing Systems - CHI '91*. New York: ACM (1991) 161-166

14.  Martin, D. W. *Doing psychology experiments*, 4th ed. Pacific Grove, CA: Brooks/Cole (1996)

15.  Mithal, A. K., and Douglas, S. A. Differences in movement microstructure of the mouse and the finger-controlled isometric joystick, *Proceedings of the ACM Conference on Human Factors in Computing Systems - CHI '96*. New York: ACM (1996) 300-307

16.  Silfverberg, M., MacKenzie, I. S., and Korhonen, P. Predicting text entry speed on mobile phones, *Proceedings of the ACM Conference on Human Factors in Computing Systems - CHI 2000*. New York: ACM (2000) 9-16

17.  Soukoreff, W., and MacKenzie, I. S. Generalized Fitts' law model builder, Companion *Proceedings of the ACM Conference on Human Factors in Computing Systems - CHI '95*. New York: ACM (1995) 113-114

18.  Welford, A. T. *Fundamentals of skill*, London: Methuen (1968)

## Discussion

*N. Roussel:* Have you considered two handed input devices (devices that you hold in one hand and operate with the other)? For example like the input devices used for video games? This would fit well with kinematic chain theory?

*S. MacKenzie:* Yes, but we haven't actually evaluated such devices yet. On the other hand we have evaluated systems where we use one hand for motion and the other hand for selection.

*P. Smith:* This task uses targets defined only by width. Can you conclude that one device is better than another overall given that vertical movement was not evaluated.

*S. MacKenzie:* This is a one dimensional task. There are also two dimensional tasks in the ISO standard however. Our experience suggests that performance on two dimensional tasks is even worse, especially for non-mouse pointing devices.

*S. Greenberg:* It's nice to see such careful work. However, your work seems to suggest that we should use mice rather than these other kinds of pointing devices.

That doesn't take into account contextual issues like the fact that when giving a presentation the act of pointing itself provides important social clues. Another issue: does the ISO standard look at other features of pointing devices like the ability to write or draw with the device, or to select in the presence of jitter?

*S. MacKenzie:* Yes, there are six different tasks which also include drawing, plus some biomechanical measures and an eleven-item questionaire. The standard includes much more than just Fitts law which we agree is inherently limited.

*R. Jezek:* Based on your experience with pointing devices, is there any influence of the driver software which might limit or expand the possibilities of the input device?

*S. MacKenzie:* We've done some experiments where we controlled driver parameters using standard control panel features and found that within limits it doesn't matter, but at the extremes throughput degrades quite noticeably.

*L. Nigay:* How did you address the problem of the fact that the users were experts with the mouse but not with the other devices?

*S. MacKenzie*: That can be a serious problem, but here the tasks were very very simple. Also, we observed very little learning across the total of 120 trials. We believe that the user's expertise coming into the task was not a significant factor.

# Does Multi-modal Feedback Help
# in Everyday Computing Tasks?

Carolyn MacGregor[1] and Alice Thomas[2]

[1] Dept. Systems Design Engineering, University of Waterloo
Waterloo, Ontario Canada N2L 3G1
cgmacgre@engmail.uwaterloo.ca
[2] 724 Solutions Inc., Toronto, Ontario Canada M2P 2G4
athomas@724.com

**Abstract.** A study was conducted to investigate the effects of auditory and haptic feedback for a "point and select" computing task at two levels of cognitive workload. Participants were assigned to one of three computer-mouse haptic feedback groups (regular non-haptic mouse, haptic mouse with kinesthetic feedback, and haptic mouse with kinesthetic and force feedback). Each group received two auditory feedback conditions (sound on, sound off) for each of the cognitive workload conditions (single task or dual task). Even though auditory feedback did not significantly improve task performance, all groups rated the sound-on conditions as requiring less work than the sound-off conditions. Similarly, participants believed that kinesthetic feedback improved their detection of errors, even though mouse feedback did not produce significant differences in performance. Implications for adding multi-modal feedback to computer-based tasks are discussed.

## 1    Introduction

Within the traditional framework of human-computer interaction (HCI) feedback to users for the purposes of navigation and interaction have been provided primarily via the visual mode. For example, cues for cursor position, current application status, and the user's most recent keyboard or mouse entries are generally obtained when the user looks at the screen. In our high-paced world, many computer users regularly divide their visual attention between more than one software application at a time. To assist with this visual juggling, the use of multi-modal feedback has been proposed as a means of reducing the load on the visual modality so as to allow the user to work more efficiently [1, 2]. The general thinking is that we can capitalize on our ability to process information through more than one modality without significantly impairing performance on the primary visual task. In HCI, the term multi-modal feedback is used to refer to feedback sources other than those provided visually via a computer monitor. The more common sources of alternative feedback are through touch (haptic feedback) and sound (auditory feedback).

While computing technology offers opportunities to provide feedback to users via other modes (e.g. hearing, touch), whether such feedback actually enhances human performance across computing tasks is not well understood. This research set out to

M. Reed Little and L. Nigay (Eds.): EHCI 2001, LNCS 2254, pp. 251–262, 2001.
© Springer-Verlag Berlin Heidelberg 2001

address this issue - specifically whether the addition of haptic and auditory feedback, separately or combined, enhances performance on a basic "point and click" computing task.

## 2    Providing Multi-modal Feedback in Computing Tasks

Besides the keyboard, the computer mouse is the most common input device used in HCI activities. While the user may employ the mouse to manipulate objects on the screen, he or she often must look at the screen to confirm an action due to the limited feedback provided via the mouse itself.    There are some mouse-like input devices, such as Control Advancements Inc.'s Virtual Reality Mouse™ (VRM), and SensAble Technologies' PHANToM® haptic interface, that can provide haptic feedback through the user's sense of touch. All computing mice provide some haptic feedback in that the user physically moves the cursor by touching the mouse.  Current use of the term haptic with respect to input/output devices typically refers to some physical sensation provided directly back to the user's hand or fingers via force feedback in the form of vibration or resistance.

In addition to force feedback, some haptic devices, such as the VRM, provide kinesthetic feedback to the user. The VRM provides the user with a sense of relative cursor position on the screen by physically restricting mouse and user hand movements to a planar workspace on the physical desktop corresponding to a fixed frame of reference on the computer screen.  With a traditional mouse, the user can pick the mouse body up and reposition it when it gets to the edge of the mouse pad. With the VRM, the mouse is mounted on a sliding metal arm that is attached to a rail housed in the VRM control box.  In addition to the kinesthetic feedback, the VRM can deliver force feedback via servomotors that create the sensation of increased friction and gravity wells as the mouse passes over screen widgets such as icons or window "edges" [3].    The VRM was originally developed to aid blind and low-visioned users in accessing and interpreting GUI interfaces. Given that devices such as the VRM are relatively new to the marketplace, the potential impact they can have in terms of user performance for sighted users is not well understood.

Akamatsu and Sate report that the addition of tactile and force feedback shortens reaction time in a target selection task [4].  However, in their study the sole task of the user was to perform the target selection task.   The benefits of tactile and force feedback as a way to reduce visual load were not specifically tested.  More recently, Oakley and colleagues investigated the use of the PHANToM® haptic device as a potential interface for reducing visual load [5].    They hypothesized that by augmenting a regular desktop interface with haptic effects participants would be able to perform scrolling and reading tasks faster, with fewer errors, and with less workload than without the haptic effects.  While they did not find any significant differences in terms of their objective measures (time and error), their participants did report experiencing a reduced workload with the haptic effects.  In both of the studies cited, the focus was on the effects of haptic feedback and additional modes of feedback, such as auditory cues, were not included in the investigation.

While haptic interfaces are still in their infancy, we have had the ability to include auditory feedback in computer-based tasks for some time. The fact that one can attend to auditory feedback without having to look at the screen suggests that strategically used auditory cues should allow a user to visually monitor one task while attending to the auditory feedback of another. Surprisingly, there has been little in the way of fundamental research which clearly demonstrates that the addition of auditory feedback significantly enhances everyday computing tasks. It has been suggested that simple musical sounds (e.g. tones) can be used as non-speech "earcons" to help users navigate through computing space [7]. While it is certainly possible to provide auditory feedback in most computer-based tasks, it is not clear whether adding sound will necessarily enhance or hinder performance, especially if it is combined with haptic feedback. Brewster and his colleagues investigated whether sound could be used to help users overcome the common problems of overshooting or slipping off of GUI buttons in point-and-select tasks [8]. They found error recovery to be significantly faster when the sonically enhanced buttons were used over standard GUI buttons. In addition, measures of perceived workload suggested that participants preferred the sonically enhanced buttons to standard ones. While this would suggest that auditory feedback does enhance computing tasks, Brewster did not include haptic feedback as part of the investigation.

Based on a review of the literature, there is some evidence that haptic and auditory feedback may enhance user performance. However, this evidence is provided by studies that examine either one or the other mode for providing feedback, and not both. And yet, the pundits suggest (without clear empirical evidence) that combinations of multi-modal feedback will enhance user performance in everyday computing [1].

Currently, the combination of haptic and auditory feedback is more likely to be used in gaming and simulation applications, or as sight-substitution aids for visually impaired users. Combining haptic and auditory feedback has been shown to improve selection performance for blind and 'visually occupied' users [6]. However, such studies tend to focus on the use of multi-modal feedback as a substitution for visual feedback rather than as a supplement to existing visual feedback. In gaming and simulation applications, contributions made by haptic and auditory feedback to actual user performance may be of less importance than the user's perception of improvement (often driven by the «WOW» factor). For example, Microsoft's Sidewinder™ joystick provides force feedback to supplement computer-graphic action games that already come with realistic sound effects, and Logitech's Wingman™ Force Feedback Steering Wheel allows players to «feel» the road and obstacles in PC-based driving simulation games. Empirical studies demonstrating the improvement in gaming performance due to combined auditory and haptic feedback are lacking.

However, not all computing is gaming and many computer-based jobs carried out by visually impaired and sighted users require multi-tasking. From a software and interface design perspective, we would like to know that the inclusion of haptic and auditory feedback features actually serves to enhance user performance rather than to degrade it. It is the potential enhancement that haptic and auditory feedback can make to everyday computing that was the motivation for this research. As a preliminary step in tackling such design issues, an empirical study was carried out.

# 3    Selecting an Everyday Computing Task

The first challenge for this study lay in deciding upon a task that would be representative of a subset of everyday computing – as opposed to more dynamic gaming or simulation activities.  By «everyday» we mean tasks that incorporate the basic actions that are part of most workplace computing activities (e.g. transcribing, pointing, selecting, monitoring, acknowledging, etc). In order to maximize the opportunities for alternative modes of feedback to enhance performance and minimize the user's sense of workload, we needed a task that would require a fair amount of visual and mental attention. At the same time, the task needed to be fairly basic so that the users would not require a great deal of familiarization or training.

Capitalizing on the work of Brewster and others, we decided to focus on point-and-select tasks, as they tend to depend heavily on visual feedback to ensure that the cursor is positioned over the appropriate target before it is selected.  Point-and-select tasks have been shown to benefit from auditory feedback [7] and haptic feedback [5]. One activity that allows for a common point-and-click interface is the entering of calculations on an online calculator.  The reading and transcribing of basic mathematical formulas is not language dependent, and thus reduced concerns one might have about participant reading levels and typing speeds that would be associated with a text transcription task.

In order to provide opportunities for auditory and haptic feedback, a "Feedback Calculator" application was developed in C++ [9]. The calculator interface was built to resemble the basic calculator interface found in most Windows-type applications. It was programmed so that the calculator buttons could be felt when using the VRM with the force feedback effects turned on.  Passing over a calculator button gives the sensation of being drawn into a gravity-well.  The user then has to push against the force to move the mouse off of the button.  In addition, each button was programmed to produce a distinct sound when the auditory feedback was turned on.

The basic task was to have participants read a mathematical equation from a separate window located on the screen and then to enter that equation into the calculator as quickly and accurately as possible. Monitoring of time to complete task and errors, in terms of incorrect selections, were automated through the Feedback Calculator program. To lessen reaction time bias the stimuli, in the way of 12-character mathematical formulae, could be randomly generated such that users could not predict consecutive character sequences as one might be able to with the reading and typing of words or sentences.

Since the calculator task itself was fairly rudimentary, we recognized that alternative feedback might not significantly enhance performance. We hypothesized that the auditory and haptic feedback may play a greater role in enhancing performance if the user must visually monitor more than one task. In order to simulate the situation when a user is trying to monitor more than one application at a time, we drew upon work by Sellen and her colleagues [10]. They compared the effectiveness of kinesthetic feedback via a foot pedal with visual feedback and asked participants to monitor numbers on an adjacent computer screen while performing a text-editing task on the main screen. Participants were to type in the number that appeared on the adjacent screen whenever they heard a randomly timed beep.  As we were investigating the effects of auditory as well as haptic feedback, we were unable to provide an auditory cue for secondary monitoring.  Instead, we had participants monitor a secondary screen and acknowledge when they noticed the letter on the screen change to one of the three possibilities (a, b, or c). Arguably, this made the dual task situation more difficult than the one used by Sellen, as the participants had to remind themselves to multi-task – which is the more typical case in the workplace. Further details of the empirical study are described below.

# 4    Method

## 4.1    Participants

Thirty students (16 males, 14 females) were recruited from the University of Waterloo to participate in the study. The criteria for inclusion in the study was self-reported experience with any Windows 3.1 applications or higher, proficiency with a mouse-input device, having good vision with or without corrective lenses, and no hearing difficulties. Three of the participants were left-handed and the rest were right-handed. None of the participants had any previous experience with the VRM. All participants were paid $10 for participating in the study that lasted approximately 1 hour.

## 4.2    Experimental Design

This study used a 3X(2X2) mixed design with Mouse Feedback Group (Control, Kinesthetic Feedback, Kinesthetic and Force Feedback) representing the between-group factor, and with Sound (On, Off) and Task Level (Single, Dual) representing repeated measures. All participants were randomly assigned to one of three mouse-feedback groups: standard Logitech™ mouse (Control C non-haptic condition), haptic VRM with no force feedback (Kinesthetic K condition); and haptic VRM with force feedback  (Kinesthetic and Force Feedback KF condition). While the specifics of mouse feedback varied between groups, the basic approach and methodology remained the same. All participants experienced both task levels (Single Task, and Dual Task) and for each task level there were sets of trials with auditory feedback (Sound On) and without auditory feedback (Sound Off). All participants did the Single Task trials first followed by the Dual Task trials. In each Mouse Feedback group, the order of auditory feedback trials was balanced with half of the participants experiencing the auditory feedback first followed by the Sound Off condition.

## 4.3    Equipment, Procedures, and Measures

The "Feedback Calculator" Application and the secondary monitoring tasks were presented on two separate monitors set side-by-side and linked to two IBM compatible PCs. The auditory feedback was provided through a head set and the volume was adjusted to a level of comfort for the participant. The Control (C) group used a regular three-button Logitech™ mouse, while the other two groups used a Virtual Reality Mouse™ (VRM).   In the kinesthetic and force (KF) feedback condition, the force feedback feature of the VRM was turned on, and it was turned off in the kinesthetic feedback (K) condition. All participants used their preferred hand to manipulate the mouse. The Feedback Calculator interfaced with the VRM such that force feedback was activated when the mouse cursor passed over the calculator buttons. In the same way, the Feedback Calculator interfaced with the VRM and regular mouse such that auditory tones representing a tonal scale were also mapped to the calculator buttons. To suggest numerical increments (i.e. 0 to 9) a tone equivalent to middle C on a piano was selected for the number 0 and was increase by a full tone for each successive number up to 9. To reduce any advantages of musicality, the tones associated with the number keys were roughly equivalent to the sound produced by rubber bands stretched taut and then plucked. The sounds associated with the basic arithmetic function keys were more distinct (e.g. a ding for the Equals key). In the Sound On condition, the appropriate tone was played whenever the cursor passed over a button.

Participants performed four blocks of trials, one for each of the repeated conditions (Sound by Task Level). In the Single Task blocks, participants were asked to enter a series of mathematical expressions and find the answers using the Feedback Calculator. Six different equations were included in each trial. For the Dual Task condition, participants were asked to perform a secondary-monitoring task concurrently with the equation-entering task. Three target letters (a, b, or c) were presented one at a time on a second monitor to the left of the primary task monitor. Three directional arrow keys on the keyboard were assigned to each of the letters. Participants were instructed to press the appropriate key with their non-preferred hand whenever they noticed the target letters change.

Speed and accuracy on the primary task were the performance measures recorded for each participant. Total errors per trial were calculated by summing the number of incorrect button clicks and the number of mouse clicks that were made over non-button areas of the Feedback Calculator window. After each condition, participants were administered the NASA TLX index for subjective mental workload. The NASA TLX consists of six linear rating scales: Mental demands, Physical Demands, Temporal Demands, Effort, Own Performance, and Frustration. The scales display a value from 1 to 100 (or percent value) for low to high respectively (with exception for rating of Own Performance that was from 1 to 100 for good to poor respectively). An overall composite score for mental workload is created from the six individual scales [10]. Once all four conditions were completed participants were asked to complete a usability questionnaire. The usability questionnaire used a series of 6-point scales to assess how well the mouse and the sounds helped in detecting the buttons on the calculator and in detecting any mistakes made when entering the equations.

## 5    Results

One concern with timed trials is that participants will trade off accuracy for speed. To check for such trade-offs, linear correlations were calculated for time and errors. None were found to be significant. While the performance measures are not clearly independent, we can view them as contributing different pieces of information to the understanding of the effects of feedback on performance. Table 1 presents the mean and standard deviations for trial times and error rates for each block based on Mouse-Feedback group and trial specifications.

**Table 1.** Means and *Standard Deviations* for Trial Times and Errors

| Mouse Feedback Group | Average Trial Time per Condition in seconds (*standard deviations*) | | | | Average Number of Errors per Condition (*standard deviations*) | | | |
|---|---|---|---|---|---|---|---|---|
| | Single Task | | Dual Task | | Single Task | | Dual Task | |
| | Sound | | Sound | | Sound | | Sound | |
| | Off | On | Off | On | Off | On | Off | On |
| Control N = 10 | 36.6 (4.0) | 37.7 (3.6) | 53.8 (9.4) | 50.3 (9.9) | 1.5 (1.9) | 1.8 (2.7) | 2.8 (2.8) | 3.2 (4.7) |
| Kinesthetic N = 10 | 38.9 (7.7) | 41.0 (5.2) | 54.9 (9.2) | 56.8 (9.4) | 1.7 (1.5) | 1.7 (1.4) | 2.2 (1.8) | 2.0 (1.3) |
| Kinesthetic & Force N = 10 | 37.9 (4.3) | 38.7 (4.3) | 48.6 (5.0) | 52.3 (9.1) | 0.9 (1.0) | 0.8 (1.0) | 1.4 (1.7) | 1.9 (1.4) |

For average time to complete trials, a significant three-way interaction (Mouse-Feedback Group by Task by Sound) was found [F (2,26) = 3.91, p < 0.03]. As expected, the dual task was found to be significantly more challenging than the single task on all measures for all Mouse-Feedback groups (p < 0.01). Thus, the tasks appeared to be successful in representing two different levels of cognitive workload. In the Single Task conditions, time to complete trials did not appear to be significantly affected by auditory feedback for any of the three Mouse-Feedback groups. It is interesting to note that in all three groups, the times to complete trials were actually slightly faster for the conditions without auditory feedback. However in the Dual Task condition, the KF group was the fastest in the absence of auditory feedback, while the control group appeared to benefit the most from having the auditory feedback.

Similar to the analyses for Average Time, a three-way ANOVA was carried out for Error rates. However, none of the analyses revealed significant results (p > 0.05). This may be due to the simplicity of the primary task such that the participants made relatively few errors under any of the conditions.

## 5.1    Subjective Workload and Usability

The subjective workload ratings were compared across Mouse-Feedback group and condition. Table 2 presents the means and standard deviations for subjective workload ratings.

**Table 2.** Means and *Standard Deviations* for Overall Workload Ratings

| Mouse Feedback Group | Overall Workload Rating 1 (low) – 100 (high) | | | |
|---|---|---|---|---|
| | Single Task | | Dual Task | |
| | Sound OFF | Sound ON | Sound OFF | Sound ON |
| Control | 68.3 (16.4) | 44.9 (14.6) | 72.0 (10.1) | 52.4 (15.7) |
| Kinesthetic Only | 62.3 (15.0) | 34.3 (10.8) | 69.8 (15.7) | 37.3 (14.0) |
| Kinesthetic & Force | 65.7 (14.6) | 41.9 (14.5) | 72.7 (11.4) | 45.7 (16.1) |

Auditory feedback had a significant effect on perception of workload [$F(1, 26) = 123.99$, $p < 0.001$]. Under both the single and dual task conditions, participants rated total workload lower when the auditory feedback was available. There was little difference between the groups in terms of workload ratings in the Dual Task Condition. All groups rated the Dual Task as having higher workload than the Single Task condition. For the Single Task condition, when auditory feedback was present the K group rated workload significantly lower than the C group ($p < 0.045$) and marginally lower than the KF group. While for the most part the groups did not differ significantly in terms of overall workload ratings in the dual task conditions, there were differences when the physical demand scale was examined. When auditory feedback was turned off, the ratings for physical demands of the task were comparable across the mouse feedback groups. However, an interesting pattern of results emerged when auditory feedback was turned on (See Table 3).

In both the single task and dual task conditions, the KF group had the highest ratings for physical demands. This was most likely due to the physical experience of overcoming resistance when moving the mouse. The physical demands ratings were significantly higher than the K group when the sound was on in the single task ($p < 0.01$) and the dual task ($p < 0.02$). As well the physical demands ratings were significantly different from the C group when the sound was on in the single task ($p < 0.04$). When the sound was off, both the C and K groups increased their ratings of the physical demands of the task. It is not clear why the presence of sound would lower the perception of physical demands.

At the end of the study, all participants were given a final questionnaire in order to evaluate their ability to detect mistakes because of the feedback experienced, as well as to provide subjective and qualitative feedback about the impact of the equipment and the sounds used. Participants were asked to rate their ability to detect mistakes they made because of the feedback from the mouse. The K group (Mean = 3.5) rated their input device slightly higher than the KF group (M = 3.2) and significantly higher than the C group (M = 2.4) [$t(9) = -2.51$, $p < 0.03$]. Participants were also asked to rate their ability to detect mistakes because of the feedback from the sounds. Participants in the KF group (3.7) generally rated the sounds as being more useful when compared to the ratings of the K group (M = 2.8) and C (M = 2.9) groups [$t(17) = -2.55$, $p < 0.02$]. Not surprising, there was a significant difference between the groups in terms of their perception of ease of movement of the mouse [$F(2,26) = 16.2$, $p < 0.001$]. The C group gave the highest rating for ease of movement (M = 4.4), followed by the K group (M = 3.9), and the KF group (M = 2.3).

**Table 3.** Means and *Standard Deviations* for Physical Demands Ratings

| Mouse Feedback Group | Physical Demands Rating 1 (low) – 100 (high) | | | |
|---|---|---|---|---|
| | Single Task | | Dual Task | |
| | Sound OFF | Sound ON | Sound OFF | Sound ON |
| Control | 44.5 (19.2) | 37.0 (24.7) | 48.5 (16.7) | 33.0 (26.3) |
| Kinesthetic Only | 45.0 (19.6) | 28.0 (18.1) | 42.5 (23.8) | 30.5 (18.0) |
| Kinesthetic & Force | 60.0 (22.6) | 57.8 (27.4) | 62.8 (27.4) | 60.6 (26.8) |

# 6    Discussion

The significant three-way interaction for time to complete trials along with the results from the participant's self-reports on workload and usability presents an interesting pattern of results.    To address the question of whether the addition of auditory, kinesthetic, and force feedback enhanced the primary task (i.e. the calculator task), we will look at the feedback modes separately and in combination, and for each of the single and dual task conditions.

## 6.1    Multi-modal Feedback and Single Task

When participants only had to focus on the primary task, the addition of auditory feedback helped to significantly lower perceived workload, even though it did not improve performance in terms of time to complete trials or error rates.    In fact the means for trial time and errors actually increased slightly for all three mouse feedback groups – although not significantly.    It may be that the sound served as a source of confirmation that the correct button was about to be selected. While such auditory confirmation might reduce cognitive load of visually checking correct selection, it may serve to increase trial time as the user processes and acknowledges the confirmation before moving on to the next entry.

The addition of kinesthetic feedback alone and in combination with force feedback did not improve performance over the traditional mouse condition.    Although not significantly different, time to compete trials took slightly longer with this additional feedback over the non-haptic feedback condition.    However, in the absence of auditory feedback, ratings of overall workload and physical demands were significantly less for the kinesthetic condition when compared to those for the traditional mouse condition.    In the kinesthetic condition, the mouse movements map directly to the cursor position on the screen.    Thus, participants did not have to lift, reposition, and reorient the mouse as one might with a traditional mouse when the edge of the mouse pad or desk is reached.    This may have contributed to the perception of reduced workload for the kinesthetic condition.    This advantage may have been reduced in the kinesthetic and force feedback condition, as those participants had to exert additional physical energy to overcome the force effects.

## 6.2    Multi-modal Feedback and Dual Task

As was the case with the Single Task, the addition of auditory feedback reduced the perception of workload for all mouse feedback conditions in the Dual Task.    While the groups did not differ significantly in terms of mean performance times, the pattern of results for the dual task is worth discussion. In the single task condition, the group in the non-haptic condition had the fastest mean times on the point-and-select task (with and without auditory feedback).    However, once participants were required to monitor a second visual display, the group receiving kinesthetic and force feedback had the faster mean times in the absence of auditory feedback.    This pattern changed once auditory feedback was added.    With the sound effects turned on for the calculator task, the group using the traditional mouse had the faster mean times (improving upon their mean time from the sound-off condition), while both haptic feedback groups showed a degrading of mean trial time.

The group receiving both kinesthetic and force feedback rated the physical demands of the mouse condition as being significantly greater than the other groups when auditory feedback was present. It is curious that the same group gave higher ratings to the auditory feedback in terms of enhancing error detection in the primary task than did either of the other two groups.

## 6.3     Does Multi-modal Feedback Help?

Including kinesthetic, force feedback in a single task did not seem to enhance or degrade performance on the point-and-select task when compared to that achieved with a regular mouse. However, including kinesthetic feedback alone seems to lessen the perception of workload and enhance perception of error detection for the point-and-select task.   Including kinesthetic and force feedback appears to enhance performance in tasks that require high levels of visual monitoring, especially if auditory feedback is not present.  Correspondingly, the groups that did receive the kinesthetic and force feedback rated their devices as being more useful in helping them detect errors than did those who used the regular mouse.  This is interesting given that the group that experienced the force feedback consistently gave higher physical demands ratings than did the other groups.  These findings are similar to that reported by Oatley et al [4].  In their study, including haptic effects did not enhance performance above that of the non-haptic condition.  However, as in our study, their participants reported that the presence of the haptic effects made it seem like the workload was reduced.

From a user's perspective of workload, adding sound makes both a single and dual task seem easier and adding kinesthetic and force feedback makes error detection seem easier.   From a task performance perspective, it would seem that adding auditory feedback to a dual visual task situation enhances performance if the user has a regular mouse.  If auditory feedback is not present, then including kinesthetic and force feedback enhances performance on visual tasks.  The design challenge lies in deciding which should take precedence, the performance outcomes or the users' subjective assessments.  From a consumer's perspective is it worth it to have feedback included that makes you think you are doing better on a task even when you may be doing worse?  From an industrial health perspective, if the operator believes that the device is reducing workload then it is may be worth it.

Empirical research into the true enhancement effects of multi-modal feedback on human performance needs to continue.  We know that a user's subjective assessment of the effectiveness of non-visual feedback may not match the user's actual performance on the task.  We know that subjective assessment may be more favorable towards the feedback source than might be warranted given performance outcomes. Some of the positive assessments of the auditory and haptic feedback may be due to the relatively short periods of time that the participants experienced the feedback, and the fact that they did not have direct knowledge of their performance in terms of trial times and errors.  We do not yet know how influential such favorable impressions may be if users are given the opportunity to self-select the combination of multi-modal feedback they believe enhances their performance, especially over extended periods of time. In the gaming industry, a consumer's perception that a multi-modal feedback device will enhance performance is more likely to drive the purchasing of a device than whether it actually enhances performance. It is not clear whether the same adoption of multi-modal feedback technology would occur with the business or home computing industry without empirical evidence of performance enhancement. Further research needs to explore whether users would elect to have multi-modal feedback included in demanding tasks if they believe it will enhance performance, even if it could be to the detriment of overall task performance.

To date, few published studies have reported empirical investigations into the effects of haptic feedback in conjunction with auditory feedback for single and dual visual tasks. It should be kept in mind that this research is limited in terms of the generalizability of the task. While it did involve components of transcription, selection, and secondary monitoring, it did not involve menu selection, navigation, and text or database entry, which are also basic components of everyday computing tasks. Thus it is not clear if computing tasks that go beyond simple target selection will truly benefit from the addition of auditory and haptic feedback. Work needs to continue to explore and build upon research involving the combination of modes of feedback if we are to realize the potential of multi-modal feedback interfaces for everyday computing tasks.

# References

1. Buxton, W. The three mirrors of interaction: a holistic approach to user interfaces. Proceedings of Friend21'91 International Symposium on Next Generation Human Interface. Tokyo, Japan, Nov. (1991) 25-27.
2. MacKenzie, I.S. Input devices and interaction techniques for advanced computing. In W. Barfield & T.A. Furness III (Eds.) Virtual environments and advanced interface design. Oxford, UK: Oxford University Press (1995) 437-470.
3. Madill, D.R. Modelling and control of a haptic interface: a mechatronics approach. Unpublished Ph.D. thesis, Department of Electrical and Computing Engineering, University of Waterloo (1998)
4. Akamatsu, M. and Sate, S. Multimodal mouse with tactile and force feedback. International Journal of Human Computer Studies, 40 (3), (1994) 443-453.
5. Oakley, I., McGee, M.R., Brewster, S. and Gray, P. Putting the feel in 'look and feel'. Proceedings of CHI'2000, Conference on Human Factors in Computing Systems "The Future is Here", April 1-5, The Hague Netherlands, (2000) 415-422.
6. Dufresne, A., Martial, O., and Ramstein, C. Multimodal User Interface System for Blind and "Visually Occupied" Users: Ergonomic evaluation of the Haptic and Auditive Dimensions. In K. Nordby, P. Helmerson, D.G. Gilmore and S.A. Arnesen (Eds.) Human-Computer Interaction Inteact'95. London: Chapman and Hall. (1995), 163-168.
7. Brewster, S.A. Using non-speech sound to overcome information overload. Displays, 17, (1997) 179-189
8. Brewster, S.A., Wright, P.C., Dix, A.J. and Edwards, A.D.N. The sonic enhancement of graphical buttons. In K. Nordby, P. Helmersen, D. Gilmore & S. Arnensen (Eds.) Proceedings of Interact'95. Lillehammer, Norway. (1995) 471-498.
9. Thomas, A.S. The effects of integrating kinesthetic feedback, force feedback, and non-speech audio feedback in human-computer interaction. Unpublished MASc thesis, Dept. of Systems Design Engineering, University of Waterloo (1998).
10. Sellen, A., Kurtenbach, G., and Buxton, W. The prevention of mode errors through sensory feedback. Human Computer Interaction, 7 (2), (1992), 191-164.

11. Hart, S.G. and Staveland, L.E. Development of the NASA task load index (TLX): Results of empirical and theoretical research. In P.A. Hancock & N. Meshkate (Eds.), Human mental workload. Amsterdam: North-Holland, (1988) 139-183.

# Discussion

*P. Gray:* Your audio feedback had a particular semantic component. Did your force feedback component?
*C. MacGregor:* No, although the VRM gives relative position information.

*P. Gray:* It seems your users got a better sense of their performance from the audio feedback but more actual help from the haptics. Could it be this difference influenced the difference in performance and subjective results?
*C. MacGregor:* Yes. Work needs to be done to understand what it is about the audio feedback that generates positive ratings from users.

*J.Höhle:* At last year's British HCI, someone recorded stress measures using galvanic skin response. Did you consider this?
*C. MacGregor:* No. We chose to use the NASA TLX because it was easily available and has been widely used in other studies.

*H. Stiegler:* I can imagine many tasks where I would like to have some feedback to help me prevent errors. And since everything is flat, you get no feedback. Perhaps your task is too simple to get good results?
*C. MacGregor:* It may be that as the task becomes more complex that some of these feedback features will be of more help; this research is really just a first step.

*K. Schneider:* You suggest that your research has raised many open questions. Which area would you think is most important to address next? Have you considered working with visually impaired people?
*C. MacGregor:* We have considered visually impaired, although we're not planning on working on that right now. We're more interested in whether novelty was a significant issue, and in how users will behave if we allow them to self select whether they want audio or haptic feedback.

*S. Greenberg:* Is there a danger of 'throwing the baby out with the bath water' because of these somewhat neutral results? I can think of many tasks that are not target acquisition that might benefit from force feedback.
*C. MacGregor:* I'm currently working with a company in Waterloo that builds immersive virtual reality systems like walk-in caves, etc. The company needs to suggest input devices to its customers. We're working on an experimental protocol to figure out which of these devices makes the most sense under what conditions.

*L. Nigay:* You're using multiple modalities, but all are conveying the same information. Have you considered using different modalities for different tasks.
*C. MacGregor:* Yes. For example we've used auditory progress bars in dual task capabilities with some interesting results. In this experiment for the dual task effect we had expected that the force auditory feedback would allow the users to more effectively split their attention.

# Information Sharing with Handheld Appliances

Jörg Roth

University of Hagen, Department for Computer Science
58084 Hagen, Germany
Joerg.Roth@Fernuni-hagen.de

**Abstract.** Handheld appliances such as PDAs, organisers or electronic pens are currently very popular; they are used to enter and retrieve useful information, e.g., dates, to do lists, memos and addresses. They are viewed as stand-alone devices and are usually not connected to other handhelds, thus sharing data between two handhelds is very difficult. There exist rudimentary infrastructures to exchange data between handhelds, but they have not been designed for a seamless integration into handheld applications. Handheld devices are fundamentally different from desktop computers, a fact that leads to a number of issues. In this paper, we first analyse the specific characteristics of handheld devices, the corresponding applications and how users interact with handhelds. We identify three basic requirements: privacy, awareness and usability. Based on these considerations, we present our own approach.

## 1    Introduction

Currently, there exists a growing market for handheld devices such as PDAs, mobile phones, electronic pens etc. Upcoming communication technologies like UMTS and Bluetooth promise new functionalities for human communication. Currently, the mostly accepted way for communication is still verbal communication, whereas symbolic or textual media such as SMS or Email are still hard to use with handheld devices.

Handheld devices may already store reasonable amounts of data, e.g. appointments, holidays, addresses and agendas. On one hand, it should be possible to exchange these data between users without too much effort, on the other hand, handheld devices are basically of private nature and thus not intended for sharing data with other people. These contradictory characteristics may be one reason, why handhelds are currently being viewed as autonomous systems without any sophisticated facilities for inter-handheld communication.

In this paper, we present requirements, issues and problems of communication-oriented, distributed applications for handheld appliances. A lot of research has been done about the design of distributed applications in desktop environments. However, due to the fundamental different nature of handheld devices, these results can hardly be adapted to handheld scenarios.

This paper is structured as follows: first, we present the fundamental differences between handheld devices and traditional desktop computers. Based on these considerations we identify three key requirements, a communication-oriented distributed handheld application has to meet: *privacy*, *awareness* and *usability*. We then present a

M. Reed Little and L. Nigay (Eds.): EHCI 2001, LNCS 2254, pp. 263–279, 2001.

framework, which allows a designer to create successful information sharing applications for handhelds.

## 2    Handheld Computing Characteristics

The notion of *handheld device*, *palmtop*, *PDA* and *organiser* is often interpreted in different ways. An older interpretation distinguishes between *pen-based devices* and *palmtops*, where the latter have keyboards. In contrast, Microsoft divides Windows CE devices into *handheld PCs* (H/PC) with a keyboard, *palmsize PCs* (P/PC), which are controlled by a pen and *handheld PC Pro* devices, which are subnotebooks [3]. To get completely confused, Microsoft calls the new pen-based devices based on Windows CE 3.0 *Pocket PC*.

In the following, we understand by *handheld devices* mobile devices with small displays, without or with just rudimentary keyboards and an autonomous power supply. Examples are:

_ 3Com's Palm devices (e.g. Palm III or Palm V),
_ Casio's Cassiopeia or
_ Electronic pens such as the C-Pen.

In particular, we do not summarise notebooks or laptops under the notion of handhelds.

Handheld computing is closely related to so-called *ubiquitous computing*. Mark Weiser introduced the concept of ubiquitous computing often called "ubicomp" [19]. His vision was a huge number of invisible and "calm" computers surrounding people in their everyday life. In contrast, handheld computing keeps the device in the foreground, whereas ubicomp devices should work in the background [6]. People using handhelds are aware of using computers and adapt their activities to the device, e.g. learn a specific kind of handwriting. However, studying handheld computing may be the right step towards ubicomp, since a number of problems is identical. Handhelds, e.g., should be suitable for everyday tasks, easy to handle and fail proof.

Table 1 shows some hardware characteristics of popular handheld devices. Compared to desktop computers, handheld devices have small memories, low computational power, limited input and output facilities and usually do not contain a mass storage. Persistence data have to be stored in the battery-buffered RAM, and thus decreases the available runtime RAM for dynamic data and runtime stack. Even worse: some CPUs (e.g. Palms CPU) allow only addressing small pieces of memory (64k) as a whole. The reduced capabilities have two major reasons:

*1. Size:* The display is limited to palm size (approx. 10cm x 8cm); chips (e.g. memory and CPU), even when highly integrated, need space for connectors and circuit boards. Due to these limitations, it is impossible or at least cost intensive to integrate high-resolution displays or a big number of electronic parts into a device.

*2. Battery life:* Fast CPUs, big memories and high-resolution displays (particularly coloured ones) consume a big amount of valuable battery power. If battery technology will not significantly improve in the future, handheld computers will always be far behind the capabilities of desktop computers.

**Table 1.** Some hardware characteristics of handheld devices

| Device | Processor | RAM | Screen | Battery life |
|--------|-----------|-----|--------|--------------|
| Palm IIIxe (3Com) | 16Mhz MC68EZ328 | 8MB | 160x160 (16 grey) | 1.5-2 months (normal use) |
| Cassiopeia E-15 (Casio) | 69MHz NEC VR4111 | 16MB | 240x320 (16 grey) | 25 hours (continuously) |
| C-Pen 800 (C-Technologies) | 100Mhz Intel StrongARM | 8MB | 200x56 (b/w) | 2-3 weeks (normal use) |

The reduced equipment of handheld devices has a big influence on software development. Handheld applications are usually not developed 'from scratch' but make use of their operating system's services. Popular operating systems like PalmOS [2], Windows CE [3], EPOC [17] or ARIPOS [5] provide the following services:

_ starting, stopping and switching applications, managing the memory and the user interface;

_ special device-dependent services like handwriting recognition, OCR, time-dependent alarms;

_ supervising the battery power, performing auto-power-off;

_ managing persistent data in the battery-buffered memory;

_ managing communication to other devices (usually to the host PC).

Software development kits allow the software designer to code and compile handheld applications on desktop computers, usually in C. Some kits provide emulators for testing applications before downloading them to the specific device.

Compared to desktop operating systems, handheld operating systems do not offer the same variety of services. Major shortcomings are:

*Limited user interface capabilities:* Usually the so-called WIMP paradigm (Windows, Icons, Menus, Pointers), which is very common to desktop computers, is, if at all, supported in a very reduced way only because of the limited display and input capabilities.

*Limited support for persistent data:* Since handhelds have no mass storage system, all persistent data have to be kept in the battery-buffer RAM. Windows CE emulates a hierarchical file system inside the RAM area. Other systems like ARIPOS and PalmOS store persistent data in so-called *databases* [2] (not to be confused with traditional databases). A database is a persistent collection of records. Each record has a unique identifier; its content is opaque to the operating system. Constructing and interpreting records solely depends on the corresponding application. Databases can be viewed as flat file system with a record-oriented structure.

*Limited or no parallel execution capabilities:* Most handheld operating systems do not support threads or processes for background tasks, a common technique for desktop computer applications. As a work-around, some systems offer so-called *timers*, which can periodically call a predefined procedure. Unfortunately, a call is only performed, when no other instruction is being executed, thus a timer does not provide real background operations.

*Limited support for communication:* Compared with desktop computers, handhelds only support a small set of communication capabilities. Handheld operating systems usually support one specific way of communication determined by the peripheral equipment. ARIPOS, e.g., only supports IrDA communication, since the C-Pen only

contains an infrared transceiver for communication. PalmOS supports serial communication and TCP/IP, but does not support TCP server sockets, essential for reacting
on incoming communication requests.

*Networking*: In addition to the limited communication support, the underlying
network itself has some drawbacks:

_ Wireless communication infrastructures currently have low bandwidths (GMS,
e.g., only provides 9600 Baud) and tend to have high error rates and abnormal ter-
minations.

_ Handhelds as communication end points are mobile in the network, i.e. often
change their network addresses.

_ Mobile devices are rarely connected, i.e. are most of the time not available in the
network because of network failures or just because the device is turned off.

Emerging technologies as UMTS, IPv6 and Bluetooth will change the way handhelds
can be used inside a network. UMTS, e.g., allows a device to be permanently connected to the network with high bandwidth. IPv6 offers with MobileIP the possibility
to keep the same IP address, even when a device moves inside the network. However,
such technologies are not yet widely available and cannot be used inside current concepts.

All topics mentioned above have big influence on the application development
process. Usually, because of the limited handheld capabilities developing handheld
applications is very cost intensive. This includes testing and debugging; since handhelds do not offer the same debugging facilities like desktop computers, testing and
debugging is quite cumbersome.

# 3    Handheld Applications

## 3.1    Application Types

Interactions with handheld applications and interactions with desktop applications
follow different usage paradigms. First, handheld applications have to respect the
limitations mentioned above; much more important: users request a different kind of
availability from such applications: handhelds do not 'boot up'. When needed, applications have to immediately appear on screen. In turn, when the handheld device is
deactivated, an application has to immediately save its state and disappear. In general,
handheld applications are being used for entering and retrieving small pieces of
information rather than processing data.

On handhelds, only a small set of application types can reasonably be used. We
analysed 32 Palm applications currently available as shareware. In order to get a
representative set of applications, we took a shareware collection of a popular German
journal [4]. Table 2 shows the results.

Five applications are utilities and so-called *hacks*, which extend the operating
system. This kind of application is not used to store information, thus not taken into
further consideration.

**Table 2.** A selection of Palm applications

| Applications | Application type | Data type | No. |
|---|---|---|---|
| Launcher III, SwitchHack, German Chars, Hackmaster, Eco Hack | Utilities | mixed | 5 |
| Brainforest, dNote, HandyShopper, PocketMoney, HanDBase | Textual notes, ideas, shopping lists, bank accounts | textual documents, tables | 5 |
| Feiertage, Yearly, DateBk3, Palm Planner | Dates, appointments, holidays | dates | 4 |
| Abacus, TinySheet, MiniCalc | Spreadsheet tools | spreadsheets | 3 |
| Desktop to Go, Documents to Go, TealDoc | Documents | textual documents | 3 |
| ptelnet, MultiMail, HandWeb | Internet tools | mails, web pages | 3 |
| DiddleBug, TealPaint | Graphical notes, freehand | graphical data | 2 |
| Parens, Currency Calculator | Calculators | numbers | 2 |
| PocketChess, TetrisV | Games | game states | 2 |
| Secret! | Security | texts | 1 |
| Route Europe | Route planner | geographic data | 1 |
| Timer | Clock | time | 1 |

Except for two programs, which allow graphical input, all applications store well structured and record oriented data, often text based. Typical data types for handheld applications are

_ texts and lists of texts,
_ date entries,
_ numbers,
_ tables or spreadsheets.

None of the applications above deals with multimedia data such as audio or video, which require a considerable network bandwidth, sufficient output devices and a huge amount of battery charge. Currently multimedia data are not suitable for handheld devices.

To summarise, most of the applications deal with simple data types such as strings or numbers, joined together to lists or tables. Only few applications support graphical data.

## 3.2   Privacy

Data stored inside a handheld device are usually viewed as *private data*. Even more than desktop computers, handhelds are viewed as personal devices [16] for storing personal data such as telephone numbers, birthdays and leisure-time activities. Some handheld operating systems protect private data. PalmOS, e.g., allows a user to mark entries as *private*, i.e. a password is needed for viewing them. In addition, a handheld device as a whole can be locked via a password.

Connecting handheld devices for data exchange may jeopardize privacy. In case of an untrusted network, applications have to offer mechanisms for guaranteeing data. No private data should be transferred across a network, other people must not be able to break into a handheld and to spy out private data.

To gain acceptance by end-users, an infrastructure has not only to ensure privacy; but also convince the user that her or his private data are kept private. This is perhaps the most crucial issue related to privacy.

## 3.3    Awareness

Mobile devices can be connected to a network at different places. Depending on the location, different information may be available. Users should be aware of their current location, including the geographic location as well as the location in the network. This information is part of so-called *context awareness*. Abowd and Mynatt list different kinds of context awareness, defined by the "five W's": *Who*, *What*, *Where*, *When* and *Why* [1]. The *Who* context, e.g., is based on information about other people in the environment, especially for looking at activities.

Sharing information between people leads to the area of groupware and CSCW (computer supported collaborative work). Collaborative applications significantly differ from single-user applications. Many users provide input (often simultaneously), output has to be processed for many users, and shared data have to be kept consistent. Groupware applications have to provide a 'feeling' of working together in a group, called *collaboration awareness*: users have to be aware of other users involved in the collaborative task.

Context awareness as well as collaboration awareness require special components inside the applications' user interfaces called *awareness widgets*. Similar to users, we call an application *aware* of something, if it explicitly takes care of a special situation, otherwise we call it *transparent*. *Collaboration aware* applications, e.g., are especially designed for supporting a group, i.e. they contain special code for group functions. *Collaboration transparent* applications originally are single-user applications, which, with help of a groupware toolkit, can be used by many users simultaneously. Collaboration transparent applications do not offer awareness widgets. A similar notion can be applied to the mobility aspect: *mobility aware* applications contain code to handle mobility, e.g. react on unstable network connections and changing network locations. *Mobility transparent* applications cannot handle such problems explicitly, but rely on an underlying platform.

## 3.4    Usability

Usable applications support users in carrying out their tasks efficiently and effectively. For handheld applications, usability may be even more important than for desktop applications. When an application is designed in isolation from the intended users, the result is all too often an application which does not meet their needs and which is rejected by end-users. An application should meet the following requirements:

*Respect hardware and software limitations:* Usable applications take into account the hardware and software limitations of their hosting handheld devices. Heavy computational tasks are not suited for handhelds; user interfaces should be designed for small displays with minimal text input; communicating across a network should consider the small bandwidth and high error rate.

*Software quality*: Handheld applications should be more fail proof than desktop applications. A locked or crashed application blocks the entire device. An application trapped in an infinite loop prevents some devices from being switched off. Cold starting a handheld often results in loosing all stored data. The problem becomes even worse, if an application communicates with other devices. A blocked device may interrupt the entire group communication. To improve software quality, design guidelines may help a developer to build well-formed applications. Such a guideline can, e.g., be found in [2]. A platform or application framework, that encapsulates standard solutions for a specific application domain, helps a developer to meet these guidelines: She can rely on a set of services and only has to code application-specific functions.

*Respect everyday requirements:* Handheld applications are used every day. Abowd and Mynatt introduced the term called *everyday computing* [1]. They state that daily activities rarely have a clear beginning or end and often are being interrupted. This issue is especially important when considering communication-oriented applications. The strict distinction between asynchronous and synchronous groupware [7] hardly applies to everyday tasks. This leads to the notion of *relaxed synchronous collaboration* when group members collaborate synchronously, but may sometimes be disconnected from the network for short periods of time [15]. In addition, everyday tasks require spontaneous, unplanned communication. Exchanging data between handhelds should be as easy as a phone call. Especially, user-driven central co-ordination or administration should be avoided.

# 4    Related Work

*Mobile phones*: Mobile phones offer simple mechanisms to transfer textual data. SMS (short message service) [13] is a protocol, which allows sending up to 160 characters to another mobile phone. It can slightly be compared to the email service on the Internet, but is based on the mobile phone infrastructure GSM (global system for mobile communication). WAP (wireless application protocol) [20] allows browsing special Internet pages on small displays. WAP provides a one-way only information channel, i.e. it is not possible to send page contents from one device to another.

*Beaming*: A simple technique for exchanging data between handhelds, so-called "beaming", comes along with the Palm device [2]. Manufacturers of other devices, e.g. of C-Pens or Windows CE devices adapted the technology. Beaming can be viewed as de-facto standard for short range data exchange between handheld devices. It is infrared based and allows exchanging single records of data, e.g. one address or one memo. The distance between communicating devices should not exceed approx. one meter. Beaming requires human interaction, i.e. each time an entry is to be transferred, the sender as well as the receiver have to interact with their devices. Beaming is only suitable for small amounts of data.

*PIMs*: Personal Information Managers (PIMs) are important tools for handhelds. They conveniently allow entering data by keyboard and then downloading them to the handheld. A popular PIM is Microsoft's Outlook [11]. In addition to synchronising data with a handheld, Outlook allows scheduling appointments in a team, i.e. small amounts of data can be exchanged between a group of people.

*Coda*: Several research platforms have been developed for data distribution and consistency in mobile environments. Coda [9] provides a distributed file system similar to NFS, but in addition supports disconnected operations. Applications based on Coda are fully mobility transparent, i.e. run inside a mobile environment without any modification. Disconnected mobile nodes have access to remote files via a cache. Operations on files are logged and automatically applied to the server when the client reconnects. Coda applications can either define themselves mechanisms for detecting and resolving conflicts or ask the user in case of conflicts.

*Rover*: The Rover platform [8] supports mobility transparent as well as mobility aware applications. To run without modification, network-based applications such as Web browsers and news readers can use network proxies. The development of mobility aware applications is supported by two mechanisms: *relocated dynamic objects* (*RDOs*) and *queued remote procedure calls* (*QRPC*). RDOs contain mobile code and data and can reside on a server as well as on a mobile node. During disconnections, QRPCs are applied to cached RDOs. As in Coda, after reconnection operations are logged and applied to server data.

*Bayou*: Bayou [18] provides data distribution via a number of servers, thus segmented networks can be handled. In contrast to Coda, replicated records are still accessible, even when conflicts have been detected but not resolved. Bayou applications have to provide a conflict detection and resolution mechanism. Ideally, no user intervention is necessary. Bayou is not designed for supporting real-time applications.

*Sync*: Sync [12] supports asynchronous collaboration between mobile users. It provides collaboration based on shared objects, which can be derived from a Java library. As in Bayou, data conflicts are handled by the application. Sync applications have to provide a *merge matrix*, which for each pair of possible conflicting operations contains a resulting operation. With the help of the merge matrix, conflicts can be resolved automatically.

*Lotus Notes*: Lotus Notes [10] has not primarily been designed for mobile computers, but allows replicated data management in heterogeneous networks. Nodes can be disconnected and merge their data after reconnection. Data in Lotus Notes have a record structure. Fields may contain arbitrary data, which are transparent to Notes. Records can be read or changed on different nodes simultaneously. When reconnecting, users resolve conflicting updates. With the help of the Notes extension Mobile Notes, it is possible to access databases via Palm devices and mobile phones.

*Discussion*: Mobile phone protocols are designed for very simple data and not practical for structured data. It is difficult to adapt application specific data with an internal record structure to these protocols. A good solution for small amounts of data provides beaming, since any application can use this communication mechanism to exchange records with other handhelds. Due to the record-by-record concept, beaming is not suitable for bigger amounts of data.

Outlook has been designed especially for office environments and can hardly be adapted to other everyday tasks. It is not possible to add new applications to Outlook. In addition, Outlook requires a considerable amount of central administration.

Most of the research toolkits above request their mobile clients to be notebook computers with, e.g., hard disks. The focus of these platforms is to maintain data consistency in a weakly connected environment. Problems related to handheld devices such as small memory and reduced computational power are not handled satisfactorily. Automatic conflict detection and resolution need a considerable amount of resources on the handheld devices. We believe that such mechanisms are (currently) not suitable for handheld scenarios.

Concepts, such as the Rover toolkit, which require mobile code and marshalling/unmarshalling mechanisms, currently cannot be adapted to handheld devices, since they are significantly different from their servers. The concept of mobile code requires platform independent code and identical runtime libraries on both platforms involved. Even though languages such as Java are running on many platforms, handheld portings will provide other runtime libraries, thus mobile code mechanisms will fail.

The platforms above leave many problems described above unsolved. Especially privacy and awareness are still open issues.

# 5    The QuickStep Approach

The QuickStep platform [15] allows developing mobility aware and communication-oriented handheld applications. Developers can use communication and collaboration primitives provided by the platform and can concentrate on application-specific details. A set of predefined awareness widgets can be integrated into an application with a few lines of code. The QuickStep approach can be described as follows:

- QuickStep supports applications with well-structured, record oriented data. It explicitly has not been designed for supporting multimedia data, graphical oriented applications or continuous data streams.
- QuickStep provides awareness widgets for collaboration awareness as well as context awareness.
- QuickStep applications are fully collaboration and mobility aware.
- QuickStep comes along with a generic server application, which allows supporting arbitrary client applications without modifying or reconfiguring the server.
- The QuickStep architecture ensures privacy of individual data.

Before describing the QuickStep platform itself, we present two sample applications developed with QuickStep.

## 5.1    Sample Applications

The first sample application allows a group of users to exchange date information (e.g. about vacations or travellings). This tool is useful in meetings, in which members want to schedule appointments for future meetings. Each member owns a handheld device, which already contains a list of appointments as well as entries indicating the time one is unavailable. The problem is to find a date, when all members are available. Figure 1 presents an application that can help to find such a date.

The figure shows the view of two users on their personal handheld device. The upper half of the window displays the days of a month. Each range of dates when someone is unavailable is indicated by a bar. To get a better overview, the view can be switched to a two-months display. The lower half of the window serves as the legend for the upper half.

a) Joerg's handheld      b) Stephan's handheld

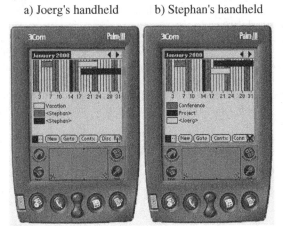

**Fig. 1.** A collaborative calendar tool

The two users Joerg and Stephan can see their own bars and the bars of each other. Foreign bars are labelled by the user name rather than the local label. For other users only the date range is of interest, not why someone is unavailable. Each user can make new entries which are distributed to the other user in real-time. With the help of this application, it is very easy to find dates, where all members are available.

The second example, the business card collector (figure 2), is a useful application for conferences. The application shows a list of all users assembled at a specific location. A user can view these cards and collect interesting cards in a persistent area. If the user permits, the business card collector publishes the card automatically, when entering a location.

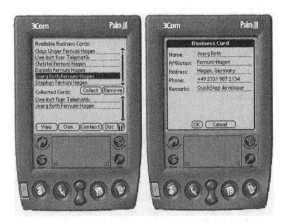

**Fig. 2.** A business card collector

To develop such an application 'from-scratch', a developer has to implement many tasks, e.g., communication protocols have to be integrated, shared data have to be managed. The application should offer awareness widgets. All these services have to be developed in addition to the main task. Developing all these functions would overwhelm a developer. QuickStep helps a developer to concentrate on the application-specific details. Data primitives as well as predefined awareness widgets can be used from the platform.

In the following, we present the QuickStep platform. After describing the basic concepts, we discuss QuickStep with the help of the three key requirements *privacy*, *awareness* and *usability*.

## 5.2    The QuickStep Infrastructure

As described above, handheld operating systems offer only limited support for communication. Most systems cannot handle communication in the background. If the handheld device is always the initiating part of the communication, we need an additional computer, which acts as a communication relay between handhelds. This computer, the *QuickStep server*, contains a generic server application, which is able to serve arbitrary QuickStep applications.

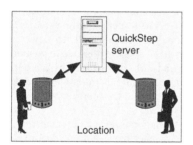 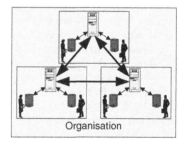

**Fig. 3.** QuickStep communication infrastructure

Figure 3 shows the QuickStep communication infrastructure. A QuickStep server operates in so-called *locations*. A location links all handheld devices together which are 'in range', i.e. which can be accessed by the specific communication technology. This can be the range of an infrared transceiver or a Bluetooth sender. Locations linked together form an *organisation*. Organisations may connect locations, which are in the same company, building, conference or public place. Table 3 shows typical examples for locations and organisations.

**Table 3.** Examples for locations and organisations

|              | Train       | Company                | Conference                |
|--------------|-------------|------------------------|---------------------------|
| **Location**     | wagon       | meeting room, hallway  | presentation room, foyer  |
| **Organisation** | whole train | company building       | whole conference          |

Connections between handhelds and QuickStep servers are usually wireless, where the QuickStep servers are connected via traditional local area networks. The QuickStep server can be viewed as 'inventory' of a specific location. Once installed, it normally has not to be reconfigured or administered. The server runs without an operator and does not need a user interface, thus can, e.g., work invisibly behind a panel.

## 5.3    Underlying Data

As mentioned above, most handheld operating systems offer an entity called *database* to handle application-specific data. The database is a common programming abstraction in handheld applications, thus the ideal abstraction for communication-oriented applications as well. QuickStep follows the same paradigm for collecting and distributing data. The QuickStep application programming interface (API) has similar database functions as the database API. An application developer can use well-known services to handle application specific data. Data stored in QuickStep databases are automatically distributed among a group by the QuickStep platform. Similar to native database services, the actual content of records is not of interest for the distribution mechanism and can only be interpreted by the application. Especially, the QuickStep server does not know the record structure.

*Conflicts*: Concurrent updates on shared data sometimes cause conflicts. Many platforms described above have complex mechanisms to detect and resolve conflicts. In our opinion, such mechanisms cannot be used inside handheld devices. Our concept for solving conflicts is simply to avoid them: it is not possible to concurrently manipulate data. For this, each record of data can only be changed by the handheld device, which originally created the record; copies residing on other handheld devices can only be viewed. To modify data which were created by another user, one has to make a private copy, which is treated as a new record.

*Mirroring and Caching*: Due to the low computational power of handhelds, heavy processing tasks should run on the server. On the other hand, with respect to the low network bandwidth, it is not possible to transfer a large amount of processing results between server and handheld. To reduce network traffic and to perform as many computations as possible on a server, we developed a combined mirroring and caching mechanism. Each handheld has its *local database*, which stores its user's data. A *cache database* stores all data of other users, the local user has currently in view. E.g., the cache database in the calendar tool stores dates of other users in a specific month. Finally, the QuickStep server has a copy of each local database, the *mirror database*. The mirror and cache databases are incrementally updated, each time a handheld device is connected to the server. The application developer has not to worry about the cache and mirror databases; they are completely set up and maintained by the QuickStep runtime system.

## 5.4    Developing with QuickStep

Figure 4 shows the environment, in which a QuickStep application is embedded.

Applications developed with QuickStep use the QuickStep API as well as the API offered by the corresponding operating system. QuickStep is built upon the database communication and user interface APIs. QuickStep does not use the operating system's communication API directly, instead it uses an intermediate layer, called the *network kernel framework*. This layer, developed by the DreamTeam platform [14] offers a generic interface for communication services such as starting and stopping connections, transferring data etc. With the help of the network kernel framework, it is possible to exchange the underlying communication API without changing the QuickStep platform. E.g., one can exchange a TCP/IP communication by a direct serial or infrared connection and only has to adapt the network kernel framework.

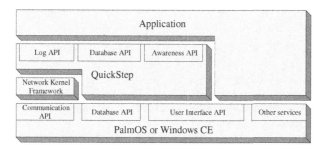

**Fig. 4.** The QuickStep programming environment

## 5.5   Privacy

To ensure privacy, QuickStep does not transfer any private data across the network. Every record can be marked as *private* (the default value). Private records reside only on the handheld and will not be transferred in any case.

Non-private records are not transferred until an anonymising process relieves them from personal information. Since the record structure is opaque to the underlying system, the anonymising function has to be provided by the application. In the calendar application, e.g., the anonymising function blanks out the labels of appointments and transfers the date range only. The entry

```
May 11-13: "Jörg is on the EHCI"
```

will be transformed to

```
May 11-13: "Jörg is away"
```

since others do not have to know anything about the reason of absence.

As an additional concept, each record has a 'time to live' entry, after which a record is deleted automatically from the QuickStep server and other handheld devices. This is done because a user wants to be sure that her or his data are not available forever on other computers (even in anonymised form). The time to life entry can be one of *session, min, hour, day* and *forever*. If the value is *session*, the corresponding record will be immediately removed from the server and handheld caches after the corresponding handheld is disconnected. The other values indicate the time, a record will reside after disconnection. The lifetime is controlled by special tasks inside the platform, the *lifetime supervisors* which exist on the handheld devices and on the QuickStep servers.

A similar concept applies to the space property. Each record has a 'space to live' entry, which defines to which servers a corresponding copy is transferred. Space to live entries can be one of *location, organisation* and *everywhere*. If the value is *location*, only the QuickStep server which serves the current location gets a copy in the corresponding database. If the value is *organisation*, the record is transferred to all servers inside the organisation. The value *everywhere* is for future use and currently not supported. We work on a concept, which allows linking multiple organisations together for exchanging data. Currently, mirroring and group management requires a tight coupling between servers, thus is only suitable for local area networks. Transferring data between organisations requires completely different mechanisms.

## 5.6    Awareness

Context information is important for users as well as for the application, which may make decisions based on contextual data. A user who collaborates may want information about the context she or he is currently working in. For this, a user can open a frame as presented in figure 5.

The context frame is the central instance for all context-related information:
_ What is the current connection state (connected or disconnected)?
_ To which server is the handheld currently connected (server name, organisation)?
_ What is my current location?
_ Who can be called in case of problems (e.g. network failures)?
_ Which other users are currently in the same location or organisation?
_ What are other users' connection states?

**Fig. 5.** The context frame

Location information is important when a user enters an unknown location. Consider a scenario where a huge building is equipped with a number of QuickStep servers (e.g. one per floor). Each QuickStep server provides information about the current location and thus can be used as a beacon for navigating inside the building.

For collaborating users, the connection state is important. If a user is disconnected, data changes of that user cannot be viewed by other users. Thus, information about the connection state should be available on the main window of an application. We designed an integrated button and state indicator (figure 1, lower right button). This widget allows connecting and disconnecting to a QuickStep server and via a small icon indicates the current state.

The button/state indicator as well as the context frame are predefined awareness widgets and can be integrated in an application via the QuickStep library. In addition, an application can retrieve state and context information via the QuickStep API and can react on events (e.g. disconnecting from a network). So, an application developer can create his or her own awareness widgets.

### 5.7    Usability

QuickStep explicitly has been designed for handheld scenarios and respects hardware and software limitations. As described above, heavy computations are avoided and network limitations are considered. Using a well-formed and tested platform a developer can rely on stable services. The database abstraction offers a suitable application framework. All services related to communication are embedded into the platform. To easily find errors in the application itself, a developer can use the log API (see figure 4). Handheld logs are stored on QuickStep servers, thus problem analysis is possible, even when an application or the entire handheld device crashes.

Everyday tasks are often run in an unexpected way. In order to encourage spontaneous communication between users, central administration has to be avoided. For this, groups of interacting users are not defined explicitly in QuickStep. All users connected to a specific QuickStep server at the same time and using the same QuickStep application automatically form a collaborative session. This concept allows running a server without defining groups centrally. It is possible for a user to join a group without having explicit permission from other users. Since a mechanism for anonymising data is integrated into the platform, a user cannot spy out other users' private data. QuickStep does not provide services for leaving a collaborative group. When a user disconnects, the server first assumes a temporary disconnection, which happens frequently. Only if a user is disconnected for a longer time (e.g. an hour), the server removes that user from the session. The period of time, a user has to be disconnected until a leave operation is performed, is defined by the corresponding application. When a user leaves, the corresponding mirror database is deleted from the server.

## 6    Conclusion and Future Work

Handheld applications require approaches fundamentally different from desktop applications. If, in addition, applications have to exchange information between users, additional issues have to be considered. We identified three properties, an ideal communication-oriented handheld application has to meet: *privacy*, *awareness* and *usability*.

The QuickStep approach meets these properties: it allows a designer to develop mobility and collaboration aware applications and has been especially designed for handheld devices. The generic QuickStep server relieves the handheld devices from heavy tasks and keeps data during disconnection. The QuickStep server operates without human intervention and can serve arbitrary QuickStep applications without modification. A server offers contextual information, which can be used by handheld applications. Data distribution is handled by a caching and mirroring mechanism.

In the future, we will follow two directions. First, we want to include traditional computers into the approach, because currently handheld computing relies on both, handheld and desktop applications. As data input is more convenient on desktop computers, an appropriate concept has to support both kinds of computers.

Second, we want to extend QuickStep to a global communication infrastructure. With this, two or more users operating at different places in the world could exchange data. Since wide area connections are considerably slow compared to local area networks, we have to develop new concepts. New technologies such as UMTS may help to address this problem.

## References

1.   Abowd G. D., Mynatt E. D.: Charting Past, Present and Future Research in Ubiquitous Computing, ACM Transactions on Computer-Human Interaction, Special Issue on HCI in the new Millennium, Vol. 7, No. 1, March 2000, 29-58
2.   Bey C., Freeman E., Mulder D., Ostrem J.: Palm OS SDK Reference, 3Com, http://www.palm.com/devzone/index.html, Jan. 2000
3.   Boling D.: Programming Windows CE, Microsoft Press, 1998
4.   Brors D.: Software Highlights für Palm-Rechner, C'T Vol. 7, Apr. 2000, 138-141
5.   C-Technologies, ARIPOS Programming, http://www.cpen.com
6.   Demers A. J.: Research Issues in Ubiquitous Computing, Proc. of the thirteenth annual ACM symposium on Principles of distributed computing, Aug. 14-17, 1994, L.A., 2-8
7.   Ellis C. A., Gibbs S. J., Rein G. L.: Groupware - some issues and experiences, Communications of the ACM, Vol. 34, No. 1, Jan. 1991, 39-58
8.   Joseph A. D., Tauber J. A., Kaashoek M. F.: Mobile Computing with the Rover Toolkit, IEEE Transactions on Computers, Vol. 46, No. 3, March 1997 337-352
9.   Kistler J. J., Satyanarayana M.: Disconnected Operation in the Coda File System, ACM Transaction on Computer Systems, Vol. 10, No. 1, Feb. 1992, 3-25
10.  Lotus Development Corporation: Lotus Notes, http://www.lotus.com/home.nsf/welcome/lotusnotes
11.  Microsoft Outlook, http://www.microsoft.com/outlook
12.  Munson J. P., Dewan P.: Sync: A Java Framework for Mobile Collaborative Applications, special issue on Executable Content in Java, IEEE Computer, 1997, 59-66
13.  Point-to-point short message service support on mobile radio interface, http://www.etsi.org, Jan. 1993
14.  Roth J.: DreamTeam - A Platform for Synchronous Collaborative Applications, AI & Society (2000) Vol. 14, No. 1, Springer London, March 2000, 98-119
15.  Roth J., Unger C.: Using handheld devices in synchronous collaborative scenarios, Second International Symposium on Handheld and Ubiquitous Computing 2000 (HUC2K), Bristol (UK), 25.-27. Sept. 2000
16.  Stabell-Kulø T., Dillema F., Fallmyr T.: The Open-End Argument for Private Computing, First International Symposium on Handheld and Ubiquitous Computing, Karlsruhe, Germany, Sept. 1999, Springer, 124-136

17. Tasker M., Dixon J., Shackman M., Richardson T., Forrest J.: Professional Symbian Programming: Mobile Solutions on the EPOC Platform, Wrox Press, 2000
18. Terry D. B., Theimer M. M., Petersen K., Demers A. J.: Managing Update Conflict in Bayou, a Weakly Connected Replicated Storage System, Proceedings of the fifteenth ACM symposium on Operating systems principles, Copper Mountain, CO USA, Dec. 3-6, 1995, 172-182
19. Weiser M.: The computer for the Twenty-First Century, Scientific American, 1996, Vol. 265, No. 3, Sept. 1991, 94-104
20. Wireless Application Protocol Architecture Specification, WAP Forum, http://www.wapforum.org/, April 30, 1998

## Discussion

*N. Graham:* A lot of design decisions were based on limitations of the current technology. How many of those limitations are fundamental?

*J. Roth:* Some characteristics of handhelds will never change, e.g. display size. Display sizes will not improve since they have to fit in a hand. Another problem is battery life. This is only improving about 10% per year. So battery life will always be a limiting factor. There will certainly be faster CPUs in the future. Also wireless technology will also continue to be low bandwidth and higher error rates.

*P. Van Roy:* Your current design seems to rely on a server. 2 people who meet and would like to share information cannot if there is no server in the network environment. What if you went to a peer-peer style? How would you guarantee coherence? What would you do if there was no server; would people not be able to communicate? There exist protocols, e.g. Ginsella, that are serverless.

*J. Roth*: It is impossible to create a network that is always on because the devices are often switched off. So we need a server.

*P. Van Roy:* ... but gnutella doesnít require that devices are always on.

*J. Roth*: If two devices want to communicate then both devices need to be on.

*S. Greenberg:* Handheld devices are a real pain. I wish I had your stuff a few years ago. How do you handle how information propagated to other devices can be modified, that is, are people modifying the original data, or a copy of it? We ask this because in our own work with propagration of information over PDAs we noticed that people were sometimes confused over whether they were modifying someone elseís personal information or if they were modifying a public copy of this information. Unlike a piece of paper, you canít easily tell if it is a copy or the original. This is also a privacy issue. Have you come across something like this?

*J. Roth*: We avoid this problem. Persons have their own records. Only the originator of a piece of data is allowed to change it. Anyone can create or change their own information, but the ownership always remains with the originator and everyone can know who the owner of the information is.

*F. Paterno:* What kinds of building blocks/support do you have for developing applications, e.g. widgets, platform encapsulations?

*J. Roth*: We are only building an infrastructure, not a complete toolkit for creating applications. We donít offer UI widgets.

# Dynamic Links
# for Mobile Connected Context-Sensitive Systems

Philip Gray and Meurig Sage

Department of Computing Science, University of Glasgow, UK
{pdg, meurig}@dcs.gla.ac.uk

**Abstract** The current generation of mobile context-aware applications must respond to a complex collection of changes in the state of the system and in its usage environment. We argue that dynamic links, as used in user interface software for many years, can be extended to support the change-sensitivity necessary for such systems. We describe an implementation of dynamic links in the Paraglide Anaesthetist's Clinical Assistant, a mobile context-aware system to help anaesthetists perform pre- and post-operative patient assessment. In particular, our implementation treats dynamic links as first class objects. They can be stored in XML documents and transmitted around a network. This allows our system to find and understand new sources of data at run-time.

## 1 Introduction

Early interactive systems just had to deal with mapping input from input devices onto application operations and system changes to output devices. However, current interactive systems exhibit a much higher degree of potential sensitivity to change in surrounding environment. In addition to changes in input they must also handle changes to sensors and incoming information from distributed information sources.

The modelling and implementation of this change-sensitivity is generally ad hoc. Attention has been given to the forms of change that are usefully exploited, to the methods of storage and communication of such changes in distributed systems, but little attention to the general mechanisms for mediating application-oriented links between change in source data and its desired consequential effect.

We propose a general software mechanism to support a variety of related types of change sensitivity. Given their genericity of structure and the dynamic environments in which they have to operate, we have also designed them to be highly configurable and able to respond to changes in their run-time environment.

In Section 2 we provide a general introduction to the notion of dynamic links and discuss their relationship to distributed context-sensitive interactive information systems. Section 3 describes the setting in which our work was carried out, the Paraglide Project, identifying the domain-related challenges that we have tried to meet via our generic configurable link structure. Section 4 presents our model and its implementation in the Paraglide Clinical Assistant. Finally Section 5 offers some conclusions and directions for further work.

M. Reed Little and L. Nigay (Eds.): EHCI 2001, LNCS 2254, pp. 281–297, 2001.
© Springer-Verlag Berlin Heidelberg 2001

# 2    Dynamic Links

## 2.1    What Is a Dynamic Link?

The notion of dynamic link is based on the concept of constraints as value dependencies. That is, some data element e is constrained by condition c if the value of e depends on the value of c. Constraint-satisfaction systems and related constraint specification languages have proved useful for a variety of applications in which value dependencies are volatile and subject to change. A number of successful user interface development environments, for example, have been implemented using constraints to specify the interactive behaviour of graphical elements ([2],[7]).

A dynamic link, in our sense, is a reified constraint. That is, it is a value dependency represented by an object in the run-time system itself that defines a relationship that may result in changes to the state of the link destination based on changes to the link source over the lifetime of the link.

## 2.2    Dynamic Link Structures

In the domain of user interface software, constraint-based mechanisms have been used for at least twenty years, although in their early incarnations the constraints were not always instantiated as links. Smalltalk's MVC provides a notify/update mechanism for creating constraints between view and model components. However, the mechanism is not intended to be visible; it implements constraints as implicit links. It is possible, but not necessary, to create specialised model components that interpose between the source data and the dependent view. Such models can be viewed as dynamic links. Other approaches have made the links explicit and hence configurable [6].

The Iconographer and Represent er systems treat the link as a central configurable element, with special visual programming tools for the configuration ([4], [5]). Little recent work has revisited this issue and we are now confronted with user interface components with complex interactive structures with only poorly configurable interfaces between linked components.

Similarly, dynamic links in hypermedia systems offer the potential to make the usually fixed document associations dynamically configurable, so that they reflect different potential views onto the document or so that they can change to accommodate changes in the remote resources to which the links can point [3].

Modern distributed systems architectures provide mechanisms for the implementation of distributed link structures (i.e., those in which source or destination of the link resides in a remote environment). For example, Elvin [8] and JMS[1] offer facilities for establishing subscription-based notifications and data delivery from remote servers. However, while this supplies enabling technology for the link, it is not sufficient to create the link itself, which often requires access to application-oriented data and operations.

---

[1] JMS is a messaging standard defined for Java by Sun Microsystems (http://java.sun.com/). We are using the iBus//MessageServer implementation of JMS, produced by SoftWired Inc. (http://www.softwired-inc.com).

## 2.3    Dynamic Links in Context-Sensitive Interactive Systems

Our concern in this paper is not with constraints or even dynamic links in general, but in their application to distributed interactive systems, particularly those in which client services are mediated via small, mobile context-aware devices such as PDAs and wearable devices. In this domain, links can serve a number of roles, particularly relating local data elements to

- other local data
- data from remote services
- data from external sensors.
  By generalising over these different forms of link we can
- hide from the link ends the nature of the link, improving reusability via information hiding and
- centralise relevant domain knowledge for use across different link sources and destinations.
  Link effects may vary according to the aspect of the target data that is affected. Thus, the linked source may cause a change in the value of the target (the most common relationship). However, it might also cause a change in the likelihood of certain values being appropriate.

We can distinguish between link update effects. The link may actually cause a change to the value of the destination object or simply notify the destination that an appropriate change has taken place in the link source and let the destination object take appropriate action.

Actual link behaviour is also variable. In some cases, the link is governed by a set of constraints, as in the case of constraints among graphical elements or between multiple views onto the same data. In other cases, complex domain knowledge may be needed to resolve the link relationship. Some links cannot be resolved without the involvement of a human agent, resulting in user-assisted links. Finally, some links depend on contextual information for their resolution; that is, they behave differently depending upon the context in which they are resolved.

Links may have to perform additional work to establish relationships and to maintain them. Thus, if a link has a remote component as a source or target, then it may have to communicate with that remote component, perhaps via middleware, to create a communication channel for transfer of data. New sources and types of data that are relevant as link sources may become available during the lifetime of a context-sensitive system; it is important, therefore, that such a system be able to discover new resources and either configure its links or create new ones to handle these resources [1].

As shall be described later in this paper, taking this high-level conceptual view of links offers potential advantages in the flexibility of software structures to support them. We are unaware of any systems, apart from the one described here, that provides this level of generality in approach.

# 3     Paraglide – An Anaesthetist's Clinical Assistant

## 3.1     Overview

The work reported here has taken place in the context of the Paraglide project which is developing a mobile, wireless context-sensitive system for pre- and post-operative assessments by anaesthetists. The Paraglide system consists of a set of clinical assistants that hold information about current cases requiring assessment, along with associated data. Clinicians use a clinical assistant to collect additional data to record their assessments and to develop plans of drugs and techniques that will be used during the operation.

Clinical assistants communicate with a set of remote services, opportunistically requesting information from these services. This information falls into two general categories: *data* and *task* oriented. The first category covers relevant medical data, such as records of previous anaesthetic records and current laboratory test results. The anaesthetist can view this data and decide whether to incorporate it into their current anaesthetic record. The second category of *task* oriented data covers information that the system can use to help the anaesthetist. Much anaesthetic work is very routine at least for any given anaesthetist. For instance, in some cases only a small set of drugs and techniques are appropriate. The type of operation, along with a few key aspects of the patient's medical record, determine the technique that will be used. The Paraglide system can offer plan templates based on the anaesthetist's previous work that match the current patient's history and the surgical procedure.

## 3.2     Types of Context & Change Sensitivity in the Domain

There are two basic forms of context- and change-sensitivity that the Paraglide system must be able to handle:

- changes to local data on the clinical assistant
- changes to the accessibility of case-relevant data from Paraglide servers (i.e., the generation of requests to such servers and the subsequent arrival of responses to these requests)[2].

Paraglide clinical assistants operate in a wireless environment with intermittent connections. Additionally, they can operate outside the normal clinical environment (e.g., in the anaesthetist's home) where the nature of the connection may be very different, without access to sensitive data that must remain inside the hospital's LAN. Thus the system must also be sensitive to changes in the connection status.

Paraglide users perform tasks that require changes in the organisation of the activity depending upon the context. Thus, a clinician operating on an emergency case will have very different demands on data than one dealing with general or day surgery. Our system must be responsive to these changes in context.

---

[2] Sensor-derived data can be modelled as a form of either local or remote data, depending upon how it is captured and communicated.

## 3.3   Scenarios and Use Cases

A Paraglide clinical assistant usually runs on a small handheld computer to allow anaesthetists to access and enter data on the move. Such systems have a number of input problems. For instance, data entry is often slow and cumbersome. In general the aim is therefore to allow users to select data rather than enter it. The system must therefore be able to predict sensible values for this to work.

When a document, such as a set of blood results, arrives, a title summary is presented in the relevant document interactor. The anaesthetist can open the document in a reader panel and view its contents. They can then choose to paste the details into their current anaesthetic record, or to delete it if not relevant.

Certain elements within the system can be automatically generated. For instance, when the anaesthetist enters the height and weight of the patient, the system can generate the "body mass index" which is used to determine if a patient suffers from obesity.

Mutual dependencies exist between a great deal of the data within the system. These frequently can only be expressed as predictions rather than actual definite changes. For instance, if a patient has a given complaint such as asthma we can predict that they are likely to be on one (possibly more) of a limited set of asthmatic drugs.

Predictions can also depend on remote data. A user could select the operating surgeon from a set of surgeons. This prediction set will change if the staff list changes and a new list is sent out. Anaesthetic plans come from remote servers; the choice of plans is affected by data on the handheld assistant (i.e., the patient's medical record) and on a remote server (i.e., the set of plans available in a database). Matching can take place externally and then the predicted set of plans can be updated.

There are therefore three dimensions of change that we must consider: predictions vs values, distributed vs local data, and implicit vs explicit update (see figure 1). In general, we wish to handle prediction updates implicitly. The user does not want or need to be informed every time the system changes its set of predictions. They need to find out when they attempt to set a value on which a prediction is based. It is also useful to highlight a field if a prediction changes such that the system believes the new value is not valid. The system can do this in a visible manner without interfering in the current activity of the user. In general, we wish to handle value updates explicitly so that the user knows what is in their system. For instance, when a set of blood results arrives the anaesthetist will look at them. The anaesthetic report acts both as an aide-memoire and a legal record. It would therefore be very unhelpful if there was data in the record that the anaesthetist had not explicitly looked at and affirmed as true. Also there is a need to examine where data came from (to see what caused a value update). There can, however, be times when value updates should be implicit; for instance the calculation of "body mass index". These implicit updates generally depend on local data changes. It is important to note that these local changes may have been propagated by distributed updates. For instance, the height and weight could come from a document; once those values have been updated, the body mass index will be calculated.

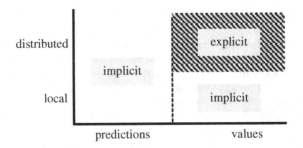

Fig. 1. Dimensions of Change Sensitivity

# 4    The Paraglide Dynamic Link Architecture

## 4.1    Overview of the Architecture

A Paraglide Clinical Assistant consists of four components: a set of interactors (user interface), a set of resources managers, a library of documents and a communications subsystem (a broker). The interactors (or widgets) are used to communicate with users and the broker mediates communication with other Paraglide Services. The relationship among the components is shown in Figure 2.

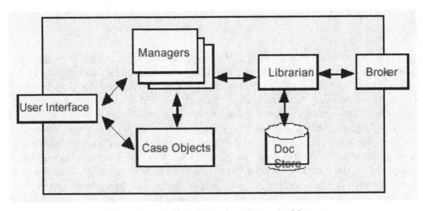

Fig. 2. The Paraglide Clinical Assistant Architecture

Many links depend on domain-related data, e.g., the link between a surgeon's specialty and the surgical procedure for a given case. Such information is held and/or mediated via resource managers. Each manager is responsible for supplying domain-related information to the system as a whole and also for creating links associated with its domain.

Managers may record history about a given topic. For instance, the scene manager can record navigation paths used by the anaesthetist to suggest possible future routes.

They can also preload and provide access to lists of data. For instance, the worker manager can load a set of surgeon records at start up.

Resource managers maintain data used within the system. There are four different types: the scene manager maintains the collection of scenes and handles and records navigation; the case manager produces new cases and provides access to them; the technical environment manager maintains data about the system such as battery power and network connectivity; and domain knowledge managers maintain data about drugs, procedures, staff etc. The case manager produces case objects that represent an object graph maintaining all data about a given case within the system.

We must therefore handle two sorts of links: those that maintain consistency between different elements of a case object; and those that import document data. Our fundamental approach is to try to unify the link framework, so that we can cope with change-sensitivity in a principled way. Therefore all links are viewed as associations that relate a source object to a destination object with respect to an aspect, or operation, via a link function:

```
link = <source, destination, operation, link_function>
```

A link goes from a given source to a given definition, applying some form of link function to transform the data from the source, and then performs some operation on that destination. For instance, a link could go from a document to the blood test results set, extracting all blood results from the document and transforming them into Java objects, before adding them to the blood test results collection.

Because links depend on potentially variable relationships and because they must be created at run-time, we also include link specification as an explicit element in the architecture. A link specification is an object that holds the information necessary to create a link of a specified type, defined in terms of the types of its arguments.

```
linkspec = <source_type, destination_type,

            operation_type, function_type>
```

As we shall see, it is occasionally useful to have multiple sources and frequently useful to have multiple destinations. For example, a pre-operative examination document (a link source) might have information relevant to patient medical history, current medications and other clinical issues, all of which are contained in different parts of the anaesthetic record and that appear in different parts of the user interface (viz., several link destinations). We can therefore think of a link as a *set* of sources, and a *set* of tuples of destination, operation and function.

Furthermore, to enable configurability, link specifications are written in XML. They can therefore be stored in documents and transferred around a network. New links types can be added without the need for recoding. They can, in fact, even be added while the system is running, thus enabling *dynamic* reconfigurability. For instance, if a new document type were to be added in a hospital, an update could be sent out to allow all Personal Clinical Assistants to interpret it, without any need to disrupt the users of the system.

## 4.2    Link Sources

There are two sorts of link source, document and value sources. It is important to note
that these sources are not simple documents or values but sources that will *provide*
new documents or values over the duration of the program.

### 4.2.1    Document Sources

In the Paraglide system, information is transmitted between services as documents,
i.e., structured text. These documents may contain information relevant to a number
of local links. This information must be extracted from incoming documents in order
to resolve the links dependent on that document.

Managers talk to brokers (through the librarian) in order to request documents.
These requests are made through the following interface. A manager generates a topic
that says what documents to get. This topic specifies which service to go to for the
data, and what to say to the service. For instance, it may specify an SQL query to
extract data from a database service.

```
public void addDocumentRequest(PgTopic topic,

                          DocumentListener l);
```

The manager provides a document listener that is used to consume relevant
documents. The document listener interface contains one method handleDoc which
consumes a document and performs some action with it. The document is also stored
in the document library until released by the consumer.

```
public void handleDoc(Document d);
```

This interface is inspired by, and built on top of the Java Messaging Service (JMS).
iBus//MessageServer, the JMS implementation we are using, guarantees document
delivery under conditions of intermittent connectivity, supporting the use of wireless
connected mobile systems.

It is significant to note that JMS uses a JavaBeans style model for interaction. We
provide a listener which is a callback function defining what to do with the given
input. This JavaBeans model is the same one used by Java user interface components.
This mapping makes it possible to unify information gathered from remote and local
sources.

To use a source we must generate a specific request from a given service. We can
specify a document source in XML in two parts: a *from* attribute specifying the
service source and a *request* attribute specifying the query to be made of the remote
service. For instance, the following source specification queries the *LabResults*
service for all blood results for a given patient.

```
<PgDocSource

  from="LabResults"

  request="select BloodResults where

              subject={/case/pgSubject/hospitalNumber}"/>
```

Note that the above example highlights two important aspects of link specifications Firstly, we can use a simple path-based syntax for referring to the elements that make up the link, including documents and the attributes of local values (further details of both context paths and document descriptors will be given below). Secondly, the specification can refer to current data within the system, i.e. references that will be resolved at run-time. Here we specify the hospital number of the patient in the current case.

### 4.2.2   Value Sources

The other link source is a local value within the data structure. For instance, we may have a link between a client and her specialty.

Paraglide uses a general JavaBeans model for all data structures. All mutable objects are *active values*, i.e., we should be able to listen for and react to changes in a value. Paraglide distinguishes two sorts of active value: *Attributes* and *Collections*. Attributes contain simple object values that satisfy the Java Beans PropertyChangeListener interface. We can add a listener to hear about changes to the value.

```
public void addChangeListener(PropertyChangeListener
l);
```

The second form is DynamicCollections. Consider a dynamic list of items. We may not want to hear about the whole change, but only be notified about incremental changes to the collection. We can therefore add listeners to hear about changes such as additions and deletions from a list.

```
public void addListDataListener(ListDataListener l);
```

Based on these two source types we have two forms of source specification. Value sources specify a particular attribute; collection sources specify a collection and an operation (we can do something when an item is either added to or deleted from a collection). Again note that we use context paths to specify a route to a given object in the current case data structure.

```
<PgValSource from="/case/client"/>
```

```
<PgCollSource op="add" from="/case/regularMedication"/>
```

A simple model might assume only a single source for any given link. In fact, it is sometimes useful to have multiple sources. For instance, the "body mass index" calculation depends on two sources: the height and weight attributes.[3]

### 4.3   Link Operations

Once we have a source we require an operation to perform with that source result. There are two general types of operation:

- updates – operations which explicitly update some data structure,
- notifications - operations which notify a data structure about a set of updates.

---

[3] In practice, we have found only a few instances where the use of multiple sources was really necessary. None of these involve document sources.

### 4.3.1    Updates

The simplest of the two forms of operation is update: an operation can explicitly update some data. For instance, when the surgeon changes we update the specialty, by setting the value. Again we have a distinction between collection and attribute destinations. With an attribute destination we can set it with a given value. In contrast, with a collection destination we can reset the collection, add or delete one or more items to or from the collection.

### 4.3.2    Notification

An operation can notify some destination object about a set of changes. Recall that when an anaesthetist receives a new document, such as a set of blood results, it is not immediately integrated into the system. The anaesthetist has the opportunity to review the document, and decide whether it is indeed accurate and relevant. The anaesthetist can then either accept the contents and paste them into the anaesthetic record or reject them, thereby deleting the document.

This form of activity is supported by notification. A notification operation contains a summary function that specifies how to summarise the document. It also contains a set of sub-operations that specify what to do if the notification is accepted. Each of these sub-operations will extract some data from the source. We can generate a summary based on these extracted elements.

For instance, we can have a message that notifies the relevant data structure with a given summary and an operation that adds a new element for each blood investigation to the system.

```
<PgNotify to=...

    summary="Blood Results {/PgBloods/@datimPublished}">

    <PgOp op="add" mode="collection"

        to="/case/bloodInvestigations">

    . . .

    </PgOp>

</PgNotify>
```

Note that the summary is a combination of static text and extractions. Here the source is a document so the extraction rule is an XML query. In contrast, if the source were a local value then the query would be a context path query.

What do we do with a notification? We send a *NotifyEvent* to the destination. This provides a summary, and two methods *accept* and *reject*. The summary method provides a *PgSummary* with a title summary (which can, for instance, be viewed in the relevant document list) and a list of child summaries that summarise the data that is extracted from the source for each child operation. The *accept* method accepts all the notification. If some of the child operations fail to work, we throw an exception summarising all the failures. The *reject* method rejects the notification. If this were a notification in response to a document, it would delete the document from the library document store.

```
public interface NotifyEvent {

  public PgSummary summary();

  public void accept() throws Exception;

  public void reject();          }
```

In practice so far we have used notification updates only with document sources. We have generally opted for explicit updates only with distributed data. However, the mechanism is available here if necessary.

While we can have one operation associated with a link function it is more helpful to have multiple operations, both for ease of specification (we can write the source only once) and efficiency (we can add only one listener that does several things, perhaps eliminating some of the common work).

### 4.4    Link Destinations

#### 4.4.1    Value Destinations
With a value destination we are changing the actual value, such as setting the surgical specialty. As outlined in the previous section there are several possible operations that can be performed on the destination. If the operation is a notify operation then we will simply notify the destination. In this case it must be notifiable object (i.e. able to accept a *NotifyEvent*). Otherwise if the destination is a collection then we will have a collection operation, if an attribute the operation will simply set the value.

```
<PgOp mode="value" op="set" to="/case/specialty">
```

#### 4.4.2    Predictions
Changes to the probability of values in the destination object have been implemented via a *predictions* component. A prediction identifies three subsets of the value-set for a destination object:
- a default
- a likely subset
- the entire value-set.

Although primitive, this provides a potentially useful way of offering alternatives to the user, especially where there is a large set of enumerated alternatives.

More sophisticated prediction models, and more sophisticated ways of utilising the predictive information, are possible. For instance, one can give probability values to destination alternatives. Also, the source of the prediction can be identified where several prediction-changing links are active on a single property.

A prediction update will generally change only the likely subset. We could imagine situations in which the system could attempt to forbid a particular value. For instance, an expert system might predict that a patient could not be on two drugs simultaneously. However, such an approach is very heavy-handed and assumes the system is always accurate. We have chosen a lightweight approach in which updates affect the likely subset by adding or deleting suggestions from it. The operations used here are therefore collection operations.

The default is then simply the most likely element in the likely subset.

```
<PgOp mode="prediction" op="add"

      to="/case/regularMedication">
```

## 4.5    Link Functions

We may wish to perform some arbitrary transformation on the data before applying the operation. To do this we use link functions. A link function is a function that performs a simple *apply* transform to a piece of data.

```
public interface PgLinkFunction {

  public Object followLink(Object context,Object obj);}
```

It takes a context object (described below) and a value and generates a new value based on this input. For reasons of efficiency, it is important that the *link-function* is a *pure* function. That is, it transforms the data without any side-effecting updates. If applied at any given time in the program to the same value it should therefore return the same result. Given these conditions we can precompile link functions in advance so that the difficult work is done at start-up, not each time the *followLink* function is called.

We support several types of link function, based on the nature of the link. These include property queries, xml queries, maps and ranges, constructors, and predefined functions. A link function can also be a composition of these functions.

### 4.5.1    XML Queries
The first two forms of link function are both types of query that extract data from the argument value. As we have seen so far there are two sorts of data that we may wish to query: local data and incoming documents.

An extract query is the first type of query. It contains two parts: an actual query and a result type. The result type can be either *collection* or *value*. An arbitrary *XMLQuery* will generally return a set of results. However, sometimes we only wish the first result form a query that we know will return a result. In this case we can use the *value* result type to return only one (i.e., the first) result.

An *XMLQuery* is based on the developing *XQL* query standard[4]. In our initial work we have only used and implemented a subset of this query mechanism. The format of a query is based on a UNIX path structure, consisting of a set of entity names separated by backslashes, e.g., "/PgBloods/PgBlood". The query can start at the root "/" or within the current context "./".

---

[4] http://www.w3c.org/

What is the current context? Remember that we're applying this query to a value. This value may be a document or it may be the result of some earlier query. This happens often in Constructor link-functions, detailed below. The context parameter in the link-function argument provides access to the root document that the source provided. We can precompile all query link functions for a given document source. We can preprocess a document when it arrives, extracting all necessary data, allowing us to parse a document once. This is particularly useful if several link-functions extract the same data.

A query can return one of four results: an attribute or a set of attributes (e.g. return the docId attribute of PgBloods entity is /PgBloods/@docId); the character data residing under an entity (eg return the text string child of /Name/-); an entity (e.g. return the PgBloods entity and attributes /PgBloods); or a whole document subtree (e.g., return the whole PgBloods entity, attributes and children).

The following extract function extracts all the PgBlood subtrees and returns a collection of them.

```
<Extract type = "collection"

          query="/PgBloods/PgBlood#">
```

### 4.5.2    Property Queries

A property query extracts a value from a local data structure. The JavaBeans model introduced the notion that all values in a bean have a String property name through which they can be accessed. A property query is based on this idea. For instance, the following property link function extracts the specialty field from its argument.

```
<PgProperty value="specialty"/>
```

A property query is in fact a ContextPath. We can apply a context path to a Context object to yield a result. Every object within the Java case object implements this Context interface.

```
public class Context {

 public Object find(ContextPath p);

}
```

A ContextPath is in fact more complex than a simple field name. It does in fact have similarity to an XMLQuery, involving a descent path which is a set of field names separated by a backslash e.g., ./subject/age. The descent path can begin either at the root or at the current value. Context paths into collections require some extra handling. We can specify either the location of an item in a list (e.g. ./1) or a query on the items within a collection (e.g. ./findings/[type='weight']).

### 4.5.3    Maps

One very useful type of function is a map from keys to values. These occur commonly in prediction links. For instance, consider the case where the procedure prediction depends on the specialty of the surgeon. We might have a map from specialty names to workers. Every time we change the specialty value we look up the new value in the map and generate a new set of likely values.

It would, however, be very tedious to have to specify all of these maps by hand. For instance, we have an XML data file containing data on the list of surgeons. Each entry in the file contains a specialty field. We would like to be able to generate the map from this file. This requirement becomes even more important with medical history data with a number of different dependencies. There are links between *issues* such as asthma, *drugs*, *findings* and *measurements*, and sometimes *operations*. For instance, a coronary bypass operation implies one of a set of serious heart conditions and likely drugs.

Our link functions allow maps to be generated from data files. We specify a file type, which provides access to the data; a *root* query to apply to the file and a set of *from* and *to* queries. Each of these queries is an XMLQuery. The *root* query extracts a set of document objects. The *from* query then extracts the map keys from each object. The *to* query extracts a result type. We may have one or more *from* and *to* queries. For instance, the following link function says: "extract from the PgWorkers xml source, the Worker entities; then generate a map where the keys are the specialty values of the worker entities, and the values are lists of worker entities themselves".

```
<MapExtract file="PgWorkers"

        root="PgWorkers/Worker"

        from="./@specialty" to="./">
```

The use of pure functions is particularly important here. We can precompile the link-function, reading in the data once and then applying all following transformations to the results. This means that when a change actually occurs we perform a simple hash-map lookup, which is cheap to perform.

### 4.5.4    Ranges

Ranges are very similar to maps. Given a value, we calculate which of a set of non-overlapping ranges it lies within and generate a set of likely results. For instance, if a patients "body mass index" is greater than a given value, they are likely to be obese. We can specify these using ranges. The measurement entity has two important attributes *minValue* and *maxValue*. We can therefore generate a list of ranges from measurement ranges to issues. We can generate a list of ranges and then simply perform a binary search.

```
<MapExtract file="PgIssues"

        root="PgIssues/Issue"

        from="./Measurement"

        to="./"

        op="range">
```

### 4.5.5    Constructors

A *PgConstructor* object converts data extracted from an XML file to a Java object, taking as parameters the target object's class name and additional parameters as necessary. These additional parameters specify Extract queries. If we're generating the object from an XML source then these will be XML Extract queries. Each of these queries generates one parameter. These queries may have children, i.e., we may have a query that extracts a collection of values, and applies a further constructor to that result.

For instance, consider the following constructor. It generates a *PgInvestigationBlood* object. It extracts the date attribute as its first parameter and then generates a collection of blood results, one for each result value for its second parameter.

```
<PgConstructor ref="PgInvestigationBlood">

<Extract type="value" query="./@date" />

    <Extract type="collection" query="./Result/@value" >

      <PgConstructor ref="PgInvestigationBloodResult"/>

    </Extract>

</PgConstructor>
```

For this all to work, we need a static Java method which generates instances; this resides in the data manager.

```
public static Object getInstance(String name,

                                Object[] params);
```

### 4.5.6    Predefined Functions

Sometimes the set of functions defined above is not enough. In this case we can call preprogrammed Java link functions. This is most common for arithmetic calculations. For instance, we can calculate the age based on the date of birth. We define a calcAge method in Java and then call it.

```
<PgPredefined value="calcAge"/>
```

The data manager also provides access to such predefined methods.

```
public static PgLinkFunction getLinkFunction(String
name);
```

# 5    Conclusions and Future Work

The generic link structure described in this paper is still in its infancy. We have implemented a restricted prototype for the Paraglide system and intend to test it in field trial-based setting in which the links will reflect the anticipated information needs of clinicians.

There remain a number of features that require further development, the most important of which are:

- bidirectional links
- mechanisms to identify and cope with pathological links (e.g., circular link sets)
- tools for specifying links and link functions
- more sophisticated predictions, including predictions from multiple sources.

We also envisage our link architecture offering additional functionality to the application, such as context-sensitive help.

**Acknowledgements**

This work was supported by EPSRC under the Healthcare Informatics Initiative (Grant GR/M53059). We wish to thank our colleagues on the Paraglide project (Chris Johnson, Gavin Kenny, Martin Gardner and Kevin Cheng) for their feedback on the ideas expressed in this paper and for their contribution to the design of Paraglide system. We also thank the referees for their helpful comments.

# References

1. G. Abowd and E. Mynatt. Charting Past, Present and Future Research in Ubiquitous Computing. Transactions on Computer Human Interaction 7,1 (March 2000), 29-58.
2. A. Borning. Thinglab - A Constraint-Oriented Simulation Laboratory. Ph.D. thesis, Stanford University, 1979.
3. L. A. Carr, D. DeRoure, W. Hall and G. Hill. The Distributed Link Service: A Tool or Publishers, Authors and Readers. Proc. 4th International World Wide Web Conference. Pp. 647-656.
4. P.D. Gray and S. Draper. A Unified Concept of Style and its Place in User Interface Design. Proc HCI '96. Springer-Verlag. pp. 49 -62.
5. P. D. Gray, "Correspondence between specification and run-time architecture in a design support tool," in Bulding Interactive Systems: Architectures and Tools, P. D. Gray and R. Took, Eds.: Springer-Verlag, 1992, pp. 133-150.
6. R. D. Hill, "The Abstraction-Link-View Paradigm: Using Constraints to Connect User Interfaces to Applications," Proc. CHI '92, 1992.
7. B.A. Myers., et.al. Garnet: Comprehensive Support for Graphical, Highly-Interactive User Interfaces. IEEE Computer 23, 11 (Nov. 1990), 71-85.
8. Bill Segall, David Arnold, Julian Boot, Michael Henderson and Ted Phelps, Content Based Routing with Elvin4, (To appear) Proceedings AUUG2K, Canberra, Australia, June 2000.

# Discussion

*J. Hohle:* It sounds like you are talking about consistency issues. Have you investigated reason or truth maintenance systems, where you can model information dependencies?

*P. Gray / M. Sage:* We are trying to build a lightweight system. There are heavy systems that will do this? Our system needs to work on handheld devices. The power of these links is the fact is that they can be connected to local and remote services. Decision support systems are far too heavy for handheld devices. Perhaps you could îplug inî queries to a remote server which could do further processing.

*C. Yellowlees:* How do you resolve the desire to employ these high-powered back-end systems with the constraint that the users in this application domain do not have persistent connections to the network.

*P. Gray / M. Sage:* The resource manager may be adapted to perform predictions locally when no network is available, and query a more powerful prediction tool on the network if it detects that such a resource is available.

*F. Paterno:* In your application, do you really need a mobile device? Or would it be sufficient to just have a computer in the patientís room connected by a LAN?

*P. Gray / M. Sage:* The hospital that we are working with is very large, so it is more useful, and cheaper, to have them on their handheld device (which the anesthetists already have and already carry). It is essential for this kind of system, given the number of places where the doctos move (including by the bedside) to always have access to the information, so this is a cheaper, more practical solution.

*J. Roth:* Are your handheld devices really mobile? Are they actually notebooks or something like a Palm?

*P. Gray / M. Sage:* We are currently using tablets and are moving to a Compaq IPAQ.

*J. Roth:* How is the design influenced by using mobile computers?

*P. Gray / M. Sage:* The information from services in a hospital will be consistent, but the doctor can use the information anywhere in the hospital.

*K. Schneider:* Given the goal of a dynamic lightweight application, how are you ensuring that the data is reliable and trustworthy? How successful has the application been? Are the doctors able to trust the info? Are they using it? What about the system changing over time?

*P. Gray / M. Sage:* The full trial has only been running for a month. So far the doctors seem to trust the information. The doctors are using our system and then printing a paper backup later. Also, it is important to note that right now their data is often not very timely (paper records transcribed by a secretary, and not delivered in a timely fashion). It is important that any changes in the system over time are visible to doctors.

# Mobile Collaborative Augmented Reality: The Augmented Stroll

Philippe Renevier and Laurence Nigay

CLIPS-IMAG Laboratory, IIHM team
BP 53, 38041 Grenoble Cedex 9, France
{Philippe.Renevier, Laurence.Nigay}@imag.fr

**Abstract.** The paper focuses on Augmented Reality systems in which interaction with the real world is augmented by the computer, the task being performed in the real world. We first define what mobile AR systems, collaborative AR systems and finally mobile and collaborative AR systems are. We then present the augmented stroll and its software design as one example of a mobile and collaborative AR system. The augmented stroll is applied to Archaeology in the MAGIC (Mobile Augmented Group Interaction in Context) project.

## 1 Introduction

One of the recent design goals in Human Computer Interaction has been to extend the sensory-motor capabilities of computer systems to combine the real and the virtual in order to assist users in interacting with their physical environments. Such systems are called Augmented Reality (AR) systems. There are many application domains of Augmented Reality (AR), including construction, architecture [26] and surgery [8]. The variety of application domains makes it difficult to arrive at a consensus definition of AR: i.e. different people, having distinct goals are using the term "Augmented Reality". In [8], we presented an interaction-centered approach for classifying systems that combine the real and the virtual. By considering the target operations (i.e., in the real or virtual world), we made a clear distinction between Augmented Reality (AR) and Augmented Virtuality (AV):

1 In AR, interaction with the real world is augmented by the computer.
2 In AV, interaction with the computer is augmented by objects and actions in the real world.

In this paper, we focus on AR systems as defined above and we describe our MAGIC (Mobile Augmented Group Interaction in Context) system. The application domain is Archaeology. MAGIC supports fieldwork carried out by archaeologists such as taking notes, sketching objects and taking pictures on site. The target of the task is clearly in the real world, the archaeological site: MAGIC is therefore an AR system. Moreover archaeologists are mobile within the archaeological site and they need to collaborate with each other. Thus, MAGIC is additionally a mobile and collaborative AR system.

M. Reed Little and L. Nigay (Eds.): EHCI 2001, LNCS 2254, pp. 299–316, 2001.

The structure of the paper is as follows: We first define what mobile AR systems, collaborative AR systems and finally mobile and collaborative AR systems are. We present related works and characterize existing systems highlighting the power and versatility of mobile collaborative AR systems. We then describe our MAGIC system and we emphasize one user interface component, the augmented stroll, and its software design. The augmented stroll is related to the mobility of the user as well as to the collaborative aspects of the AR system.

## 2  Augmented Reality: Mobility and Groupware

### 2.1  Augmented Reality and Mobility

First AR systems were designed for a specific use in a fixed environment such as the digital desk [27]. Progress made in wireless networks (RF, Radio Frequency and IR, InfraRed, signals) in terms of quality of services make it possible to build mobile augmented reality systems [11]. We believe that mobile AR has a crucial role to play for mobile workers, bringing computer capabilities into the reality of the different workplaces. Let's envision an augmented reality system that will help in deciding where to dig in the streets to access the gas tubes. Similar systems already exist such as the Touring machine system of the project MARS (Mobile Augmented Reality Systems) [10] or the NaviCam system [18]. The user, while walking in a building such as a museum, in the streets or in a campus, obtains contextual information about the surrounding objects or about a predefined path to follow.

> Definition: A mobile AR system is one in which augmentation occurs through available knowledge of where the user is (the user's location and therefore the surrounding environment).

Even though the user's location has an impact on the augmentation provided by the system, the latter does not necessarily maintain this location. Indeed, as explained in [11], on the one hand, the user's location and orientation are generally known by outdoor systems such as the Touring machine system, the position being tracked by a GPS. On the other hand, for indoor AR systems, objects and places identify themselves to the system (RF, IR or video based tags): hence the system does not maintain the user's location. Going one step forward, in [11], objects are not only tagged for identification but also contain a mobile code that for example describes the virtual object, i.e. augmentation of the real object.

### 2.2  Augmented Reality and Groupware

Several collaborative AR systems have been developed. We focus on collaborative systems that enable a group of users to perform a task in the real world, as defined above. Systems such as the StudierStub [20] that allows multiple collaborating users to simultaneously study three-dimensional scientific visualizations in a dedicated room is not part of our study because the task of studying a virtual object, is not in the real world. The shared real environment of the two users is augmented by the computer but the task remains in the virtual world.

In a collaborative AR system, augmentation of the real environment of one user occurs through the actions of other users. As a counterexample, let us consider the system NetMan [2]. One user, a technician is mobile fixing the computer network of a University. While looking at cables, s/he is able to perceive more information displayed in the head-mounted display, including what the cable is connected to. Netman is therefore an example of mobile AR, as defined above. In addition the technician can orally communicate with an expert, the expert seeing what the technician is seeing thanks to a camera carried by the technician. But the real environment of the technician is not augmented by information defined by the expert. Netman is therefore a mobile AR system that also provides human-human communication services, in other words, a mobile AR and collaborative system but it is not a collaborative AR system if we refer to the following definition:

> Definition: A collaborative AR system is one in which augmentation of the real environment of one user occurs through the actions of other users and no longer relies on information pre-stored by the computer.

A review of the literature enables us to classify the collaborative AR systems into three categories, as schematized in Figure 1. We first consider the classical distinction in groupware [9], that is the distance between users. We also take into account the distance between one or several users and the object of the task. Because the object and/or its environment is augmented, at least one user must be next to the object or else the system is no longer an augmented reality one and falls into the collaborative tele-operating class.

The first category, namely remote collaboration in one augmented reality, includes systems in which at least one user is physically next to the object of the task and some users are distant. For example in [12] and in [13], two systems dedicated to repairing a physical object, the user is next to the object while an expert is at distant position. The real environment of the user is augmented by information provided by the expert.

The second category, namely remote collaboration in augmented realities, encompasses systems where there are several objects of the tasks, remotely linked together and physically present in different sites. Each user performs actions on their own physical object of the task. This is for example the case of a collaborative augmented whiteboard [22]. The real environment (the office with whiteboard) of each user is augmented by information provided by others. Such a system implies that the real environment of each user share common attributes or objects.

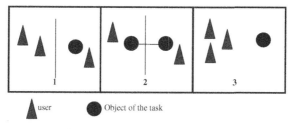

**Fig.1.** Three categories of collaborative AR systems: (1) remote collaboration in one augmented reality, (2) remote collaboration in augmented realities and (3) local collaboration in one augmented reality

The last category, namely local collaboration in one augmented reality, represents systems where all the users are positioned together next to the object of the task. The Shared-Space System [4] and the collaborative Mah-Jongg game [23] are such examples. The users are all together, and the shared physical environment is augmented by information and action from the users. Although they are all together (same physical environment), the key point is that their augmented environments are different.

## 2.3  Augmented Reality: Mobility and Groupware

Mobile collaborative systems are rapidly finding widespread use due to the recent progress in networking technology. For example, a new protocol of continuous real time transport between a wireless network and a fixed network such as Ethernet is presented in [17]. This protocol is compatible with the quality of service of the current wireless networks. Moreover the studies carried out by the UMTS consortium [25] foresee, in the short run, flows of data of about 2Mbit/s. Finally hierarchically structured networks combining total networks with local area networks such as Bluetooth [5] constitute a promising approach for providing quality of service in developing CSCW for mobile users. An example of existing collaborative systems is ActiveMap [14] that enables users to locate on a map all participants who carry infrared badges. Another very different example is RAMSES [1], in the archaeology domain. Each archaeologist in the field takes notes on a Palmtop connected to a radio frequency (2 Mb a second) network so that notes can be shared by the group of archaeologists working in the same field.

Although mobile collaborative systems are now possible and systems already exist, and while some existing AR systems are mobile and some are collaborative, few AR systems combine the mobile and collaborative aspects. Mobile and collaborative AR systems combine the characteristics of mobile AR and collaborative AR, in other words:

> Definition: A mobile and collaborative AR system is one in which augmentation occurs through available knowledge of where the user is and what the other users are doing.

The only existing system as far we know is the collaborative augmented reality game, called WARPING [21]. Instead of recreating a virtual world, the game is based in the real world, the system only adding the magical possibilities related to the game. Nevertheless the current version of WARPING [21] is not completely mobile - one user being in front of an augmented desktop while the second one makes movements (hand gestures and head) but is assumed to stay in about the same place in front of the walls of the room. But imagine a collaborative AR game, where a participant acquiring new magical powers because of certain actions can see through the walls and discover other participants looking at him. Clearly a role playing game is a great application domain for mobile collaborative AR because the designer can focus only on the magical possibilities, related to the game, that augment the physical world.

In the following sections, we will present MAGIC, a mobile collaborative AR system that supports fieldwork carried out by archaeologists. This is another suitable application domain for mobile collaborative AR because:

- The archaeological site is not simulated by the system so we only focus on its augmentation.
- Archaeologists are working in groups in an archaeological site.
- Archaeologists need extra information from databases, from their colleagues working on the site and from experts (such as an expert on water-tanks).
- Archaeologists need information about found objects in the archaeological site. It is important to note that found objects are removed from the site, before starting a new stratum. The exploration of a site is organized according to stratums or levels: the main assumption is that objects, found within a given stratum, are more recent than ones found deeper.

We first give an overview of the MAGIC system, its equipment and infrastructure. We then focus on one user interface component the augmented stroll and its software design.

The design of the MAGIC system is based on a study of the tasks of archaeological fieldwork, from literature, interviews and observations in Alexandria (Egypt). We organized the identified tasks according to the functional decomposition of the Clover Model: Coordination, Communication and Production [19]. The archaeological fieldwork in Alexandria is time-constrained because the archaeological site must be explored in less than three months (rescue archaeology). Tools that can make such fieldwork more efficient are therefore important. To do so the main idea of the MAGIC project is to allow analysis of data directly on the site.

## 3  MAGIC: Equipment and Graphical User Interface

### 3.1  Equipment

We base the system on the paper metaphor. We therefore chose a computer whose use resembles that of a note pad or a paper sheet. The pen computer Fujitsu Stylistic was selected. This pen computer runs under the Windows operating system, with a Pentium III (450 MHz) and 196 Mo of RAM. The resolution of the tactile screen is 800x600 pixels outdoor and 1024x768 pixels indoor. In order to establish remote mobile connections, a WaveLan network by Lucent (11 Mb/s) was added. Connections from the pen computer are possible at about 200 feet around the network base. Moreover the network is compatible with the TCP/IP protocol.

As shown in Figure 2, Augmented Reality needs dedicated devices. First, we use a head mounted display (HMD), a SONY LDI D100 BE. Its semi-transparency enables the fusion of computer data (opaque pixels with a 800x600 resolution) with the real environment (visible though transparent pixels). Secondly, a Garmin GPS III plus is used to locate the users. It has an update rate of one per second. The GPS accuracy is of one meter at the University of Grenoble (France) and of 5 centimeters in the Alexandria archaeological site (Egypt, International Terrestrial Reference Frame ITRF).

**Fig. 2.** A MAGIC user, equipped with the HMD and holding the pen computer

On the tactile screen, the location of each user is displayed on top of the site map, allowing coordination between users. The GPS is also useful for computing the position of a newly found object and removed objects. Finally, capture of the real environment by the computer (real to virtual) is achieved by the coupling of a camera and a magnetometer (HMR3000 by Honeywell that provides fast response time, up to 20 Hertz and high heading accuracy of 0.5° with 0.1° resolution). The camera orientation is therefore known by the system: thus the latter can automatically add the orientation (an arrow to the North) on the pictures taken. (Archeologists used to put a physical rule on the floor showing the north before taking a picture.). As shown in Figure 2, when the magnetometer and the camera are fixed on the HMD, in between the two eyes of the user, the system is then able to know the position (GPS) and orientation (magnetometer) of both the user and the camera.

### 3.2  User Interface on the Pen Computer

The description of the user interface is organized according to the functional decomposition of the Clover Model: Coordination, Communication and Production. Nevertheless it is important to note that the user interface is not designed according to the Clover Model: indeed some components of the user interface are used for several purposes, such as coordination and production. Figure 3 presents the graphical user interface displayed on the tactile screen of the pen computer.

**Coordination**: Coordination between users relies on the map of the archaeological site, displayed within a dedicated window (at the top of Figure 3). The map of the archeological site is the common place of interactions. It represents the archaeological field: site topology, found objects and archaeologists location. Each type of information can be displayed thanks to magic lenses [3] ("grid" magic lens for topology, "objects" magic lens and "users" magic lens). This map is shared by all archaeologists and allows them to coordinate with each other.

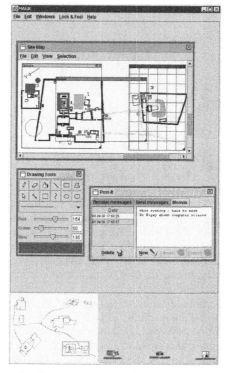

**Fig. 3.** User interface on the pen computer

The site is displayed in two distinct zones on screen: the detailed site map at the top of Figure 3 and a radar view at the bottom left of Figure 3. The radar view shows a non detailed complete site map. On top of the view, a green frame represents the part of the site detailed in the site map window. Visual consistency between the two parts is guaranteed:

- By scrolling in the site map window (small precise movements), the frame is moved in the radar view.
- By dragging the frame, the part of the site displayed in the site map window is modified. By resizing the frame, the view displayed in the site map window is zoomed in/out.

Resizing the site map window is another way of zooming in/out, the same part of the site being displayed in a bigger/smaller window. We define three levels of zoom for displaying the objects. First the most detailed view enables the user to see all the objects. At the intermediate level, small objects, such as a coin or pottery, are not displayed. Finally, the less detailed view only displays very big objects such a wall or water-tank.

The current position of each user is displayed on the map. A user is represented by a circle. The color of a circle translates the degree of accessibility of the corresponding user. We base the choice of colors on the traffic lights metaphor (green = available for others, orange = can eventually be disturbed and red = do not disturb).

The accessibility window of Figure 4 enables the user to set her/his degree of accessibility.

**Fig. 4.** Accessibility window

In addition the system must manage the cases of lost network connections. This is for example the case of an archaeologist going inside a water-tank. On the screen of other users still connected, the circle corresponding to the user having lost the connection is stationary (last known location) and blurs progressively. On the screen of the user having lost the connection, all the other users are represented by stationary gray circles, immediately informing the user of the lost connection.

**Communication**: Three communication tools are implemented. The first one is a post-it enabling a user to post a note on the screen of another user. Text can be typed in using a transparent virtual keyboard, as shown in Figure 5. Speech recognition is another way of specifying text that we have not yet integrated. A voice recording also supplements the textual post-it so that users can send a voice post-it. The selection of a user (recipient) is made by dragging her/his corresponding circle in the post-it window or by typing her/his name if the user is not inside the site such as for the case of a distant expert. The second tool is a chat room as shown in Figure 5. Again, the specification of the participants can be performed by dragging their corresponding circles from the sitemap window to the chat room window. The last communication tool enables users to share pictures, and discuss them. One example is shown in Figure 5. We use sounds and textual messages displayed on the HMD to keep the user informed of the arrival of a new post-it or request for communication. All the services provided by the mediaspace developed in our team [6] can be incorporated in MAGIC (video and audio communications).

**Production**: Production covers two complementary activities: describing found objects and analyzing them. For each found object, archaeologists fill a form describing the object, draw some sketches or very precise drawings and take pictures. In particular, the object must be located in the site (stratum, location and orientation). When an archaeologist describes an object, the description is maintained locally on the pen computer. Using a toolglass, s/he can then validate the new object. After validation, the object is then added to the shared database and is visible on the map of each user. Analyzing objects mainly consists of dating them: dating an object enables the archaeologists to date the corresponding stratum and then continue the fieldwork. Analysis of objects relies on comparisons with known objects from other archaeologists or reference manuals and on discussions with other archaeologists in the site or with a distant expert.

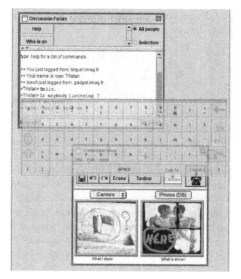

**Fig. 5.** Human-Human communication tools

The next section describes the links between the real and the virtual in order to assist the archaeologists in describing objects as well as analyzing them.

# 4   Augmented Stroll

The virtual world corresponds to all the information maintained by the computer: database of objects, reference manuals etc. The real environment is the archaeological site. In order to smoothly combine the virtual and the real, we create a gateway between the two worlds. This gateway has a representation both in the virtual world on the screen of the pen computer (bottom right part window in Figure 3) and in the real environment, displayed on the HMD.

Information from the real environment is transferred to the virtual world thanks to the camera carried by the user. The camera is positioned so that it corresponds to what the user is seeing, through the HMD. The real environment captured by the camera can be displayed in the gateway window on the pen computer screen as a background. Information from the virtual world is transferred to the real environment, via the gateway window, thanks to the HMD. For example the archaeologist can drag a drawing or a picture stored in the database to the gateway window. The picture will automatically be displayed on the HMD on top of the real environment. Moving the picture using the stylus on the screen will move the picture on top of the real environment. This is for example used by archaeologists in order to compare objects, one from the database and one just discovered in the real environment. In addition the user can move a cursor in the gateway window that will be displayed on top of the real environment. The ultimate goal is that the user can interact with the real environment as s/he does with virtual objects. Based on this concept of gateway between the real and the virtual, we implemented the clickable reality and the augmented stroll. We first describe the clickable reality and the augmented stroll and then focus on their software design.

## 4.1  Clickable Reality

Based on the gateway window, we allow the user to select or click on the real environment. The camera is fixed, on the HMD, in between the two eyes. Before taking a picture, the camera must be calibrated according to the user's visual field. Using the stylus on screen, the user then specifies a rectangular zone thanks to a magic lens. The specified rectangular zone corresponds to a part of the real environment. As shown in Figure 6, the lens is both displayed in the gateway window on the pen computer and on the HMD. Inside the lens, there is a button for transferring the selected part of the real environment to the virtual world as a picture. A short cut for this action would be a speech command "take a picture" when the speech recognizer will be integrated. The picture is then stored in the shared database along with the description of the object as well as the location of the object. The next step is then to restore this picture in the context of the real environment: we call this action augmented stroll.

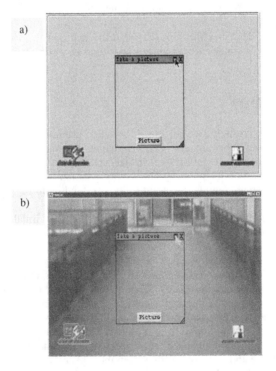

**Fig. 6.** Clickable reality: (a) Magic lens displayed on the pen computer (b)View displayed on the HMD

## 4.2 Augmented Stroll

Because a picture is stored along with the location of the object, we can restore the picture in its original real context (2D location). When an archaeologist walks in the site, s/he can see discovered objects removed from the site and specified in the database by colleagues. As schematized in Figure 7, based on the current position and orientation of a user, the system is able to determine the available objects stored in the database whose locations belong to the visual field of the user. The system indicates on the HMD that a picture is available. S/he can then see the object as it was before being removed from the site. The augmented stroll is particularly useful to see objects belonging to a stratum higher than the current one, because by definition the objects have all been removed. We envision to let the user specify what is the stratum of interest, so that the user will only see the objects of a given stratum while walking in the site. The augmented stroll is an example of asynchronous collaboration and belongs to the mobile collaborative AR class as defined in paragraph 2.3.

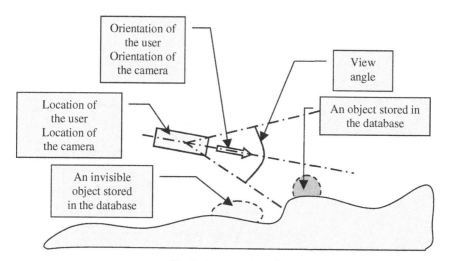

**Fig. 7.** Augmented stroll

## 4.3 Software Design

Our software design solution for implementing the augmented stroll draws upon our software architecture model PAC-Amodeus [15]. PAC-Amodeus uses the Arch model [24] as the foundation for the functional partitioning of an interactive system and populates the key element of this organization, i.e., the Dialogue Controller, with PAC agents [7]. Although the software architecture model is not new, we show here how the model, dedicated to the software design of multimodal systems [15], can be applied to the design of an AR system. In particular we highlight the software design of the gateway between the real and the virtual.

### 4.3.1    Overall Architecture of MAGIC

PAC-Amodeus incorporates the two adaptor components of Arch, the Interface with the Functional Core (IFC) and the Presentation Techniques Component (PTC), to insulate the keystone component (i.e., the Dialogue Controller, DC) from modifications occurring in its unavoidable neighbors, the Functional Core (FC) and the Low Level Interaction Component (LLIC).

Each component of the model briefly exposed before, appears in the design of MAGIC. One instance of the five PAC-Amodeus components is replicated for every user. The PAC-Amodeus architectures communicate with each other via their Interface with the Functional Core (IFC). Indeed all PAC-Amodeus architectures are linked together via a unique shared Remote Functional Core (FC) that communicates with the IFCs. The IFC maintains links between the Local Functional Core (Local FC) and the Remote Functional Core (remote FC), as shown in Figure 8. In addition the IFC operates as a translator between the database and the data structures used in the Dialogue Controller (DC). The IFC is split into two parts: one that bounds the Local FC with the DC and one that gathers network functionality, and consequently the connection with the remote FC.

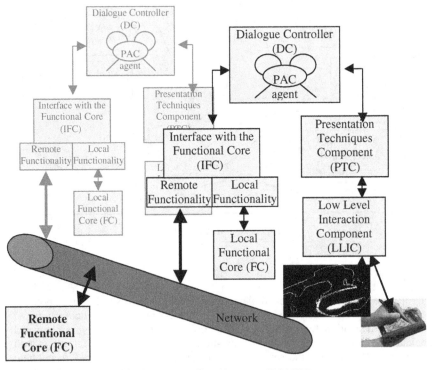

**Fig. 8.** The overall architecture of MAGIC

At the other end of the spectrum, the Low Level Interaction Component (LLIC) is represented in terms of several components: (1) The JAVA event handler and graphic machine, (2) The augmented reality glasses driver (3) The GPS interface and (4) The electronic compass interface. In turn, the Presentation Techniques Component (PTC) is split into several parts: the PTC is no longer dependent on devices, but translates information from the drivers in terms understandable by the Dialogue Controller. For example, (x, y) positions of the pointer on the graphic tablet are translated into the selection of an object.

The Dialogue Controller (DC) is responsible for task sequencing on each user's workstation. As explained above, the DC is independent of the Functional Core and of the network, as well as of the underlying software and hardware platform, such as the AWT toolkit and interaction devices. The model is geared towards satisfying the flexibility and adaptability software quality criteria. This component is refined in terms of PAC agents. In the following section, we only describe the PAC agents dedicated to the augmented stroll.

### 4.3.2    Software Design of the Augmented Stroll

The hierarchy of PAC agents implementing the augmented stroll has been devised using the heuristic rules presented in [16]. In particular we apply the following rules:

> *Rule 1: Use an agent to implement an elementary presentation object.*
> *Rule 2: Use an agent to implement a group object.*
> *Rule 4: Use an agent to maintain visual consistency between multiple views*

Rule 4 is illustrated by Figure 9.

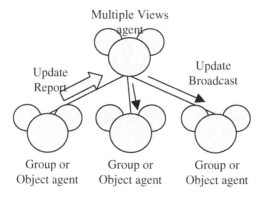

**Fig. 9.** The Multiple Views agent

Applying these rules, the augmented stroll is implemented by three agents, presented in Figure 10. One agent called "Gateway" corresponds to the Multiple Views agent and is introduced to express the logical link between the two representations of the gateway: the virtual representation and the real representation. The virtual representation corresponds to one agent, called "Virtual Representation agent". Its Presentation facet implements the gateway window displayed on the pen computer that is presented in Figure 3. The real representation is another agent, namely "Real Representation agent" and its Presentation facet manages the information displayed on the HMD. Any action with visual side effect on a view (Virtual Representation agent or Real Representation agent) is reported to the Gateway which broadcasts the update to the other siblings. To better understand the roles of each agent, we explain the message passing through the hierarchy of agents in the context of the two scenarios: the clickable reality and the augmented stroll.

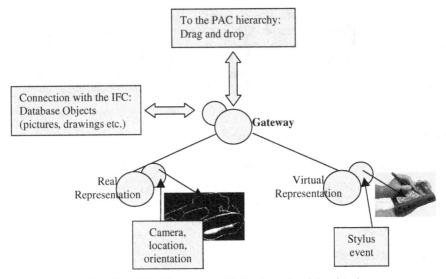

**Fig. 10.** PAC architecture: combining the real and the virtual

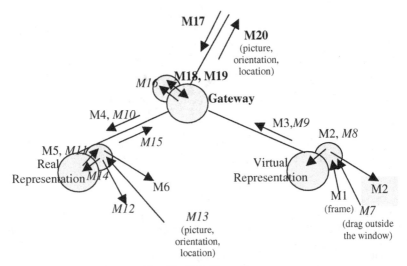

**Fig. 11.** Clickable reality

Figure 11 shows the message passing through the hierarchy of agents in the context of the first scenario: clickable reality. To take a picture, the user has to select what s/he wants to capture using the stylus on the gateway window. It corresponds to message M1. A frame is displayed in the gateway window (M2) and in parallel on the HMD (messages M2, M3, M4, M5 and M6). When the user using the stylus drags the frame outside the gateway window (M7), the Virtual representation agent is asked by the Gateway agent to provide the corresponding picture, orientation and location (messages M8, M9, M10, M11, M12, M13, M14, and M15). The picture is saved in the Abstraction facet of the Gateway agent (M16). The agent receiving the drop event then asks the Gateway agent for the picture (M17, M18, M19 and M20) in order to display it via its Presentation facet.

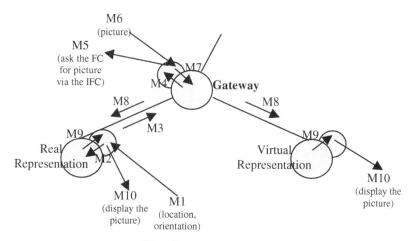

**Fig. 12.** Augmented Stroll

In the second scenario, we consider the augmented stroll. While the user is walking, messages specifying the location and orientation (M1) are sent to the Presentation facet of the Real Representation agent from the Presentation Techniques Component (PTC), as shown in Figure 12. Data are passed to the Gateway agent (M2, M3 and M4). In turn the Abstraction facet of the Gateway agent asks the Functional Core if it contains a picture at this point (location and orientation) (M5). If a picture is available, it is received by the Abstraction facet of the Gateway agent (M6). The latter then dispatches the picture to the Real representation agent and the Virtual representation agent (M7 and M8). The picture is then displayed on the HMD as well as in the gateway window on the pen computer (M9 and M10).

# 5 Summary and Future Directions

We have presented our definitions and classifications of mobile AR systems as well as collaborative AR systems. These definitions highlight two main characteristics of mobile and collaborative AR systems: "augmentation of the real environment of one user occurs through available knowledge of where the user is and what the other users are doing". We have then described our MAGIC system whose application domain is Archaeology, focusing on the mobile and collaborative aspects: the augmented stroll and its underlying concept of clickable reality. For the two concepts, we explained the software design based on the PAC-Amodeus software architecture model. We focused on the software architecture implementing the gateway between the real and the virtual, a generic software design solution to smoothly integrate the two worlds, without being dependent on the interaction devices.

The next step in our work is to experimentally test MAGIC in order to evaluate the usability of the clickable reality and the augmented stroll. The first tests by archaeologists in Alexandria are planned in June 2001. Although our application domain is Rescue Archaeology, we believe that the clickable reality and augmented stroll are generic concepts that can be applied in other domains such as game: a group of players can for example search for objects that have been virtually placed in the real environment by other users.

### Acknowledgements

This work is supported by France Telecom R&D, Laurence Pasqualetti being the supervisor. We wish to thank our partners from the University of Toulouse, P. Salembier and T. Marchand who are currently studying the archaeological fieldwork from a social point of view and the CEA (Centre d'Etude d'Alexandrie in Egypt) for welcoming us. Special thanks to the reviewers for their constructive comments and to G. Serghiou for help with style and English Grammar.

## References

1. Ancona, Dodero, Gianuzzi: RAMSES: A Mobile Computing System for Field Archaeology. LNCS 1707, Handheld and Ubiquitious Computing First International Symposium, HUC'99 (1999) 222-233
2. Bauer, Heibe., Kortuem, Segall: A Collaborative Wearable System with Remote Sensing. Proceedings of the Second International Symposium on Wearable Computers, ISWC'98 (1998) 10-17
3. Bier, Stone, Pier, Buxton, DeRose: Toolglass and Magic Lenses: The See-Through Interface. Proceedings of Siggraph'93, Anaheim, Computer Graphics Annual Conference Series, ACM (1993) 73-80
4. Billinghurst, Weghrst, Furness: Shared-Space: An Augmented Reality Interface for Computer Supported Collaborative Work. Workshop Proceedings on Collaborative Virtual Environment (1996)
5. Bluetooth SIG. http://www.bluetooth.com
6. Coutaz, Crowley, Bérard, Carraux, Astier: Using Computer Vision to Support Awareness and Privacy in Mediaspaces. Computer Human Interaction 1999, CHI99 Extended Abstracts Proceedings, ACM, (1999) 13-14
7. Coutaz: PAC: an Implementation Model for Dialog Design. Proceedings of Interact'87 (1987) 431-436
8. Dubois, Nigay, Troccaz,, Chavanon, Carrat: Classification Space for Augmented Surgery, an Augmented Reality Case Study. Conference Proceedings of Interact'99 (1999) 353-359

9. Ellis, Gibbs, Rein: Groupware : Some issues and experiences. In Communications of ACM, 34(1), (1991) 38-58
10. Feiner, MacIntyre, Höllerer, Webster: A touring machine: Prototyping 3D mobile augmented reality systems for exploring the urban environment. Proceedings of the First International Symposium on Wearable Computers, ISWC '97 (1997) 74-81
11. Kangas, Röning: Using Code Mobility to Create Ubiquitous and Active Augmented Reality in Mobile Computing. Mobicom' 99 Seattle, ACM (1999) 48-58
12. Kraut, Miller, Siegel: Collaboration in Performance of Physical Tasks: Effects on Outcomes and Communication. Conference Proceeding of Computer Supported Cooperative Work, CSCW'96, ACM, (1996) 57-66
13. Kuzuoka, Kosuge, Tanaka: GestureCam: a Video Communication System for Sympathetic Remote Collaboration. Conference Proceeding of Computer Supported Cooperative Work, CSCW'94, ACM, (1994) 35-43
14. McCarthy, Meidel: ActiveMAP : A Visualization Tool for Location Awareness to Support informal Interactions. LNCS 1707, Handheld and Ubiquitous Computing First International Symposium, HUC'99 (1999) 158-170
15. Nigay, Coutaz: A Generic Platform for Addressing the Multimodal Challenge. Proceedings of Computer Human Interaction, CHI'95, ACM (1995) 98-105
16. Nigay, Coutaz: Software architecture modelling: Bridging Two Worlds using Ergonomics and Software Properties. Book chapter of Palanque, P., Paterno, F.: Formal Methods in Human-Computer Interaction. Springer-Verlag: London Publ. (1997) 49-73
17. Pyssyalo, Repo, Turunen, Lankila, Röning: CyPhone – Bringing Augmented Reality to Next Generation Mobile Phones. Proceedings of Designing Augmented Reality Environments 2000, DARE'2000 (2000) 11-21
18. Rekimoto: Navicam: A Magnifying Glass Approach to Augmented Reality. Presence vol. 6, n° 4 (1997) 399-412
19. Salber: De l'interaction homme-machine individuelle aux systèmes multi-utilisateurs. Phd dissertation, Grenoble University, (1995) 17-32
20. Schmalstieg, Fuhrmann, Szalavari, Gervautz: Studierstube: An Environment for Collaboration in Augmented Reality. Workshop Proceedings on Collaborative Virtual Environment (1996)
21. Starner, Leibe, Singletary, Pair: MIND-WARPING: Towards Creating a Compelling Collaborative Augmented Reality Game. International Conference on Intelligent User Interfaces, IUI 2000 New Orleans, ACM (2000) 256-259
22. Streitz, Geißler, Haake, Hol: DOLPHIN: Integrated Meeting Support across Local and Remote Desktop Environments and LiveBoards Conference Proceeding of Computer Supported Cooperative Work, CSCW'94, ACM (1994) 345-358
23. Szalavari, Eckstein, Gervautz: Collaborative Gaming in Augmented Reality. Proceedings of the Symposium on Virtual Reality Software and Technology (1998) 195-204
24. The UIMS Workshop Tool Developers 1992. A Metamodel for the Runtime Architecture of an Interactive System, SIGCHI Bulletin, 24, 1, 32-37
25. UMTS Forum. http://www.umts-forum.org
26. Webster, Feiner, MacIntyre, Massie, Krueger: Augmented Reality in Architectural Construction, Inspection, and Renovation. Proceedings of Computing in Civil Engineering, ASCE, (1996) 913-919
27. Wellner: The Digital Desk calculator : tangible manipulation on a desk top display. In Proceedings of the fourth annual ACM symposium on User interface software and technology (1991) 27-33

## Discussion

*L. Bergman:* In your augmented stroll you are displaying a hypothesized model of the site. This reflects some uncertainty in the 3D models. Is this a problem when you are presenting the information?
*P. Renevier:* We are not constructing 3D models, simply showing found and removed objects in place.

*L. Bass:* I can generalize this question: how will you represent the uncertainty in the situation.
*P. Renevier:* There is no uncertainty. We only show pictures, not 3D models.

*J. Williams:* I realize that it is not necessary to use calibration as in 3D graphics (augmented reality), and virtual environments, but it seems there must be some methods of locating the userís position. How is this achieved?
*P. Renevier:* We only match virtuality and reality by using position (GPS and compass). We can have a precision of 5 cm.

*N. Graham:* Did you develop these systems by doing an external design and then applying the PAC modelling rules? Is this a good way of doing it in general? Is it really this algorithmic? How did it work in practice?
*P. Renevier:* I made the architecture before the implementation, using the rules.

# Modelling and Using Sensed Context Information in the Design of Interactive Applications

Philip Gray[1] and Daniel Salber[2]

[1] Dept of Computer Science, University of Glasgow,
Glasgow G12 8QQ, Scotland
pdg@dcs.gla.ac.uk
[2] IBM T.J. Watson Research Center, 30 Saw Mill River Road
Hawthorne, NY 10532, USA
salber@acm.org

**Abstract.** We present a way of analyzing sensed context information formulated to help in the generation, documentation and assessment of the designs of context-aware applications. Starting with a model of sensed context that accounts for the particular characteristics of sensing, we develop a method for expressing requirements for sensed context information in terms of relevant quality attributes plus properties of the sensors that supply the information. We demonstrate on an example how this approach permits the systematic exploration of the design space of context sensing along dimensions pertinent to software development. Returning to our model of sensed context, we examine how it can be supported by a modular software architecture for context sensing that promotes separation between context sensing, user interaction, and application concerns.

## 1 Introduction

Until very recently, most applications were used in a static setting using little more than user input to define what could be done and to drive the interaction forward. This situation has been transformed by the explosion of portable machines, embedded computation, wireless communications, distributed networks and cheap, plentiful sensors.

Hardware and software resources are running ahead of design and engineering models, tools and architectures. While it is easy to envision endless uses of context in interesting new applications, it is much harder to identify the issues involved in designing systems that use information sensed from the environment and also hard to incorporate sensors into applications. There is little support for systematic exploration of the design space of a context-aware application and for the evaluation of the consequences of design choices on architecture and implementation.

In this paper, we (1) propose a model of sensed context information that accounts for the complexity of sensing, as opposed to traditional user input; (2) present an approach to the design of context-aware applications that deals explicitly with properties of sensed context; and (3) introduce a preliminary software architecture model that captures typical operations on sensed context and its properties. In section

M. Reed Little and L. Nigay (Eds.): EHCI 2001, LNCS 2254, pp. 317–335, 2001.

2, we propose a definition of sensed context and analyze its constituents. Section 3 outlines our design approach and illustrates with an example the systematic exploration of a design space that it supports. In section 4, we propose a software architecture model for sensed context information. We illustrate its use with the example of section 3. Section 5 examines some related work, particularly in software architecture. We conclude and outline plans for future work in section 6.

## 2    A Model of Sensed Context

Context sensing in interactive applications refers to the acquisition of information from the surrounding environment. We first define sensed context in terms of properties of real world phenomena. We then analyze the features of sensed context and propose a model that captures its most relevant characteristics.

### 2.1    Defining Sensed Context

We are interested in the sort of information that:
- can be accessed via sensors,
- capture properties of real-world phenomena, and
- can be used to offer application functionality or to modify existing functionality to make it more effective or usable.

In each case the information is sensed from the physical context in which the application is being used. This is part of the overall context of use, which can also include information that is not sensed (typically, the user's emotional state, the social organisation of the artifacts, etc.). To reflect this relationship, we make a distinction between context and that subset of it that is capable of being sensed, viz., sensed context.

The term 'context' suffers from an embarrassing richness of alternative definitions. Dey, Salber and Abowd [1] provides a useful review and offers a version that is a useful starting point for our definition of sensed context:

**context** $=_{def}$ any information that can be used to characterize the situation of an entity, where an entity is a person, place, or object that is considered relevant to the interaction between a user and an application, including the user and the application themselves. Context is typically the location, identity and state of people, groups and computational and physical objects.

This definition needs some refinement to capture our notion of sensed context. Sensed context refers to that part of context that comes from the physical environment; i.e., that part of context that is accessible via sensors, in other words, the properties of phenomena. The term 'phenomenon' refers to "an occurrence, a circumstance or a fact that is perceptible by the senses." [2] This term comes close to expressing that set of things we wish to include as the subjects of sensed context information if we take the "senses" to include non-human sensing devices.

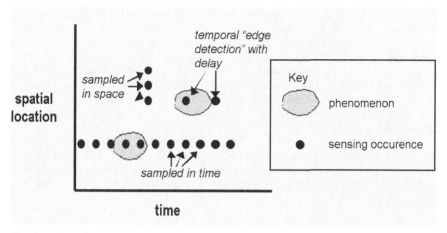

**Fig. 1.** Relationships of phenomena and sensing activities in space-time. The sensing activity at the bottom is performed by a fixed sensor that provides samples at regular intervals. The phenomenon at the top is sensed twice, but there's a delay in detecting the beginning and the end of the phenomenon, for instance because sensing occurs at fixed intervals.

The relationship of sensors to phenomena can be related in space-time as shown in figure 1. That is, phenomena "occupy" spatial and temporal locations. Often, occurrences of sensing provide "samples" of the phenomena or special events, such as the boundary of the phenomena in space or time. Many of the interesting design and implementation questions about sensed context relate to the sensor-phenomenon relationship, including the timeliness and accuracy of the sensed information and whether identity of entities across phenomena can be determined.

The notion of 'interaction' in the original definition is ambiguous; it is not clear whether this refers to what is achieved via the interaction (viz., tasks) or the means by which the tasks are performed (viz., concrete user interface, dialogue, etc). We wish to be non-restrictive and thus explicitly include both as legitimate aspects of human-computer interaction that sensed context may influence. Finally, it may be difficult or impossible to determine at a particular time if some sensed context is or is not actually relevant. What is important is its *potential* relevance to the interaction, whether or not someone considers it to be such.

Therefore, we propose the following definition:

**sensed context** $=_{def}$ properties that characterize a phenomenon, are sensed and that are potentially relevant to the tasks supported by an application and/or the means by which those tasks are performed

Sensed context is ultimately *derivable* from some sensory apparatus, or sensor(s). It doesn't follow, however, that it *is* so derived. The fact that Daniel speaks French may be acquired via some sensor (e.g., a multi-lingual speech recognition system) or via a database query, with no sensing involved. Also, sensed context may be derived from other sensed context via some transformation or interpretation. In the case of Daniel's speaking French, the input from the sensor (microphone) may have to go through sophisticated processing; nevertheless, we wish to refer to Daniel's speaking French as sensed context if the source involved one or more sensors.

Note that the same physical apparatus might serve to provide conventional input and sensed context. For example, a keyboard might provide input strings for processing by the application (user input) and the same input events could be a sensor for user fatigue, based on the average time between keystrokes. Similarly, a sensor can be exploited by a user as an input device. Consider the case of two persons who have agreed that moving past a proximity sensor at their office door will be used as a signal to the other that it's time for lunch. The difference lies in the relationship to user's intentions, conscious or otherwise, not the way the information is subsequently handled by the system.

It can be difficult at times to draw a distinction between sensed context and user-generated input. Nevertheless, it can still often be useful in thinking about the impact on the user of different design alternatives; choices of user input versus independently sensed context can be critical to the feasibility and usability of certain application types.

## 2.2    Modelling Sensed Context

According to our definition, sensed context is information and we therefore model it as such. The particular aspects captured in the model are there because we believe them to be important for the purposes of reasoning about designs and, in some cases, for implementing them.

Sensed context is propositional in nature, typically of the form "phenomenon P has property p", e.g., "this device is at location y", "this probe has temperature t", "the time is t". Relational information is also possible, such as "person p is near landmark l", "group g is meeting in room r at time t". Being propositional, sensed context information can be formulated using a formal representation such as first-order predicate logic, composed into more complex sensed context expressions and have associated with them meta-propositional properties, such as judgements of the quality of the information (e.g., its probability) or the nature of its sensory source (e.g., the operational parameters of the sensor). In this paper we will not pursue the formalization of sensed context information, but will focus on the associated meta-information.

To summarize, sensed context has several important characteristics that we will utilize in our model, including

- information content, especially
  - the sensed properties of the phenomena
  - the subject(s) of the sensing
- meta-information, including
  - information quality attributes
  - information about the source of the content

We now discuss each of these characteristics in turn.

### 2.2.1    Information Content: Sensed Context Types

Although we place no limits on the type of properties or relationships that can be expressed in sensed context, certain types are particularly prominent because

- the information offers real benefits in functionality and usability and

- current sensor technology and interpretation techniques enable the information to be captured and transmitted cost-effectively.

These information types include:

*spatial location*

We include here propositions about physical location, such as the location of a person or artifact in a building or the position of a car in a road system. It may include location in terms of global or local reference systems and absolute or relative position. Abstract spaces may also be spatially located and thus, indirectly, treated as sensed context, but only if they are marked out in physical space.

*time*

Time may include interval and point references. It is relatively easy to acquire and is useful in association with other properties, as a time stamp.

*identity*

Typically, identification is performed by a sensor capturing distinctive features of the entity to be identified (e.g., iris configuration, a facial color patch, or the id of an associated "tag"). Often, it is sufficient to be able to assert that different sensing occurrences relate to the same entity; that is, sensed context enables assertions of identity of entities across time and space.

### 2.2.2    Information Content: The Subject of Sensed Context

If one takes sensed context to be propositional, then there must be a subject or subjects for the propositions. That is, sensed context is not just a property such as location, but an assertion that something is at a location. It may be the case that a given sensor simply delivers a location value, but the data becomes useful information when implicitly interpreted as the location of the device. Typically, it is not the location of the sensor that is of interest, but of something else (e.g., the person holding the GPS), but the further inference from "This GPS is at location x" to "Person P is at location y" depends on the initial interpretation of the data. During design it can be important to identify both the ultimate subject of the information content from the application's point of view and the information provided by the sensor(s), so that the transformational requirements are clear.

### 2.2.3    Properties of Sensed Context: Meta-Information

One of the distinguishing features of sensed context information, compared to other sorts of information utilised in an application, is the importance of the way it is acquired. Sensors are subject to failure and noise. Often, they only capture samples of phenomena, hence their output is approximate only. Furthermore the process of interpretation of sensed context is also subject to ambiguity and approximation. In addition, sensors may be used in conjunction with actuators that perform actions in the physical world.

These aspects are sufficiently important that we propose as a central feature of our model a set of meta-attributes of the information, including:

- forms of representation
- information quality
- sensory source
- interpretation (data transformations)
- actuation

As shall be discussed in section 3 below, this meta-information is crucial when reasoning about sensed context during design and can be valuable as a potentially controllable aspect of information flows in the run-time system.

We shall look at each of these categories in turn.

**Forms of Representation.** Many sensing devices, such as GPS and timing devices, are capable of offering their output in different data formats. Therefore, we wish to capture this potential as a meta-property of the information itself, so that it can be described, and reasoned about, even when the sensory source is not yet specified or even known. Furthermore, transformations can serve to change the form of representation of a property, for instance transforming GPS data into building names. Expressing the desired form of representation helps identify the transformations required.

**Information Quality.** We can identify the following information quality attributes:

- coverage – the amount of the potentially sensed context about which information is delivered
- resolution – smallest perceivable element
- accuracy – range in terms of a measure of the property
- repeatability – stability of measure over time
- frequency – sample rate; the temporal equivalent of resolution
- timeliness – range of the measure in time; the range of error in terms of the time of some phenomena; the temporal equivalent of accuracy

These properties are perhaps easiest to consider in the case of spatial location. For example, a piece of sensed context might provide its information in terms of British Ordnance Survey coordinate system, covering mainland Britain, to a resolution of 100m and an accuracy of +-10%. Similarly, if the information is also temporal, we can identify its form of representation (seconds), its frequency (5 kHz) and its accuracy (or timeliness) (±100 ms).

We believe these are well-defined attributes of quality of all, or at least many interesting, properties of sensed context. Thus, we might have a way of determining the language of a speaker from his or her speech. We can describe information quality attributes for this sensed information:

*coverage*: are all languages recognized? If not, what is the subset?

*resolution*: can a distinction be made between similar languages? Thus, if the system cannot resolve the difference between various Slavic languages (Russian, Polish), this is a resolution issue.

*accuracy*: in this case, it may be difficult to distinguish between the effect of resolution and accuracy problems. However, the distinction can be drawn: accuracy refers in this case to the probability that the system will identify the wrong language. This is different from it's not being able to distinguish between them.

*repeatability:* given the exact same input, e.g., a recorded utterance, is the determined language the same over successive trials?

*frequency* might be at the level of a single utterance, with *timeliness* measured as a delay of up to 5 seconds from the end of the utterance.

**Sensory Source.** So far we have focussed on the nature of sensed context as information. However, it is often necessary to think about where that information comes from.

Among the attributes of the acquisition process by which the information is acquired, we identify the following as candidates for our model:

- reliability
- intrusiveness
- security or privacy

Just as with sensed context, sensors can be characterized by meta-information about the apparatus. The set of properties of a sensor are difficult to fix globally. They are typically closely related to the physical, behavioral and operational characteristics of the sensor. However, we would expect most sensors to at least have the following:

- cost
- operating profile

**Transformation.** We may wish to specify the transformation process by which sensor output is transformed into usable sensed context information. At the design stage it may be useful to identify that a transformation is unreliable or computationally costly. A specification of the data transformations is of obvious benefit at the implementation stage.

**Actuation.** Although context-awareness at present is more concerned with acquiring and deriving information from sensed context, we need to keep in mind the possibility of acting on the real world through effectors. Actuation can also influence the sensing processes by shutting down a faulty sensor, or modifying its operating parameters, for example by reorienting a GPS antenna.

# 3    Supporting the Exploration of Sensed Context Design Options

Understanding the nature of sensed context and the particular role of meta-data allows us to proceed further and examine how this new understanding might be used for application design and development. In particular, we are interested in considering how these requirements might be used to support systematic exploration of the design space of a context-aware application. What follows is an initial and tentative investigation of how to utilise our model. This section is intended to be illustrative of what might be done.

Our approach relies on checklists that are intended to uncover design choices and the consequences these might have in terms of software development. We will use the example of a museum context-aware tour guide throughout this section. Besides representing one of the most common types of context-aware application reported upon in the literature, the application features some clear, simple requirements and a rich design space. It is not our intention to present a solution to the design problem, but simply to indicate the kind of issues that we believe our approach can uncover.

As shall become evident as we work through this example, the design space is very rich, with many trade-offs requiring more exploration than we can manage in a relatively short paper. Consequently, we have focussed on a rather restricted view of the design process and its legitimate areas of concern. In particular, the example set out in section 3.1 below focuses on design issues related to the provision of (some) application functionality with no consideration of the wider contextual framework in which such a functional requirement might arise nor the context-related issues associated with the decisions about the method of delivery of that functionality. This may seem ironic in a paper devoted to a consideration of sensed context, but we believe this is a reasonable simplification of the design process, at least for an early investigation such as we are reporting here.

**Activities Involved in the Approach.** Our approach is based on a set of design-oriented activities that together enable issues of sensed context to be identified, related to other design considerations and explored. The activities include:

- identifying sensed context possibilities
- eliciting and assessing information quality requirements
- eliciting and assessing requirements of the acquisition process
- consideration of issues of
  - intrusiveness, security, privacy
  - transformations of the data from source to "consumer"
  - transmission and storage
- eliciting and assessing sensor requirements

The sequence of these activities in our list suggests a very rough order in which they might be addressed, but the order is not intended to be restrictive. One might well begin with any of the activities and work out to others, returning thereafter to re-assess earlier decisions.

The outcome of these activities will be a set of requirements related to sensed context and a set of design issues related to those requirements. At this stage of development our approach does not offer any particular assistance with how to resolve the set of issues raised. These have to be handled as any set of design problems that have alternative solutions and demand resolution of conflicting requirements.

### 3.1    Identifying Sensed Context Possibilities

We now consider our example of a context-aware museum tour guide. For the sake of brevity, we will concentrate on a single but representative feature, namely:

*Deliver information in the language of the visitor that describes the exhibit that the visitor is attending to.*

We will assume that the exhibits are laid out in a single exhibition space and that each exhibit has associated with it a set of descriptions, each with the same content but expressed in different languages. The set of languages supported is some subset of all natural languages. Visitors are people that wander around the museum (i.e., they are mobile), following a path from exhibit to exhibit, reading descriptions about the exhibit they are currently attending to.

Note that at this level of description, we are making no assumptions about the choice of input and output devices that are to be employed. Similarly, we are not specifying what specific sensors or sensing techniques we might use. Before we identify the sensed context involved in this case, we may make assumptions (a visitor is attending to only one exhibit at a time) or impose constraints (users should not have to carry devices). We will not introduce any design constraints in this exercise because we want to show how we can explore the space generally, although typically constraints identified at this point can limit the space and make the design process more manageable.

In this example the primary entities of interest are: visitor, exhibit. For the visitor, we are interested in her language and for the exhibit, the description(s) that are available. Also, we are interested in the relation of attending to between the visitor and the exhibit[1] she is attending to. We now have three information needs and can ask of each if it is potentially sensed:

- *visitor's language* – we can envisage using speech recognition to identify the speaker's language (assuming visitors are talking amongst themselves, or are asked to speak to a device at the beginning of their visit) or via a tag attached to the visitor and readable by some sensor, either at an exhibit or elsewhere.
- *exhibit's description* – this is directly sensible, e.g., via an infrared broadcast or indirectly sensible via a transmitted context identifier that can be used for content lookup
- *visitor's attention* – this is potentially sensed in a number of ways, including proximity sensing, orientation capture and eye gaze tracking, using sensors on the visitor, the exhibit or globally in the exhibition gallery.

Once the primary entities and related information of interest are identified, additional characteristics of the potential context sensing can be investigated. Two characteristics are particularly useful: spatial and temporal footprints. Interestingly, these characteristics are easily captured if we reason in terms of phenomena and use the graphical representation of figure 2, showing three exhibits and two languages. Notice that the diagram captures the language need in relation to exhibit and time; the precondition of the visitor's attention is abstracted away. The diagram makes clear that it is the linguistic need that is central; the user's attention to the exhibit is just an enabling pre-condition. Additionally, it is evident that the identity of the user is not important in this case, a condition that might have been overlooked if focusing on the possibility of sensing the visitor's location and language.

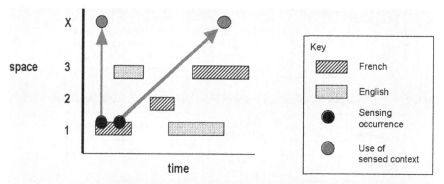

**Fig. 2.** Time-space diagram of phenomena (visitors speaking English or French attending to exhibits), sensing occurrences, and use of sensed context by the museum tour application. Locations labeled 1, 2, and 3 are locations of exhibits. In this example, sensed context is used by a centralized application running on a server whose location (X) is represented at the top of the diagram.

---

[1] The relationship might be expressed from the point of view of the visitor (the exhibit I am attending to) or the exhibit (the visitor attending to me). We shall refer to it from either point of view as appropriate.

The diagram also makes apparent the potential communications and storage requirements for sensed context. We have placed on the space-time diagram possible context sensing events, along with context use by the application. For each sensed context information that we want to acquire, if there is a spatial gap between the sensing event and the actual use of the information acquired, the information will have to be communicated. If there is a temporal gap, the context information will need to be stored.

## 3.2    Sensed Context Requirements Checklist

Context-aware application design must factor in the variable quality of sensed context data. One necessary step is to express information quality requirements for sensed context information needed to support the application functions.

Our approach relies on running sensed context information (that is, each predicate over subjects and features) through a checklist of the quality criteria of section 2.2.3. Each criterion must be considered in turn for each piece of sensed context

As an example, we will consider two of three potential sensed phenomena: the visitor's language and the visitor's attending to an exhibit:[2]

**Table 1.** Information quality criteria for context proposition "visitor's language is X".

| Coverage | All languages supported by the application |
|---|---|
| Resolution | Language (a lesser requirement would be: linguistic family) |
| Accuracy | 100% |
| Repeatability | Stable (same language for successive inputs of a pre-recorded utterance) |
| Timeliness | Before first information delivery |
| Frequency | • Once, assuming (1) visitor won't switch languages during the visit and (2) we can attach the language info to the visitor and retrieve the visitor's identity from one display to the next<br>• Else, once for each information delivery |

**Table 2.** Information quality criteria for context proposition "visitor is attending exhibit Y".

| Coverage | All exhibits |
|---|---|
| Resolution | A single exhibit |
| Accuracy | 100% |
| Timeliness | Before visitor moves on (on the order of seconds) |
| Repeatability | 100% stable |
| Frequency | Once per exhibit within timeliness requirements (on the order of seconds) |

---

[2] The third possibility, sensing the exhibit description, is a reasonable approach. One might, for example, have the exhibit broadcast all its descriptions, and have the visitor's hand-held display receive only the one in the appropriate language (the "teletext" strategy). We have decided to leave this option out of consideration solely for reasons of simplicity in our example.

Notice that several design issues are identified by this exercise. We may have to be clearer about the relationship of languages to dialects. More importantly, since we have a quality requirement that only one language needs to be sensed per visitor (per visit), we can entertain the possibility of a single sensing occurrence in order to capture this information (although we may then need to store the information for later use).

The next two activities, assessing sensor requirements, transmission and storage needs are all closely linked to particular acquisition strategies or patterns. The following seem to be reasonable candidates for sensing strategies in this example:

- Determining the language of the visitor
  - Looking it up (preferences)
  - Asking the visitor
  - Associating a sensible tag with the visitor that contains a language id
  - Listening to speech
- Visitor is attending an exhibit
  - Near it (but might fail if user is turning her back to the display) -> Proximity
  - Facing it -> Proximity + relative orientation
  - Looking at it -> Gaze
  - Asking the visitor
  - Based on history

Arising out of this exploration are a number of consequent design issues. In the case of proximity sensing (location) there are related issues of whether the user senses the display or vice versa; in the former case, we are confronted with a question of power consumption and possible privacy issues if the identity of the visitor is also made available (perhaps via the sensed tag strategy of language sensing. Considering proximity-based location as a source of attention, it becomes apparent that proximity is not sufficient if one can be attending to two different exhibits from the same position or if the proximity sensors necessarily have overlapping coverage. If there is no overlap, proximity sensing is probably acceptable; if there is overlap, then it must be either replaced or enhanced by other means. Table 3 captures the most significant design issues for the two pieces of sensed context.

Checklists, or structured questions, of the sort we have presented here are, we believe, a step in advance of the unstructured, ad hoc approach typical now. It is a form of literate specification [3], helping to identify issues, make issue coverage visible, assist in the documentation of the design process and offer a means for traceability of design decisions. While we believe our small example suggests the technique is both feasible and useful, it remains to be evaluated in a realistic setting.

# 4    Mapping Sensed Context to a Software Architecture

Software modelling of sensed context within a context-aware application must take into account the complex nature of context information. The salient characteristics of sensed context information outlined in section 2 suggest a key distinction that we want to capture in the software model: both the actual information content of sensed context and meta-information that characterizes sensed context should be handled explicitly and distinctly. In addition, section 2 introduced the need for controlling sensors and sensing processes. Finally, we want to capture the ability not only to sense context, but also to act on the real world through actuators. Our software model for context-aware applications aims at emphasizing features that are specific to sensed context and that should be given particular attention. Furthermore, the model we propose focuses on sensed context: we do not explicitly model traditional user interaction that may occur in the application. Other models exist for that purpose, and they may be used in conjunction with our model. We will show an example of dual use of our model along with existing user interface models.

**Table 3.** Combined assessment of sensing strategies. This table summarizes the issues that must be addressed. Items in bold denote major issues.

| *item* | *information quality* | *acquisition process* | *sensors* |
|---|---|---|---|
| *language* | | | |
| sensed tag | accuracy & repeatability | need stored language data | **if RFID tags: orientation of tag wrt. reader; RF interference** |
| listening to speech | **coverage & accuracy** | need multi-lingual recognition; assumes user is talking | **ambient noise** |
| ask user | | **intrusive** | n/a |
| lookup | | need stored language data | n/a |
| *attending to* | | | |
| proximity | **accuracy if overlaps between exhibits** | **privacy if linked to identity** | locus of sensing; **power if mobile** |
| orientation | accuracy | **not sufficient on its own; needs proximity** | |
| gaze | accuracy | | **awkward headgear** |

## 4.1    Requirements for a Sensed Context Architecture

Context sensing is still very much an exploratory domain. Developing efficient context sensing capabilities requires practical knowledge of sensors and associated software techniques, which might build upon disciplines as varied as signal processing or machine learning. The set of skills required is far different from that used to develop user interfaces or application logic. As a result, a paramount requirement of a context sensing architecture is to facilitate the separation of concerns between user interface, application logic, and context sensing. In addition, we want to support iterative design (typical of an exploratory domain), and thus emphasize modularity.

## 4.2    Global View of the Architecture

To achieve separation of concerns, we introduce a set of functional components that focus on bridging the gap between sensors and applications. In an architecture model like Arch [4], this set of components is organized in two layers that correspond to two primary classes of operations on context: acquisition from the physical world, usually through sensors, and transformations of sensed information to extract information meaningful to the application. These operations occur at two different levels and can be represented as a supplementary branch in the Arch model as shown in figure 3. Arrows represent flow of control, data, and meta-data. This point will be clarified when we introduce the components that constitute the two context layers.

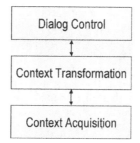

**Fig. 3.** Context handling as a supplementary branch in the Arch model. The classic user interface and application logic branches have been omitted for clarity.

This approach clearly emphasizes separation of concerns between the three branches of the extended Arch. This however, does not preclude the possibility that context-handling components may have user interfaces to allow control by or feedback to the user. A major difference with the traditional Arch model should be mentioned. As noted in Salber *et al.* [5], context components are long-lived and typically precede and survive the application they serve. Indeed, context may be required at any time by an application and the relevant context data may have been acquired long before the application requests it. Consider a handheld application that helps the user remember where she last parked her car: The location of the car when the user steps out must be recorded even though the retrieval application is not running, and maybe never will. Thus, the context transformation and acquisition layers may be active independently of any application. Connections between the Dialog Control component and the context branch are established dynamically. In addition, several applications may require the same context information, possibly simultaneously. The handheld car park payment application may be notified that the car is in a parking meter area and propose to the user to negotiate a fair price. In this sense, context services are similar in principle to operating system services such as network demons, or low-level windowing facilities.

## 4.3    Context Handling Components

Using sensors, as opposed to user input devices, entails dealing with a lot of data. Sensed context is much richer and data-intensive than user input. Thus, our model should primarily be concerned with organizing the flow of data. The model we have presented in section 2 provides an important insight into a distinguishing characteristic of sensed context. Meta-information that describes the quality and source of context data should be considered as important as the actual sensed information. In addition, context transformation mechanisms may be controllable and transformations may result in actions on the environment. We want to reflect in our software architecture the three flows we have identified: data, meta-data and control.

We populate each of the two layers identified in the previous paragraph with modular, composable, context-handling components. They all share the same structure represented as an hexagon shown in figure 4. Each lower side of the hexagon represents inputs. Each of the three sides receives one of the three flows of data, meta-data and control. Upper sides of the hexagon represent outputs and generate data and meta-data flows, along with actions on the environment. For a given component, any input or output arrow might be omitted if it is not relevant. Conversely, a component may receive multiple input flows on any one of its lower sides, or generate multiple outputs.

Each context-handling component performs transformations on context data. Although we have isolated the acquisition layer in the global view of the model, it is populated with components that share the common structure of figure 4. The only difference is that an acquisition component acquires its inputs from sensors (or from a sensing component part of a library, when they become available). In the generic model of the context-handling component, the data input flow is processed and produces the data output flow. Meta-data input may be used to influence processing (e.g., discard inaccurate data). New meta-data that describes the data resulting of the transformation is generated. Depending on the transformation involved, this may consist in updating the meta-data, or generating completely new meta-data. The control input allows other components to influence the transformation process itself. This may consist in modifying a parameter (e.g., the sampling rate of a sensor), requesting generation of the latest data (i.e., polling), starting or shutting down the component (e.g., shutting down a faulty sensor). The action output is the channel through which actions on the environment are performed. This may include turning on a light, changing the speed of a motor, driving a camera that tracks a person in a room, etc. Tight coupling between inputs and the action output is possible at the level of a single or a few components.

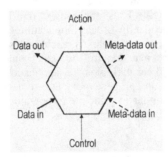

**Fig. 4.** The structure of a context-handling component.

The output of any context-handling components may be connected to the input of any other component. In general, information content outputs will be connected to the information input of another components, and similarly for meta-information and control/action. However, there may be connections from the meta-information input of a component to a second component's information input. The second component would then monitor the quality of the information provided by the first component, and possibly provide meta-information about its own monitoring function. We will give an example of this configuration in the guided tour example. In general though, control and action flows are kept separate from information content and meta-information flows. Control is also used for an important role: In a changing environment where sensors may appear and disappear, control drives the dynamic reconfiguration of the organization of components, establishing and breaking connections between a component and its peers. This issue is still being investigated but we believe it is crucial for handling context in a comprehensive way, especially when devices or sensors are mobile. Ideally, dynamic configuration, as well as transparent networking (including dynamic adaptation to the networking resources available) and other services such as service discovery and brokering between the application needs and the capabilities of the sensed context components should be handled by a middleware infrastructure, acting behind the scenes of our architecture. We know these services are required, but they fall outside the scope of our present concerns for this architecture model. We assume they are provided by the target platform. Ultimately, we aim at eliciting precise requirements for such a platform. This goal is beyond the scope of this paper.

### 4.4    Example Use of the Architecture for the Museum Tour Application

We have chosen to show how to use the architecture on a subset of the museum tour guide: the determination of the language of the visitor. Assessing the design issues of table 3, we use two complementary sensors in this example: speech recognition and RFID tag readers. Since each of these techniques has shortcomings under certain conditions, we combine the information they provide for better reliability.

The two input sensors are modeled by a context-handling component as shown in figure 5. A third component is in charge of determining the visitor language and combines the information content based on the meta-information provided, which might include indications of failures to read a tag, and confidence factors from the speech recognizer. Based on the meta-information, the combining component might decide to shutdown or reconfigure one of the two sensors. This is expressed by the "Action" output being fed to the "Control" input of the sensor components. The processes involved in selecting and combining the information content from the sensor components should of course be completely specified in a complete example.

We believe that the organization of components presented here is a potential candidate for a pattern of context sensing: combining information from several unreliable sensors and allowing on-the-fly reconfiguration of the low-level sensing components. Other potential pattern candidates include for instance the management of collections of context information or the monitoring of a single sensor by a dedicated monitoring component. We believe this is a promising area of research that our proposed architecture can support.

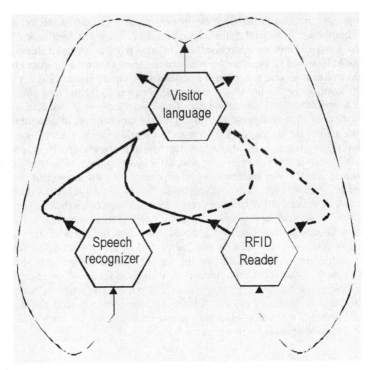

**Fig. 5.** Components involved in determining the visitor language in the museum tour guide application. Solid bold lines indicate sensed context information flow. Dashed lines indicate meta-information. Fine lines indicate control flow.

## 5    Related Work

In the field of context-aware computing, there is little work yet that aims at modelling sensors to takes into account the fact that sensed context information acquired from sensors is often of variable quality. For user input, Card, McKinlay and Robertson have devised a comprehensive model of input devices that shares interesting similarities with our approach [6]. They also felt the need to model device properties such as resolution to help designers make their requirements explicit. There have been several recent proposals of software architecture models for context-aware applications. Winograd, for example, presents a data-flow architecture based on networks of observers that abstract information from sensors [7]. Observers construct "context models", data structures that are stored in a "manager" blackboard and made available to applications. There are two interesting differences with our approach. First, Winograd's approach makes explicit the needs of applications: They provide the context model structure that observers populate. Second, the architecture doesn't account for ambiguous context information. Kiciman and Fox have a data-flow approach that introduces mediators which establish dynamic connections between components [8]. The context toolkit framework by Salber, Dey and Abowd was the basis of the architecture model presented here [5]. The context toolkit assigns specific roles to components, e.g., interpreters transform data, aggregators act as repositories. We have generalized the approach of the context toolkit by introducing a generic context-handling component that can be instantiated to play different roles. This approach is, we believe, more extensible (new classes of components can be defined) and allows for encapsulation of well-understood behaviors in a single component. But most importantly, the approach presented here is more expressive, in the sense that it captures meta-information that describes the quality of sensed context. Recent work by Dey, Mankoff and Abowd extends the context toolkit to allow manual disambiguation by the user of imprecise context data [9].

# 6    Conclusions and Future Work

We have presented a comprehensive view of how sensed context might be handled in interactive systems. In particular, we have presented a model for sensed context that accounts for the inherent uncertainty of data acquired through sensors or derived by subsequent transformations. We have examined how this model can be used to explore relevant dimensions of the design space of a context-aware application. Finally, we have introduced a proposal that incorporates the characteristics of sensed context information in terms of a software architecture model.

The model of sensed context we have described in this paper still needs refinement. We have exercised it on small examples but we need to better assess its value and its shortcomings. We plan to use it for the design of two real context-aware applications in collaboration with designers. We expect these real-world design experiences to provide some insights on how to better structure our exploration of the design space. We also anticipate to be able to address other questions, such as appropriate ways to describe aspects of interactive system functionality  relevant to the use of sensed context (e.g., using UML). Another expected result is a set of requirements for tools to support design, and documentation of the design process. A first step toward this goal will be to express our model as an XML DTD so that models of sensed context information can be used by software tools.

# References

1. Dey, A.K., Salber, D., Abowd, G.D.: A Conceptual Framework and Toolkit for Supporting the Rapid Prototyping of Context-Aware Applications. Human-Computer Interaction (to appear) (2001)
2. The American Heritage Dictionary of the English Language, 3rd ed. Houghton Mifflin Company, Boston (1992)
3. Cockton, G., Clarke, S., Gray, P., Johnson, C.: Literate Development: Weaving Human Context into Design Specifications. In: Benyon, D., Palanque, P. (eds.): Critical Issues in User Interface Systems Engineering. Lecture Notes in Computer Science, Vol. 1927. Springer-Verlag, Berlin Heidelberg New York (1996) 227-248
4. Bass, L., Little, R., Pellegrino, R., Reed, S., Seacord, R., Sheppard, S., Szczur, M.R.: The UIMS Tool Developers' Workshop: A Metamodel for the Runtime Architecture of an Interactive System. SIGCHI Bulletin 24 (1992) 32-37

5. Salber, D., Dey, A.K., Abowd, G.D.: The Context Toolkit: Aiding the Development of Context-Enabled Applications. In: Williams, M.G., Altom, M.W., Ehrlich, K., Newman, W. (eds.): Proceedings of the ACM SIGCHI Conference on Human Factors in Computing Systems. ACM Press, New York (1999) 434-441

6. Card, S., McKinlay, J., Robertson, G.: The Design Space of Input Devices. In: Chew, J.C., Whiteside, J. (eds.): Proceedings of the ACM Conference on Human Factors in Computing Systems. ACM Press, New York (1990) 117-124

7. Winograd,    T.:    Towards    a    Human-Centered    Interaction    Architecture. http://graphics.stanford.edu/projects/iwork/papers/humcent/ (1999)

8. Kiciman, E., Fox, A.: Using Dynamic Mediation to Integrate COTS Entities in a Ubiquitous Computing Environment. In: Thomas, P., Gellersen, H.-W. (eds.): Proceedings of the Second International Symposium on Handheld and Ubiquitous Computing. Lecture Notes in Computer Science, Vol. 1927. Springer-Verlag, Berlin Heidelberg New York (2000) 211-226

9. Dey, A.K., Mankoff, J., Abowd, G.D.: Distributed Mediation of Imperfectly Sensed Context in Aware Environments. GVU Technical Report No. 00-14 (2000)

## Discussion

*S. Greenberg:* What do you when the sensor you are interested in (such as a light sensor) may be affected both by natural phenonoma and user use (pick up). What do you do to distinguish these two cases.
*P. Gray:* This is a case of transformation within the process. These hexagonal components each do a transformation and uncertainty can be added at each stage of the process.
*D. Salber:* You need to know the specifics of each transformation and add a quality attribute that gives a level of certainity.

*J. Coutaz:* Sensors are spacial temporal things? The granularity of time and space varies for an individual using a particular application. How can you claim that one model can capture all of this granularity.
*P. Gray:* We don't know right now but there is going to be a workshop on location modeling at the next ubicom and maybe we will have an answer then.

*D. Damian:* How do you handle the ambiguity implicit in natural language when you use language as a sensor.
*D. Salber:* We use speech to recognize the language not the content of the speech.
*P. Gray:* It depends on the sensing strategy you use and we are not proposing a strategy but simply trying to provide a method for evaluating the strategies you might use.

*F. Paterno:* Sensed information sometimes can be misleading. For example, the position of users in a museum can be considered in order to indentify their interests: if they are close to a work of art, the system can infer that they are interested in it. However, this is not always true. In some cases the position can be determined by extreanous factors (e.g. they are meeting friends or they are tired and decide to stop wherever they are). Don't you think that we would probably need some other types of information to obtain meaningful context-dependent applications?
*P. Gray:* I agree. You almost always need other information.

*L. Nigay:* Problems addressed by content sensing (disambiguity, uncertainity) seem similar to the ones addressed by multimodal interaction.

*D. Salber:* Yes, but some problems are specific to sensing (e.g. coverage, drift).

*J. Höhle:* How does this work about XML, XQLessor queries relate to work by agent people. i.e. KQ (knowledge query and manipulation language).

*D. Salber:* XML is used for sensor context and to express specific needs. This is basically similar but less general than KQML.

# Delivering Adaptive Web Content
# Based on Client Computing Resources

Andrew Choi and Hanan Lutfiyya

Department of Computer Science
The University of Western Ontario
(choi@csd.uwo.ca, hanan@csd.uwo.ca)

**Abstract.** This paper describes an approach to adapting Web content based on both static (e.g., connection speed) and dynamic information (e.g., CPU load) about a user's computing resources. This information can be transmitted to a Web Server in two different ways. XML is used so that there is one copy of the content, but multiple presentations are possible. The paper describes an architecture, prototype and initial results.

## 1 Introduction

The users of electronic commerce applications have certain expectations about the Quality of Service (QoS) provided. By QoS, we are referring to non-functional requirements such as performance or availability. An important measurement of Quality of Service is "key to glass" response time. This refers to the time that passes after a user clicks the mouse button or presses the return key to submit a request to a WWW server to the time that the results of that request are displayed on the monitor glass. Electronic commerce retailers (e-tailers), such as Amazon.com, Chapters.ca, and other dot-coms, recognize that the reference to the WWW as being the "World Wide Wait" is a major impediment in the growth of electronic commerce applications.

To date most of the work for improving Quality of Service (QoS), has revolved around the web server and the network. There has been little work on the client side despite the research [4] showing the importance of the client computing resources especially the network link from the client machine to the Internet. This paper focuses on the client side. The approach taken is based on findings that show that web pages that were retrieved faster were judged to be significantly more interesting then web pages retrieved at a slower rate. Other findings in [2] indicates that the user prefers web pages to be progressively displayed rather than web pages that are displayed at a slower rate, but may have more graphics at a higher resolution. This provides the user with feedback concerning the web server, knowing that the downloading process is still taking place. The conclusion is that that the response time affects a user's impression of a web site.

Let us now look at the following scenario: Let us assume that we have two users: user *A* is using a slow computer with a phone modem for connection to the Internet and user *B* is on a fast computer with a cable modem for connection to the Internet.

M. Reed Little and L. Nigay (Eds.): EHCI 2001, LNCS 2254, pp. 337–355, 2001.
© Springer-Verlag Berlin Heidelberg 2001

The difference between the response time of these two users, assuming the network and web server loads are identical, relies on client computing resources. To state the obvious, user $B$ receives the content faster than user $A$.    This difference in client computing resources has forced content providers [1] to provide a compromise version of content that hopefully will not place an undue burden on clients with fewer computing resources, yet will remain satisfactory for clients with the high-end in computing resources. One approach is to have different content delivered to different clients [1]. Currently, this requires that for a single content provider would need to maintain multiple versions of the same web site. Our research addresses this problem. We provide a solution that allows a web site to have a single copy of the content, but provides a different presentation of that content based on the client's computing resources. The main advantage of this approach is the elimination of unnecessary multimedia content being delivered to clients who do not have the necessary computing resources that best supports this content.

The following sections of this paper will present the research that we have done to date in improving a client's "key to glass" response time. Section 2 provides the design architecture, followed by the implementation in Section 3.    A set of experiments in Section 4 shows the improved "key to glass" response time for clients. In Section 5 we will examine a few other techniques in improving response time. Finally, we shall provide some concluding thoughts in section 6.

## 2 Architecture

The architecture (graphically depicted in Figures 1 and 2) entails two primary components: the client and server components.    The client component (which includes the Web Browser) is used to collect client computing resource information. Computing resource information includes, but is not limited to, the type of processor, type of Internet connection, percentage of processor load and percentage of memory being used.    There are two approaches to getting this information to the web server. The first is a *push* approach where the information accompanies each web request to a web server.    The second is a *pull* approach where the information is requested by the web server when needed. Upon receipt of the client request, the server component (which includes the Web Server) is responsible for analyzing the client computing resource information to determine an appropriate presentation descriptor that is to be applied to the content being requested by the client. We will now describe these components in more detail.

### 2.1 Client Component Architecture

The Client Component of the architecture is graphically depicted in Figure 1.

The Client Component collects the client computing resource information that is needed by a Web Server for processing a client request. We refer to this as *QoS Data* and it may include the following: (i) *CPU Type*: This represents the CPU speed of the client device that is requesting the Web page. (ii) *Connection Speed*: This measures the connection speed of the client device to the Internet.    (iii) *CPU Load*: This represents the amount of processor utilization on the client device. (iv) *Memory Load*:

This represents the amount of total memory, both physical and virtual memory, being used on the client device (percentage).

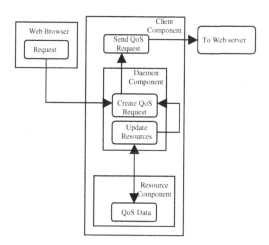

**Fig. 1.** Client Components

The two components involved in collecting *QoS Data* and including it in a web request are the Daemon and Resource components. In the push approach the Daemon converts a web request from any web browser into a request that includes *QoS Data*. In the pull approach, the Daemon process maintains (i.e., stores) the *QoS Data*. It provides an interface to other processes to retrieve this data. The *QoS Data* dynamically changes over time. The Resource Component is responsible for monitoring the computing resources to provide the information needed in *QoS Data*. It provides an interface to allow for monitoring of the necessary resources. It must be noted that this is just an interface. The actual implementation is based on the specific client device.

## 2.2 Server Component Architecture

The Server Component of the architecture is graphically depicted in Figure 2. The RequestHandler component is responsible for the initial processing of a request that is received by the web server.  It extracts the *QoS Data* (done differently for the push and pull models).  It passes the *QoS Data* to the QoS component, which encapsulates an algorithm that uses this data to determine an appropriate *classification* value. More specifically it encapsulates the order of operations (some are calls to the interface of the Classification component).  The Classification component maintains the following: information  needed to determine the classification value (described in the next paragraph) and a mapping from a classification value to a *presentation descriptor* value. The *presentation descriptor* value is applied to the web content that is to be returned to the client.

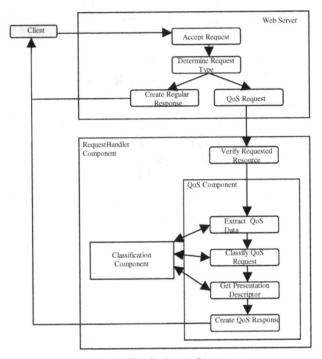

**Fig. 2.** Server Components

The algorithm used for computing the *classification* value is briefly described as follows. Each of the data items in *QoS Data* ( *CPU Type, Connection Speed, CPU Load* and *Memory Load*) are assigned weights to reflect their contribution to the "glass response time". For example, the weight assigned to *CPU Type* could be 30%, for *Connection Speed* it could be 30%, for *CPU Load* it could be 30% and for *Memory Load* it could be 10%. This weighting assumes that the type of the CPU, the connection speed and the load on the client machine contribute equally and much more than memory load. A weight is applied to a *category* value. We will use *CPU Load* to illustrate what we mean by category value. One possible categorization is based on having two categories: CPU loads that represent less than 50% utilization and CPU loads that represent 50% or more utilization. We know that if there is higher utilization then there will be fewer CPU cycles on the client device to process the response. Thus, resulting in the end result of a slower response time. We can assign a *category* value to each of these categories. The appropriate *category* value is determined and is multiplied by the weight associated with *CPU Load*. The sum of these values is the *classification* value The *classification* value is mapped to a *presentation descriptor* value which is then applied to the content. The result of this application is what gets sent back to the client.

## 2.3 Interactions

This section is intended to examine the flow of a request from the client device to the final delivery of a response from the web server. We assume that the specified resource in the request is available.

1.  The request is made in the web browser.

2.  In the *push* approach, the request is intercepted from the web browser by the Daemon component. The Daemon component adds the *QoS Data* to the request. In the *pull* approach, the request is sent through without any additional modifications (in the current prototype, a web request is considered QoS enabled if it uses the ".xml" extension). The *QoS Data* used by the Daemon Component is updated periodically (the period is relative to the speed of the processor) by the Resource component.

3.  When a request arrives at a web server, the Request component determines the nature of the incoming request i.e., whether the request is QoS enabled or not. In the *push* model, the Request component checks to see if the QoS Data is appended to it. In the *pull* model, it checks to see if the request uses the ".xml" extension. If so, the RequestHandler component will retrieve that data from the client. If the request does not include *QoS Data* or does not use the ".xml" extension, the web server will immediately process the request; otherwise the request is forwarded to the QoS component.

4.  The QoS component processes the *QoS Data* to determine the appropriate classification value based on the brief description of the algorithm presented in the previous section. It encapsulates an algorithm that has operations that make calls to the Classification component.

5.  The QoS Component takes the *classification* value and requests from the Classification component the associated *presentation descriptor* value.

6.  The URL for the web content requested by the client is processed with the returned *presentation descriptor* value. Based on this processing, the appropriate QoS enabled response is produced for the client device. This is very critical in our process. If the client device was a wireless cell phone, it would not be capable of viewing images or handling HTML content. Hence an appropriate *presentation descriptor* value would be required to convert the web content into Wireless Markup Language (WML) content, so that the client device could display the results.

7.  The response with the web content produced from the previous step is returned to the client device that made the request. The client device's web browser then processes the web content and displays it accordingly.

If there was any problem with the *QoS Data*, then default classification values are used (in our prototype we defaulted to classifications resulting in the simplest content presentations).

# 3  Implementation

In this section we describe the prototype implementation.  Open-source code (Mozilla) is available for the Netscape browser, but the source code for this browser consists of over four million lines of code. Due to this complexity, it was decided to first implement the push approach. This was done using the Remote Method Invocation (RMI) API provided by Java.

## 3.1 Client Component

This section describes implementation details associated with the Client Component.

### 3.1.1 Daemon Component

We implemented the pull approach for the retrieval of client computing resource information. The Daemon component is implemented using a Java class that provides a listen queue for monitoring incoming requests for client computing resource information from the web server. The Java RMI enables the web server to make a method call to the remote object as if the remote object were local, without the need to explicitly program the socket connection interface. The following defines the class for the Daemon component.

```
class Cl {
        static Classification classification;

        public Cl();
        public String getClClass();
        public static void main();
}
```

The *classification* object will be discussed in greater detail in the following section. The *getClClass()* method is the remote method that is invoked by the web server. Its purpose is to return the client computing resource information in a string format. Even though most of the resource values are bound to be numerical values, Java provides methods that transform numbers into strings and vice-versa rather easily. The *main()* is responsible for collecting the client device connection speed to the Internet and providing this in the creation of the *classification* object. In addition, it provides the initialization required to set up the necessary RMI server to listen for incoming RMI requests.  Once all the RMI initialization has been established the Daemon component will listen to port 1099 for incoming RMI requests.

### 3.1.2 Resource Component

The Resource component monitors the CPU utilization, memory utilization, and the type of processor. Depending on the operating system running on the hardware, each operating system provides different techniques to obtain this information, such as command line utilities, system library calls, etc. Thus, the implementation of the resource component is based on the Abstract Factory design pattern as detailed in [3]. The intent of the Abstract Factory pattern is to provide an interface for creating families of related or dependent objects without specifying their concrete classes [3].

There are two classes to note. The first is the Classification class.

```
class Classification {
        String ClientClassification;
        ResourceFactory Factory;
        long waitinterval;

        public Classification (long);
        public String getClassification();
        public void Update();
}
```

The *ClientClassification* member is used to hold the client resource information in a string format that is obtained from the *ResourceFactory* object (which will be examined in detail shortly). The *Factory* member will contain an object for either the *SolarisFactory* or *WindowsNTFactory*. Lastly the *waitinterval* will contain a value to determine the duration at which the client resource information will be refreshed.

In the Classification constructor, we use a standard Java method call *System.getProperty("OS.name")* to get the operating system name. Based on that value, we are capable of determining which type of *ResourceFactory* object needs to be created. The *Factory* variable is initialized with this object. A call to *Factory.getCPUType()* is made to get the type of processor that the client has. This is used to determine the refresh rate at which the client polls for the client computing resource information (stored in the *waitinterval* member). On slower machines the update frequency is set to a longer interval and for faster machines to quicker refresh rates. The rationale behind this is that slower machines have fewer CPU cycles, so these cycles should be used more sparingly. Lastly the *ClientClassification* member is constructed to store client computing resource information as a string. The *getClassificaiton()* method is used to return the value of *ClientClassification* to the calling object (which will be the class being used to implement the Daemon process). Finally, the *Update()* method performs the operation of making method calls to the *Factory* object (e.g., *getCPUType()*, *getCPULoad()*, *getMemory()*) in order to get the current client computing resource information. The frequency of the updates is based on the *waitinterval* that was obtained in the constructor of the *Classification* object.

```
interface ResourceFactory {
String getDeviceType();
float getCPUType();
float getCPULoad();
float getMemory();
    }
```

The ResourceFactory is an abstract class. This allows for any number of subclasses to be created for any number of operating systems for various devices. Our prototype contains only a single implementation for the Solaris operating system. Future work will entail creating a similar ResourceFactory for the Windows operating system. With our implementation for the Solaris operating system, we were able to use native Solaris system calls to obtain the necessary information in the SolarisFactory. In obtaining the value for the getCPUType() method, we used the **"sysinfo"** utility to obtain the processor speed. For both the getCPULoad() and getMemory() methods, calls were made to the **"top"** utility for up to date utilization data on the client device. In applying the Abstract Factory pattern, it would be relatively easy for new subclasses of ResourceFactory to be created and incorporated into the client component.

## 3.2 Server Component

Before detailing the server implementation for this paper, it must be recognized that portions of the Cocoon servlet from the Apache Software Foundation were used. The Cocoon software already provides the necessary infrastructure for handling incoming HTTP requests and transforming XML files with XSL stylesheets. We shall discuss only those components of the Cocoon software that we have made modifications to.

### 3.2.1 RequestHandler Component

The RequestHandler component is implemented using a Cocoon class implementation that extends the *HttpServlet* class, which provides the ability to handle HTTP requests and responses. Once the *Cocoon* object has been successfully created, it remains in an idle state waiting for incoming HTTP requests for a resource with the .XML extension. Upon receipt of such an HTTP request, the *service()* method of the *Cocoon* class extracts from the received request the resource that it is requesting. If the requested resource is a valid resource on the server, then both the HTTP request and HTTP response objects are passed to the *Engine* object (used to implement the QoS Component), otherwise an HTTP error response is sent back to client.

### 3.2.2 QoS Component

The QoS component is implemented as a class called *Engine*. The initialization of an *Engine* object requires from the *Cocoon* object information that includes the XML parser and the XSL processor to be used. These are used to initialize two objects that are encapsulated by the *Engine*: *parser* and *transformer*. It is apparent that only one instance of the parser and transformer objects will exist. Thus, the *Engine* object uses the Singleton pattern [3]. The intent of the Singleton pattern is to ensure a class only has one instance, and provides a global point of access to it. In our implementation, we use software from the Apache Software Foundation  Thus, the XML parser is *Xerces* and the XSL processor is *Xalan*.

The *Handle()* method of the *Engine* object is the heart of the entire process. The Document Object Model (DOM) is constructed for the XML content. This is done through the *parser* object. The next step retrieves the individual data items  of the client computing resource information (*QoS Data*) via a call to *Classification.GetClassData()* (the *Classification* object implements the Classification component and is discussed in the next section). The data that is returned is then passed on to the *Classifcation.VerifyClassData()* method that determines the *classification* value for the requesting client. The validity of the *classification* value is then determined with a call to *Classification.ValidClass()* and if the *classification* value is valid a call to *Classification.ReturnLocation()* for the directory location of the XSL stylesheet is made. Based on the number of classifications, there could be any number of directories for stylesheets. In this example, we assume three classifications for constructing simple web pages, default web pages, and complex web pages. Once the location of the XSL stylesheet is determined, the actual XSL stylesheet is applied to the XML DOM using the XSL processor. Once this has completed, the processed contents from the processor object are inserted into the *response* object, which is then delivered to the client device to be displayed.

### 3.2.3 Classification Component

The Classification component encapsulates that information required to assign a *classification* value to clients based on their computing resources and a mapping from *classification* values to a presentation descriptor value. This component is implemented using the *Classification* class. The initialization of the *Classification* class retrieves from files the information needed to classify the clients.

```
public class Classification implements Defaults {

    String[] classdata;
    int ClientClassification;
    float CPUTypeWeight;
    float ConnectionWeight;
    float CPULoadWeight;
    float MemoryWeight;
    float[] ClientClass;
    float[] CPUTypeClass;
    float[] ConnectionClass;
    float[] CPULoadClass;
    float[] MemoryClass;

    public Classification();
    private float LoadWeight(BufferedReader infile);
    private float[] LoadClass(BufferedReader infile);
    public String GetClassData (HttpServletRequest request);
    public int VerifyClassData (String value);
    private float CalculateWeight(float resourcevalue, float[] inputclass, float weightclass, int flag);
    private int DetermineClass (float clientvalue, float[] inputclass);
    public String ValidClass (int classdata);
    public String ReturnLocation (int locationclass);
}
```

The *Classification()* method  (used to create the *Classification* object) uses the private *LoadWeight()* method to load from a file the weight values into the following variables:  *CPUTypeWeight, ConnectionWeight, CPULoadWeight* and *MemoryWeight*. Table 1 provides an example set of weights.

**Table 1.** Resource Weights

| Member | Value |
|---|---|
| CPUTypeWeight | 30 |
| ConnectionWeight | 30 |
| CPULoadWeight | 30 |
| MemoryWeight | 10 |

The total combined weights must sum to 100. Each resource can be considered as a percentage, providing an administrator with an easy way to judge the significance of each resource.

The private method  *LoadClass()*  loads  the  variables  *CPUTypeClass, ConnectionClass, CPULoadClass* and *MemoryClass*. These variables are arrays.

Each array stores measurement values.    A measurement in an array location $i$ represents a boundary point.    Thus, for any array $a$, $(a[i],a[i+1])$ represents a categorization of values. All values less than $a[0]$ is a categorization and all values greater than $a[n-1]$ is a categorization, where $n$ is the number of categories , is a category.    These values are used to determine classifications. Table 2 shows an example.

**Table 2.** Resource Categories

| Member | Array Location | | | |
|---|---|---|---|---|
| | 0 | 1 | 2 | 3 |
| CPUTypeClass | 90 | 166 | 300 | 450 |
| ConnectionClass | 28800 | 56000 | 500000 | 1500000 |
| CPULoadClass | 75 | 50 | 25 | XXXXX |
| MemoryClass | 66 | 33 | XXX | XXXXX |

Not all of the arrays will contain the same number of entries.  In this example, the *CPUTypeClass* member has four entries and the *MemoryClass* contains two entries. The *CPUTypeClass* is used to categorize the speed of the process in the client device. As can be seen, the categories correspond to devices with less than a 90 MHz processor, 90 MHz to less than 166 MHz, 166 MHz to less than 300 MHz, 300 MHz to less than 450 MHz, and over 450 MHz. We note that the *CPUTypeClass* and *ConnectionClass* are stored in ascending order, while *CPULoadClass* and *MemoryClass* are in descending order.  The reasoning is that in the case of the CPU type and Connection speed, a higher value means that there are more computing resources and therefore, the system can more quickly respond to content from the server. With the CPU load and Memory load, a higher value represents degradation in system performance.

The initialization process also constructs an array that maps a classification level to the associated directories for the XSL files. Table 3 provides an example.

**Table 3.** Classification Categories

| Member | Array Location | | |
|---|---|---|---|
| | 0 | 1 | 2 |
| Classdata | /simple/ | /default/ | /complex/ |

The method *VerifyClassData()* is used to verify that the contents that were returned by the *GetClassData()* method is properly formatted.  If the string is properly formatted, the calculation to determine the classification of the client begins. We will use the following string in our example: *0.1 COMP 269 56000 30 40*.  The first thing checked is the version number of the incoming information.  This will be important if future work determines that other client computing resources are found to influence the performance of a system. The second item to be checked is the type of device making the request.  If the device is other than a computer running a web browser, different computations may be required.  In our work, we have accounted for the standard computer and web browser with "COMP".  However, another possible device is that of a wireless cell phone, in which case we will have the value of "WLESS".  This will require that we provide another directory for "/wireless/" that stores a unique XSL stylesheet to transform the content to be displayed on a cell phone type device using WML.  In this example, we are dealing with a computer and web browser. The next piece of information is the type of processor in the computer, which is 269 MHz. A call is made to *CalculateWeight()*. The value 269, the *CPUTypeClass* and *CPUClass* arrays and a flag value of 1 are passed to *CalculateWeight()*. If the flag value is 1 then the array passed in is in ascending order otherwise it is in descending order.

The *CalculateWeight()* method is aware that the array *CPUTypeClass* is an array of ascending values since the flag value is 1. At this point, it checks each value at location *CPUTypeClass[i]* until it finds a value of *CPUTypeClass[i]* that is greater than 269. In this example (based on Table 2) this is the third array location (and thus the third category). There are five categories in which the CPU types are separated (these include under 90 and over 450 MhZ). In our implementation the category value is computed based on the array location and is equal to $(i+1)/n$ , where $i$ is the category and $n$ is the number of categories. In our example, since value 269 is in *CPUTypeClass[2]*, the category value is 3/5. The weight for the CPU type was defined to be a value of 30, thus we arrived at a calculation 3/5 * 30 = 18 for the CPU type in our example. This value is returned to the *VerifyClass()* method and a running tally is kept, as the *CalculateWeight()* method is called three more times, once for each of the other resources of Connection speed, CPU Load, and Memory Load.

For connection speed the calculation is 3/5 * 30 = 18. This value is returned to the *VerifyClass()* method. The total grows from a value of 18 (from the CPU Type) to 36.

The next value to be calculated is that of 30 for the CPU Load. The array, *CPULoadClass*, is organized in descending order (as indicated by a flag value of –1). Instead of comparing to see if the CPU load passed in is less than CPULoadClass[i], the comparsion is a greater than or equal to. Since the CPU load falls into the third category, we obtain a calculation of 3/4 * 30 = 22.5. Once back into the *VerifyClass()* method, the total value grows from 36 to 58.5.

Once again we must perform the same calculations on the Memory Load. In performing the calculations 2/3 * 10 = 6.6666, we get a value of roughly 6.667. Once back in the *VerifyClass()* method, the total value grows from 58.5 to a value of 65.167.

In working through this example, it is apparent that the highest possible value that can be calculated in this process is 100. Every request that is received with a ".xml" extension will be processed and assigned a value ranging from 0 to 100. This design allows us to easily extend any of the resource categories, as well as adding additional resources at a future time without affecting the number of classification categories stored.

We have now calculated a value for the resources of the client computer to be that of 65.167. Still in the *VerifyClass()* method, we must determine which *classification* value is assigned to the client request. This is done by making a call to the *DetermineClass()* method. In a similar manner as above, the *ClientClass* is an array with boundary values at each of the array locations. The values of 25 and 60 are the entries in the array locations for 0 and 1 respectively. The value of 65.167 fits into the third category. In this method, the counting of categories begins at 0, so the first category would be a value of 0, the second category a value of 1, and the third category of 2. A value of 2 is returned to the *VerifyClass()* method. As this is the last item performed by the *VerifyClass()* method, this same value is returned to the object that had called *VerifyClass()*, in this case a value of 2 is returned. In this example, we were able to calculate a valid classification for the request. There are instances where an error message would be returned in place of a valid classification. For example, if during transfer the client resource information were to become corrupt and instead of receiving a numeric value for the CPU type, the server received a string of characters. The processing of the request would fail and an error code would be returned instead of a valid classification.

The *ValidClass()* method would be called after the *VerifyClass()* method to determine the nature of the validity for the calculations and the returned value. In examining the value, it is determined if it fits into a classification or if the value were an error code. If the value is an error code, the appropriate error message is constructed and an error string is returned from the method. Otherwise, a null string is returned if the value is a valid classification.

The last method for this object is the *ReturnLocation()* method, which returns the directory location for retrieval of the appropriate XSL stylesheet to be applied.

## 4  Experiments

This section briefly describes our initial experimentation.

### 4.1 Experimental Environment

The experiments were performed on an isolated network with three Sun Ultra workstations. The workstations varied in their hardware configurations. Two of the machines functioned as the clients that requested web content via HTTP. These machines are named Strawberry and Doublefudge. Strawberry is a Sun Ultra 5/10, with a 269MHz UltraSPARC-IIi processor and 128MB of physical memory. Doublefudge is a Sun Ultra 60, with dual 359MHz UltraSPARC-II processors and 1024 MB of physical memory. While the third workstation named Vanilla, a Sun Ultra 5/10, with a 269MHz UltraSPARC-IIi processor and 256MB of physical memory, was used as the host delivering the web content.

It was critical in our experimentation to simulate connection speeds equivalent to those most commonly used today. This was achieved by establishing a Point-to-Point Protocol (PPP) connection via a serial interface. In establishing a PPP connection we were able to control the connection speeds between the two client machines of Strawberry and Doublefudge to the host machine Vanilla. The maximum connection speed that was achieved was equivalent to 400 Kbps, while using the lower extreme of 38.4 Kbps.

The web server used in our experimentation was Java Web Server 2.0. This web server software provided the necessary infrastructure to handle servlet execution without requiring additional components. In addition to the Cocoon 1.6 Java servlet from the Apache Software Foundation, we also required an XML parser and XSL processor. These two were also obtained from Apache, with the XML parser being Xerces 1.01 and the XSL processor being Xalan 19.2.

As mentioned earlier, it was necessary to produce heavy CPU load and memory load situations on the client workstations. The added load on the clients was achieved by the use of two utilities that we developed. The first is a CPU load generator utility and the second is a memory load generator. These two programs allowed us to generate high and low load values for both CPU and memory usage.

We also developed a utility that allows the user to make specific HTTP requests for a specific URL from a web server at the command line. In using this utility, scripts were developed to eliminate the need for manual requests through a web browser. In using this utility, our experimental results are based on the HTTP 1.0 design, in which HTTP connections do not remain persistent, instead each object in the web page requires a completely new connection to be established.

## 4.2 Factor Analysis for Original System

The first set of experiments performed were used to determine which of the four factors: Connection speed, CPU type, CPU load, and Memory load contributed most significantly to the delivery of web content to the client device. We used a $2^4$ full factorial design with 3 replications. Each of the factors had a high and low level. Four experiments were performed, in a similar manner. The first experiment consisted of the web server delivering standard HTML web content, with a default page having contents equaling 28KB. The results from this experiment found that the factor contributing the most to the content delivery was the connection speed, accounting for over 69.747 and the CPU load for 14.815 of significance. A similar experiment was performed with a complex HTML page equaling 1500 KB. This time the connection speed accounted for almost all of the significance with a value of 99.904 and all others were insignificant.

We then performed the experiment with the same factors, but instead making requests for the XML content through the original Cocoon servlet. The default XML page, being the same size as the HTML content of 28KB, produced almost identical results as the HTML default page. The significant values were once again the connection speed with 66.033 and the CPU load at 15.919. Once again with the complex XML content, the connection speed was most significant with a value of 99.849 and all others being insignificant.

In studying these two scenarios, the HTML content delivered by the web server is similar to a file server, obtaining stored files from a storage media and delivering it to the client. While the XML content requires the web server to process the content by applying an XSL stylesheet to the XML content. It appears with our experimentation that this additional overhead of processing only increases the response time marginally.

## 4.3 Prototype Results

From examining the results for the factorial experiments, the most significant factor for "key to glass" response time for the client device is the connection speed to the Internet. Based on this, we assigned the greatest weight to the connection speed of 70%, followed by CPU load at 20%, and the Memory load and type of CPU each with 5%.

A time of over eight minutes was required to obtain the 1500KB complex web page from the web server with a slow connection in our previous experiments. With knowledge of the slow connection being the most significant factor, our prototype will only provide the complex content if the user is on a fast connection. Hence eliminating the eight minutes of wait time for the complex web content.

The comparison of the times is taken with respect to the original Cocoon implementation. Included with this timing is the duration required for the Java RMI call to the client to obtain the QoS data, based on the *pull* model. In our QoS enabled servlet implementation, we also provided a timing for the RMI call. So, we were able to record the exact time required to perform the remote call. On average the RMI call from the web server to the client took 432.014 milliseconds to complete. This value is rather small in nature and didn't contribute greatly to the overall "key to glass" response time.

# 5 Related Work

## 5.1 Replicated Servers

When examining the idea of replicated servers on the Internet, we are interested in two key characteristics to help improve "key to glass" response time. The first factor is the distance to which the web server is located from the client requesting the web content. In saying distance, we can not assume the physical locality of the web server, but more importantly we must use the network hop distance. This value represents the distance needed to travel to arrive at the web server by the client request over the Internet. It is evident if the request and response spends a large amount of time travelling on the network, the response time experienced by the user will be affected.

The second crucial factor is the time required by the web server to process the client request and ultimately deliver the appropriate response. When considering this factor we must also examine additional items that affect the web server's ability to process requests. One such item deals with the number of requests coming into the web server, as from 1998 to 2002 the number of users on the WWW are expected to triple in volume [5]. It is apparent that a web server could potentially become over burdened serving client requests. As well as the client requests, the other item of great significance is the type of requests the clients are submitting. At the onset, web servers delivering content were very similar to file servers. A client request would arrive at the web server and the web server would deliver static web content to the user. There was very little computational overhead associated with client requests. With the evolution of the WWW, web content complexity grew hand in hand. In addition to static web content, there are now web sites that dynamically create web content unique to each client request, providing a personalized experience. In these dynamic web pages, they are not just accessing static content from files, but they are accessing databases to obtain client information, as well as other web sites for up to date content. The technology behind the dynamic content creation has also advanced, with traditional Common Gateway Interface (CGI) scripts and applications, Active Server Pages (ASP) from Microsoft, Java servlets, to eXtensible Markup Language (XML) and eXtensible Style Language Transformation (XSLT). Along with these technologies comes the added processing required by the web server. Web servers are becoming more complex entities, they are no longer just file servers, but have the added responsibility of creating dynamic web content.

## 5.2 Caching

The idea of caching is similar to that of Replicated Servers, which was discussed in the previous section. The intent of caching is not to replicate the entire server's contents, but to only reproduce those portions that have been frequently accessed. In doing so, the load of the web server will be reduced.

In examining caching, there are three points at which the caching can take place. To examine these three points, let us begin by looking at the client making requests for web content. This is the very first place that caching can occur. Once a client makes a request for web content the web content can be cached at the local client. In almost all instances, the caching is the responsibility of the web browser, as the browser is responsible for the handling of requests and the returned responses. If the browser notices that a request for a particular web resource has been cached, it will retrieve it from the cache, and bypass the submission of the request to the web server. This helps in improving response time by providing the obvious advantages, first there will be no network latency as the request doesn't even need to travel onto the Internet. Secondly, the web server will not need to service the client request, hence eliminating the server load. The advantages have been shown and the disadvantages will be discussed later in this section.

The next point for caching to occur is at a proxy location. Proxies can reside at various locations, but we will refer to the location at which an autonomous system is connected to the Internet. Since all the client requests on the autonomous system need to travel through the proxy to reach the Internet, it is ideal for this proxy to cache web content responses. To understand the sequence of events that occurs in this situation, we will have two clients $C1$ and $C2$ on an autonomous system that travels through proxy $P$. $C1$ begins by making a request for www.needinfo.com, the request travels to $P$ and then travels out onto the Internet to the www.needinfo.com site. The response is returned by the web server hosting www.needinfo.com and the response travels to $P$ where the response is cached. $P$ then sends the response to $C1$ where it is processed. Later that day $C2$ also makes a request to www.needinfo.com, the request travels to $P$, but the response for $C1$ was previously cached, $P$ needs only to return the cached contents to $C2$. In this manner, the $C2$ request only needs to travel to $P$ and doesn't even need to travel onto the Internet to the www.needinfo.com web server. Similar to caching at the client only, this approach helps both reduce the amount of traffic travelling onto the Internet and reduces the server load by eliminating repeated requests.

The last point of caching can take place directly at the web server that is handling client requests. The rationale behind this caching is to eliminate the processing load from the web server. For those instances of complex web content, in which server processing is required. Instead of performing the processing for each client request, it would be beneficial to cache the web content that is produced. The client request would arrive at the web server, it would look through the cache for the web content, if it exists then it would be retrieved, otherwise the web server would process the request to deliver the web content. Even though this process doesn't alleviate the network traffic to the web server, it helps reduce the server load by eliminating repetitive processing of identical requests.

As shown through the three caching locations, caching is an advantageous technique to improve upon response time for the user. Along with the benefits there are also definite drawbacks to this technology. The first to come to mind is the need for cache consistency. Cache consistency is the requirement to keep the cache content, either being at the client cache, proxy cache, or server cache, consistent with the content that is delivered by the web server. In any of these instances, web content may be cached in these three locations, in addition there may be updates at the web server delivering the content. The cached locations must be aware of this update, so that out dated content is not delivered to the requesting client.

The second problem deals with personalization on the WWW. This deals with web sites that personalize web content based on the individual, for instance a web site that greets you by your first name in the web content that is returned. This could happen at a banking site, which requires you to log in, once logged in the web site is aware of who you are and creates content specific to you. When using caching, this personalized content is not relevant to others accessing the same web site, as the web server would create content that is specialized for that individual. In this manner the content is created dynamically, there are also other instances in which dynamic content could be created but not personalized. An example could be an e-tailer that creates web content for merchandise they sell on the WWW. This is dynamically created based on the price of merchandise that is stored in another database, but is infrequently changed. In this instance, rather than have each request to the web server processed, this content can indeed be cached. In looking at the situation from this standpoint it would be beneficial for dynamic content caching. Again we must weigh the advantages gained by caching the dynamic content and the possible problems posed if the cached contents become outdated. In this manner not all web content can and should be cached to improve response time.

## 5.3 Adaptive Web Content

At the beginning of this paper we stated the significance for "key to glass" response time. In the previous two sections we looked at the techniques of replicated servers and caching in an attempt to stay within a 10 seconds threshold. In examining these other two techniques you will notice some similarities. The first is the desire to alleviate work imposed on the web server, which is delivering the content to the client. With replicated servers, the workload is distributed among several identical servers. On the other hand with caching, attempts have been made to cache copies of the dynamically created content, which totally eliminate the processing required on the web server.

The second common characteristic is the desire to bring the content closer to the requesting client. Once again with replicated servers, the ideal situation is to strategically place identical servers at various locations on the Internet. In a similar manner with caching, if there are strategically placed cache servers, the time traveled on the network will decrease, along with the decreased time traveled on the network there will be an increase in the "key to glass" response time.

You are more than likely pondering the question, why do we need adaptive web content when replicated servers and caching appear to improve the "key to glass" response time. The techniques suggested in the previous two sections are possible solutions, but with the solutions exist drawbacks, especially with the present state of the WWW. To improve response time in both techniques requires additional servers to be placed throughout the Internet. For companies and more importantly individuals, who are restricted by limited budgets cannot afford the cost of additional servers and network connections.

As the WWW grows, the trend for web content is towards dynamic personalized content rather than the more traditional static content that is seen today. This is due in part to the increased number of retailers providing e-commerce on the WWW. The information for the e-tailers is usually stored in a database, with dynamically created content containing queried data from the database. For example, an e-tailer may sell widgets on the WWW, they already have a legacy database system that contains all the prices of the widgets they manufacture. The price of these widgets changes monthly, and the database has several thousand widgets. Instead of creating a new set of web content containing the updated prices monthly, the e-tailer instead queries the existing database for the price quotes. Hence, the only way to create the dynamic content for a price quote is for the web server to dynamically create the content for each client request.

In addition to dynamic web content, the amount of multimedia content on the WWW is also increasing. The e-tailer who is selling widgets on the WWW today is not only giving a descriptive commentary of the widgets, but is also providing graphical images of each widget they sell. The number of images and resolution for each image are also increasing. With the increased number of people accessing the WWW, the size and complexity of web content is also rising. This is where adaptive web content will benefit the client's "key to glass" response time. In an attempt to decrease the response time, web servers perform two actions when processing a client request. The first item is the actual creation of the content to be returned to the client, this includes obtaining the content to place into a web page, such items as retrieving the links to image files from disk, making the database queries, and formatting the content appropriately. The second such task is the actual retrieval of static data from hard disk and delivery of the content to the client. Of the two processes, most of the research has concentrated on the second step concerning the retrieval of static data and the delivery of that content to the client. The interest of this technique is not to reduce the total amount of content delivered, but to only reduce content when the server is not capable of delivering a level of QoS to the client, and at that point reduce the content to once again achieve the expected level of QoS.

## 6 Conclusion

As the number of users accessing the Internet and WWW increase, it is expected by the year 2002 there will be over 320 million people accessing the WWW alone [5]. That is why we see the influx of e-tailers to the WWW. In addition, the demands of the users are also increasing, with the most predominant being the QoS expectation, in particular with the response time for the delivery of web content. As mentioned earlier this is crucial for people returning to a web site, as long delays can guarantee loss of customers.

Presently, with most web servers there is no consideration for the resources of the client. All client requests are treated as if they were equal in nature, even if the client hardware and network connections differ greatly.   Our research  supports the conclusion that others have shown in that the client resources, in particular the Internet connection, can greatly affect the response time experienced by the end user.

In our implementation we were able to show that providing adaptive web content can greatly improve upon the user's QoS.

Present web sites must cater to the lowest common denominator, which are clients connecting to the Internet with a phone modem at 28.8 Kbps. This is done to make sure all the visitors to the web site are content with a certain level of QoS, especially those with the slow connection. What about the users on fast connections? This doesn't allow them to take full advantage of their connection. With our design, the web site provider has the ability to create various levels of content based on the available client resources. In using this technique, an e-tailer's web site could provide a short video clip to the user on the fast connection, but provide a single jpeg image to the user on the slow connection for the merchandise they are trying to sell. In both circumstances the users will receive the information on the merchandise. However, the e-tailer does not have to store different versions of the content. Hence, they save on disk storage and maintenance costs. Although, XSL style sheets are to be stored, these will not have the same storage costs of storing different versions of the content.

Further work includes developing the push model, to have each client HTTP request include the QoS Data. We will develop various Resource Factories for different operating systems, in particular those for Windows and Linux operating systems. There could also be additional client computing resources that affect the response time. Further experimentation will examine this as well as determine this. We will also examine the problem of optimal categorization and weights. Finally, we must further examine the effectiveness of this approach from a user's point of view. Currently, we only look at client computing resources. We would also like to include user profiles that allow us to also use priorities.

### Acknowledgements

This work is supported by the National Sciences and Engineering Council (NSERC) of Canada, the IBM Centre for Advanced Studies in Toronto, Canada and the Canadian Institutute of Telecommunications Research (CITR).

# References

1. Abdelzaher, T. F., Bhatti, N., Web Content Adaptation to Improve Server Overload Behavior, In *Proc. of the 8th International World Wide Web Conference*, Toronto, Canada, May 1999.
2. Bouch, A., Kuchinsky, A., Bhatti, N., Quality is in the Eye of the Beholder: Meeting Users' Requirements for Internet Quality of Service, *Internet Systems and Applications Laboratory, HP Laboratories*, Palo Alto, January, 2000.
3. Gamma, E., Helm, R., Johnson, R., Vlissides, J., *Design Patterns Elements of Reusable Object-Oriented Software*, Addison Wesley Longman, Inc., 1997.
4. Seshan, S., Stemm, M., Katz, R. H., Benefits of Transparent Content Negotiation in HTTP, *Proc. Global Internet Mini Conference at Globecom 98*, Sydney, Australia, November, 1998.
5. "The Internet, Technology 1999, Analysis and Forecast", *IEEE Spectrum*, January, 1999.

## Discussion

*F. Paterno:* The most important portion of your QoS data seems to be the static data. Is it true that only static information is necessary.

*H. Lutfiyya:* That seems to be the case right now but I have a hard time believing that this will hold through future experiments.

*F. Paterno:* Key to glass time is not a single number but is spread out over time. Have you thought about capitalizing on that.

*H. Lutfiyya:* We have not yet considered that.

*D. Salber:* The RMI call takes 432 milliseconds. Is that per HTTP request?

*H. Lutfiyya:* That is per HTTP request. That is using the pull model and if we used the push model the latency should be reduced.

*S. Greenberg:* I would have thought that this could be done more efficiently at the server side by the content designer. That is, embed the sequence into the page a priori rather than deciding this through inference.

*H. Lutfiyya:* Right but that might be somewhat slower. You might be slowing things down for your high speed users.

*J. Roth:* Why do you choose something like CPU load and memory load since these factors change quickly over time. What have no categories related to the graphical capabilities of the client.

*H. Lutfiyya:* I agree with you. There are other hardware characteristics that we should incorporate. We just haven't done it yet.

*L. Bass:* Often web page responses are spread out over time with the page filling in slowly. Have you thought about capitalizing on this.

*H. Lutfiyya:* No

# How Cultural Needs Affect User Interface Design?

Minna Mäkäräinen[1], Johanna Tiitola[2], and Katja Konkka[2]

[1] Solid Information Technology, Elektroniikkatie 8,
FIN-90570 OULU, Finland
Minna.Makarainen@solidtech.com
[2] Nokia Mobile Phones, Itämerenkatu 11-13, P.O.BOX 407,
FIN-00045 NOKIA GROUP, Finland
[Johanna.Tiitola, Katja.Konkka]@nokia.com

This paper discusses how cultural aspects should be addressed in user interface design. It presents a summary of two case studies, one performed in India and the other in South Africa, in order to identify the needs and requirements for cultural adaptation. The case studies were performed in three phases. First, a pre-study was conducted in Finland. The pre-study included literature study about the target culture. Explored issues included facts about the state, religions practiced in the area, demographics, languages spoken, economics, conflicts between groups, legal system, telecommunication infrastructure and education system. Second, a field study was done in the target culture. The field study methods used were observations in context, semi-structured interviews in context, and expert interviews. A local subcontractor was used for practical arrangements, such as selecting subjects for the study. The subcontractors also had experience on user interface design, so they could act as experts giving insight to the local culture. Third, the findings were analyzed with the local experts, and the results were compiled into presentations and design guidelines for user interface designers. The results of the case studies indicate that there is a clear need for cultural adaptation of products. The cultural adaptation should cover much more, than only the language of the dialog between the device and the end user. For example, the South-Africa study revealed a strong need for user interface, which could be used by non-educated people, who are not familiar with technical devices. The mobile phone users are not anymore only well educated technologically oriented people. Translating the language of the dialog to the local language is not enough, if the user cannot read. Another design issue discovered in the study was that people were afraid of using data-intensive applications (such as phonebook or calendar), because the criminality rates in South Africa are very high, and the risk of the mobile phone getting stolen and the data being lost is high. In India, some examples of the findings are the long expected lifetimes of the products, and importance of religion. India is not a throwaway culture. When a device gets broken, it is not replaced with a new one, but instead it is repaired. The expected lifetime of the product is long. The importance of religion, and especially religious icons and rituals, is much more visible in everyday life, than in Europe. For example, people carry pictures of Gods instead of pictures of family with them. Addressing this in the user interface would give the product added emotional value.

M. Reed Little and L. Nigay (Eds.): EHCI 2001, LNCS 2254, pp. 357–358, 2001.

## Discussion

*D. Damian:* What was the major change that Nokia made after this study.
*M. Mäkäräinen:* Its difficult to say. People realized that localizing the product was not just a matter of translating text.

*H. Stiegler:* Do you think that 3 years is a long time in a repair culture.
*M. Mäkäräinen:* No.

*M. Damian:* How does Nokia decide where to go to do these studies.
*M. Mäkäräinen:* It depends on where mobile phones have some penetration and a large potential market can be seen.

*H. Lutfiyya:* Do you think the UI design process will change the way that people will interact with the phone.
*M. Mäkäräinen:* The way people interact with the phone is different in different cultures no matter what the UI design process is.

*N. Graham:* What level of precision did you have about what you wanted to find out.
*M. Mäkäräinen:* Our scope was too broad we tried to cover every phase. For this year I planned to cover only the visual facets. We observed people making phone calls and reading e-mail and doing other types of communication but it should be more narrow.

*J. Willans:* I guess there is a tradeoff between Nokia's commercial interest and some of the knowledge learned in this study.
*M. Mäkäräinen:* Commercial aspects and business models were not analyzed in this study.

# Author Index

# Lecture Notes in Computer Science

For information about Vols. 1–2165
please contact your bookseller or Springer-Verlag

Vol. 2205: D.R. Montello (Ed.), Spatial Information Theory. Proceedings, 2001. XIV, 503 pages. 2001.

Vol. 2206: B. Reusch (Ed.), Computational Intelligence. Proceedings, 2001. XVII, 1003 pages. 2001.

Vol. 2207: I.W. Marshall, S. Nettles, N. Wakamiya (Eds.), Active Networks. Proceedings, 2001. IX, 165 pages. 2001.

Vol. 2208: W.J. Niessen, M.A. Viergever (Eds.), Medical Image Computing and Computer-Assisted Intervention – MICCAI 2001. Proceedings, 2001. XXXV, 1446 pages. 2001.

Vol. 2209: W. Jonker (Ed.), Databases in Telecommunications II. Proceedings, 2001. VII, 179 pages. 2001.

Vol. 2210: Y. Liu, K. Tanaka, M. Iwata, T. Higuchi, M. Yasunaga (Eds.), Evolvable Systems: From Biology to Hardware. Proceedings, 2001. XI, 341 pages. 2001.

Vol. 2211: T.A. Henzinger, C.M. Kirsch (Eds.), Embedded Software. Proceedings, 2001. IX, 504 pages. 2001.

Vol. 2212: W. Lee, L. Mé, A. Wespi (Eds.), Recent Advances in Intrusion Detection. Proceedings, 2001. X, 205 pages. 2001.

Vol. 2213: M.J. van Sinderen, L.J.M. Nieuwenhuis (Eds.), Protocols for Multimedia Systems. Proceedings, 2001. XII, 239 pages. 2001.

Vol. 2214: O. Boldt, H. Jürgensen (Eds.), Automata Implementation. Proceedings, 1999. VIII, 183 pages. 2001.

Vol. 2215: N. Kobayashi, B.C. Pierce (Eds.), Theoretical Aspects of Computer Software. Proceedings, 2001. XV, 561 pages. 2001.

Vol. 2216: E.S. Al-Shaer, G. Pacifici (Eds.), Management of Multimedia on the Internet. Proceedings, 2001. XIV, 373 pages. 2001.

Vol. 2217: T. Gomi (Ed.), Evolutionary Robotics. Proceedings, 2001. XI, 139 pages. 2001.

Vol. 2218: R. Guerraoui (Ed.), Middleware 2001. Proceedings, 2001. XIII, 395 pages. 2001.

Vol. 2219: S.T. Taft, R.A. Duff, R.L. Brukardt, E. Ploedereder (Eds.), Consolidated Ada Reference Manual. XXV, 560 pages. 2001.

Vol. 2220: C. Johnson (Ed.), Interactive Systems. Proceedings, 2001. XII, 219 pages. 2001.

Vol. 2221: D.G. Feitelson, L. Rudolph (Eds.), Job Scheduling Strategies for Parallel Processing. Proceedings, 2001. VII, 207 pages. 2001.

Vol. 2223: P. Eades, T. Takaoka (Eds.), Algorithms and Computation. Proceedings, 2001. XIV, 780 pages. 2001.

Vol. 2224: H.S. Kunii, S. Jajodia, A. Sølvberg (Eds.), Conceptual Modeling – ER 2001. Proceedings, 2001. XIX, 614 pages. 2001.

Vol. 2225: N. Abe, R. Khardon, T. Zeugmann (Eds.), Algorithmic Learning Theory. Proceedings, 2001. XI, 379 pages. 2001. (Subseries LNAI).

Vol. 2226: K.P. Jantke, A. Shinohara (Eds.), Discovery Science. Proceedings, 2001. XII, 494 pages. 2001. (Subseries LNAI).

Vol. 2227: S. Boztaş, I.E. Shparlinski (Eds.), Applied Algebra, Algebraic Algorithms and Error-Correcting Codes. Proceedings, 2001. XII, 398 pages. 2001.

Vol. 2228: B. Monien, V.K. Prasanna, S. Vajapeyam (Eds.), High Performance Computing – HiPC 2001. Proceedings, 2001. XVIII, 438 pages. 2001.

Vol. 2229: S. Qing, T. Okamoto, J. Zhou (Eds.), Information and Communications Security. Proceedings, 2001. XIV, 504 pages. 2001.

Vol. 2230: T. Katila, I.E. Magnin, P. Clarysse, J. Montagnat, J. Nenonen (Eds.), Functional Imaging and Modeling of the Heart. Proceedings, 2001. XI, 158 pages. 2001.

Vol. 2232: L. Fiege, G. Mühl, U. Wilhelm (Eds.), Electronic Commerce. Proceedings, 2001. X, 233 pages. 2001.

Vol. 2233: J. Crowcroft, M. Hofmann (Eds.), Networked Group Communication. Proceedings, 2001. X, 205 pages. 2001.

Vol. 2234: L. Pacholski, P. Ružička (Eds.), SOFSEM 2001: Theory and Practice of Informatics. Proceedings, 2001. XI, 347 pages. 2001.

Vol. 2237: P. Codognet (Ed.), Logic Programming. Proceedings, 2001. XI, 365 pages. 2001.

Vol. 2239: T. Walsh (Ed.), Principles and Practice of Constraint Programming – CP 2001. Proceedings, 2001. XIV, 788 pages. 2001.

Vol. 2240: G.P. Picco (Ed.), Mobile Agents. Proceedings, 2001. XIII, 277 pages. 2001.

Vol. 2241: M. Jünger, D. Naddef (Eds.), Computational Combinatorial Optimization. IX, 305 pages. 2001.

Vol. 2242: C.A. Lee (Ed.), Grid Computing – GRID 2001. Proceedings, 2001. XII, 185 pages. 2001.

Vol. 2245: R. Hariharan, M. Mukund, V. Vinay (Eds.), FST TCS 2001: Foundations of Software Technology and Theoretical Computer Science. Proceedings, 2001. XI, 347 pages. 2001.

Vol. 2247: C. P. Rangan, C. Ding (Eds.), Progress in Cryptology – INDOCRYPT 2001. Proceedings, 2001. XIII, 351 pages. 2001.

Vol. 2248: C. Boyd (Ed.), Advances in Cryptology – ASIACRYPT 2001. Proceedings, 2001. XI, 603 pages. 2001.

Vol. 2250: R. Nieuwenhuis, A. Voronkov (Eds.), Logic for Programming, Artificial Intelligence, and Reasoning. Proceedings, 2001. XV, 738 pages. 2001. (Subseries LNAI).

Vol. 2251: Y.Y. Tang, V. Wickerhauser, P.C. Yuen, C.Li (Eds.), Wavelet Analysis and Its Applications. Proceedings, 2001. XIII, 450 pages. 2001.

Vol. 2252: J. Liu, P.C. Yuen, C. Li, J. Ng, T. Ishida (Eds.), Active Media Technology. Proceedings, 2001. XII, 402 pages. 2001.

Vol. 2254: M.R. Little, L. Nigay (Eds.), Engineering for Human-Computer Interaction. Proceedings, 2001. XI, 359 pages. 2001.

Vol. 2256: M. Stumptner, D. Corbett, M. Brooks (Eds.), AI 2001: Advances in Artificial Intelligence. Proceedings, 2001. XII, 666 pages. 2001. (Subseries LNAI).

Vol. 2258: P. Brazdil, A. Jorge (Eds.), Progress in Artificial Intelligence. Proceedings, 2001. XII, 418 pages. 2001. (Subseries LNAI).

Vol. 2260: B. Honary (Ed.), Cryptography and Coding. Proceedings, 2001. IX, 416 pages. 2001.

Vol. 2264: K. Steinhöfel (Ed.), Stochastic Algorithms: Foundations and Applications. Proceedings, 2001. VIII, 203 pages. 2001.